Irrigation and Drainage - Recent Advances

Edited by Muhammad Sultan and Fiaz Ahmad

Published in London, United Kingdom

Irrigation and Drainage - Recent Advances
http://dx.doi.org/10.5772/intechopen.100884
Edited by Muhammad Sultan and Fiaz Ahmad

Contributors
Muhammad Sultan, Hafiz Muhammad Asfahan, Fiaz Ahmad, Marjan Aziz, Redmond R. Shamshiri, Uzair Sajjad, Muhammad Usman Khan, Muhammad Farooq, Md Shamim Ahamed, Faizan Majeed, Fahd Rasul, Hassan Munir, M. Salman Ayub, Muhammad Safdar, Sobia Shahzad, Rehan Mehmood, M. Adnan Shahid, Abid Sarwar, Umair Gull, Wajid Nasim Jatoi, Muhammad Mubeen, Summera Jahan, Shakeel Ahmed, Erion Bwambale, Felix K. Abagale, Geophrey K. Anornu, Younsuk Dong, Valdemiro Pitoro, Rodrigo Sánchez-Román, João Queluz, Tamires Da Silva, Sérgio Jane, Kevim Muniz, Pratik Ghutke, Rahul Agrawal, Huma Zia, Mariana de Oliveira Pereira, Jailton Garcia Ramos, Carlos Alberto Vieira de Azevedo, Geovani Soares de Lima, André Alisson Rodrigues da Silva, Gustavo Bastos Lyra, Luciano Marcelo Fallé Saboya, Patrícia Ferreira da Silva, Alina Buber, Yuri Dobrachev, Alexander Buber, Evgenii Ratkovich, Mohammad Hudzari Haji Razali, Abdul Quddus Puteh, Alawi Haji Sulaiman, Mohamad Hakim Mohamad Yatim, Meinarti Norma Setiapermas, Anggi Sahru Romdon, Yulis Hindarwati, Fawibe Oluwasegun Olamide, Bankole Abidemi Olalekan, Mustafa Abdulwakiil Adeyemi, Sokunbi Uthman Tobi, Joseph Oladipupo Julius, Fawibe Kehinde Oluwaseyi, António Canatário Duarte, Amparo Melián-Navarro, Antonio Ruiz-Canales, Muthuminal R, Mohana Priya R

First published in London, United Kingdom, 2023 by IntechOpen
IntechOpen is the global imprint of INTECHOPEN LIMITED, registered in England and Wales, registration number: 11086078, 5 Princes Gate Court, London, SW7 2QJ, United Kingdom

British Library Cataloguing-in-Publication Data
A catalogue record for this book is available from the British Library

Additional hard and PDF copies can be obtained from orders@intechopen.com

Irrigation and Drainage - Recent Advances
Edited by Muhammad Sultan and Fiaz Ahmad
p. cm.
Print ISBN 978-1-80356-209-4
Online ISBN 978-1-80356-210-0
eBook (PDF) ISBN 978-1-80356-211-7

For EU product safety concerns:
IN TECH d.o.o., Prolaz Marije Krucifikse Kozulić 3, 51000 Rijeka, Croatia,
info@intechopen.com or visit our website at intechopen.com.

Meet the editors

Dr. Muhammad Sultan is an associate professor in the Department of Agricultural Engineering, at Bahauddin Zakariya University, Pakistan. He obtained a BSc and MSc in Agricultural Engineering from the University of Agriculture Faisalabad, Pakistan, and a Ph.D. in Energy and Environmental Engineering from Kyushu University, Japan. He did postdoctoral research in energy and environmental engineering at Kyushu University and in mechatronic systems engineering at Simon Fraser University, Canada. He has published more than 300 articles in international journals and conferences, as well as books and book chapters. He has been a reviewer for more than 100 SCIE journals and is currently an editor for 10 others. His research focuses on developing energy-efficient temperature and humidity control systems for agricultural applications including greenhouse, fruit/vegetable storage, livestock, and poultry applications.

Dr. Fiaz Ahmad is an associate professor in the Department of Agricultural Engineering, at Bahauddin Zakariya University, Pakistan. He obtained a BSc and MSc in Agricultural Engineering from the University of Agriculture Faisalabad, Pakistan in 2004 and 2007, respectively. He obtained a Ph.D. in Agricultural BioEnvironmental and Energy Engineering from Nanjing Agriculture University, China in 2015. He completed a postdoc in Agricultural Engineering at Jiangsu University, China in 2020. He earned B.Sc. (2004) and M.Sc. (2007) in Agricultural Engineering from the University of Agriculture, Faisalabad (Pakistan). He is the author of more than sixty journal and conference articles. He has supervised ten master students to date. In addition, he is supervising five master's and two doctoral students. He completed three research projects and is currently working on two others. His research interests include various aspects of agricultural and farm mechanization.

Contents

Preface **IX**

Chapter 1 **1**
Agrovoltaic and Smart Irrigation: Pakistan Perspective
by Hafiz M. Asfahan, Muhammad Sultan, Fiaz Ahmad, Faizan Majeed, Md Shamim Ahamed, Marjan Aziz, Redmond R. Shamshiri, Uzair Sajjad, Muhammad Usman Khan and Muhammad Farooq

Chapter 2 **27**
Sustainable Irrigation Management for Higher Yield
by Fahd Rasul, Hassan Munir, Aftab Wajid, Muhammad Safdar, M. Salman Ayub, Sobia Shahzad, Rehan Mehmood, M. Adnan Shahid, Abid Sarwar, M. Danish Majeed, Umair Gull, Wajid Nasim Jatoi, Muhammad Mubeen, Summera Jahan and Shakeel Ahmed

Chapter 3 **51**
An Overview on Techniques Involved in Recharging Ground Water and Its Impacts
by Muthuminal R. and Mohana Priya R.

Chapter 4 **67**
An IoT-based Immersive Approach to Sustainable Farming
by Pratik Ghutke and Rahul Agrawal

Chapter 5 **89**
Smart Rainwater Harvesting System for Sustainable Agricultural Irrigation and Drainage System
by Mohd Hudzari Haji Razali, Abdul Qudus Puteh, Alawi Haji Sulaiman and Mohamad Hakim Mohamad Yatim

Chapter 6 **105**
Water Sustainability through Drainage Reuse in Agriculture – A Case for Collaborative Wireless Sensor Networks
by Huma Zia

Chapter 7 119
Improvement of Tertiary Irrigation Networks, Changes in Cropping Patterns, and Increasing Cropping Index at Kendal Indonesia
by Meinarti Norma Setiapermas, Anggi Sahru Romdon and Yulis Hindarwati

Chapter 8 137
Irrigation Scheduling Methods: Overview and Recent Advances
by Younsuk Dong

Chapter 9 149
Fundamentals of Irrigation Methods and Their Impact on Crop Production
by Fawibe Oluwasegun Olamide, Bankole Abidemi Olalekan, Sokunbi Uthman Tobi, Mustafa Abdulwakiil Adeyemi, Joseph Oladipupo Julius and Fawibe Kehinde Oluwaseyi

Chapter 10 169
Effect of Irrigation Depths and Salinity Levels on the Growth and Production of Forage Palm Orelha de Elefante Mexicana
by Mariana de Oliveira Pereira, Jailton Garcia Ramos, Carlos Alberto Vieira de Azevedo, André Alisson Rodrigues da Silva, Geovani Soares de Lima, Luciano Marcelo Fallé Saboya, Patrícia Ferreira da Silva and Gustavo Bastos Lyra

Chapter 11 187
Anaerobic Filters: Alternative Solution for the Treatment of Domestic Wastewater for Reuse in Vegetable Irrigation
by Valdemiro Pitoro, Rodrigo Sánchez-Román, João Queluz, Tamires Da Silva, Sérgio Jane and Kevim Muniz

Chapter 12 205
Theoretical Approaches to Water Use Optimization for Rice Irrigation Systems in the Lower Kuban
by Alina Buber, Yuri Dobrachev, Alexander Buber and Evgenii Ratkovich

Chapter 13 225
Smart Irrigation for Climate Change Adaptation and Improved Food Security
by Erion Bwambale, Felix K. Abagale and Geophrey K. Anornu

Chapter 14 243
Resilience of Irrigated Agriculture to Face the Challenges in Mediterranean Climatic Conditions (Iberian Peninsula)
by António Canatário Duarte, Amparo Melián-Navarro and Antonio Ruiz-Canales

Chapter 15 261
Seed Soaking Times and Irrigation Frequencies Affected the Nutrient Quality and Growth Parameters of *Hordeum vulgare* L. Cultivated in Hydroponics
by Ryan Anthony Smith, Muhali Olaide Jimoh and Charles Petrus Laubscher

Preface

Irrigation and drainage play a key role in the success and productivity of on-farm agriculture. Good quality water is a prime mover of agriculture and farming. Approximately more than 80% of available freshwater reserves are utilized to irrigate farmlands to produce food and fibers. The deficiency in available freshwater has intensified due to climate change and population growth, which is devastating the agriculture sector. The global commitment to the scholar-coined term "Day Zero – Water & Food" urges researchers to explore advanced irrigation techniques and/or management strategies for effectively utilizing and conserving the freshwater at farms. Consequently, this book presents state-of-the-art research and review studies related to irrigation methods, irrigation management, groundwater recharging techniques, rainwater harvesting, and computer intelligence for developing smart irrigation.

Irrigation and Drainage - Recent Advances consists of fifteen chapters that present novel concepts, recent advancements, and management strategies. Chapter 1 discusses the significance of Agrovoltaic energy for smart irrigation. Chapter 2 presents key findings on sustainable irrigation management strategies for acquiring higher yields. Chapter 3 presents a comprehensive review of groundwater recharging techniques, providing critical insights into governance and policy optimization. Chapter 4 explores advancements in the Internet of Things (IoT) for developing smart irrigation systems. Chapter 5 discusses rainwater harvesting methods for sustainable agriculture. Chapter 6 examines how to reuse agriculture drainage and improve water quality through collaborative wireless sensor networks. Chapter 7 presents a case study of Kendal Indonesia and how the city has worked to improve its tertiary drainage networks to increase the cropping index. Chapter 8 describes several smart irrigation scheduling approaches that regulate freshwater demand according to field requirements. Chapter 9 explores the impact of irrigation methods on crop production. Chapter 10 discusses the effect of Irrigation depths and salinity levels on the growth and production of forage palm Orelha de Elefante Mexicana. Chapter 11 explores the performance of an anaerobic filter for treating wastewater for reuse in vegetable irrigation. Chapter 12 provides the details of a theoretical approach for optimizing the irrigation of rice fields. Chapter 13 proposes and discusses a novel idea for utilizing existing agricultural lands for the tri-benefits of obtaining food, creating solar energy, and controlling irrigation water supplies using an artificial intelligence approach. Chapter 14 discusses the challenges encountered by the irrigation sector in Mediterranean climatic conditions. Finally, Chapter 15 discusses the impacts of irrigation frequencies on the nutrient value and growth of agricultural produce.

The editors are pleased to share the recent progress of Irrigation and drainage technologies, methods, and management strategies. We believe this book will be useful not only for agricultural engineers but also for associated agricultural and water resource scientists. Finally, we would like to thank the authors for their valuable contributions.

Muhammad Sultan and Fiaz Ahmad
Department of Agricultural Engineering,
Bahauddin Zakariya University,
Multan, Pakistan

Chapter 1

Agrovoltaic and Smart Irrigation: Pakistan Perspective

Hafiz M. Asfahan, Muhammad Sultan, Fiaz Ahmad, Faizan Majeed, Md Shamim Ahamed, Marjan Aziz, Redmond R. Shamshiri, Uzair Sajjad, Muhammad Usman Khan and Muhammad Farooq

Abstract

The present study aims to investigate the prospects and challenges that need to be encountered for the adaptation of the novel agrovoltaic irrigation system (AVIS) in Pakistan. The agro-production scenario in Pakistan is periodically declining and leading toward the high delta crops, which develop severe pressure on the conventional energy and water resources. Groundwater might be a viable water source, but its pumping requires massive energy. In addition, excessive pumping declines the water table at a higher pace as compared to the recharge rate hence leading the country toward the exploitation of the valuable reservoir. The AVIS could be an energy-efficient and reliable irrigation solution in a manner of harvesting solar energy for driving smart irrigation systems capable to pumps the metered groundwater according to field requirements. Lack of local understanding, skilled/technical personnel, dependence on local technology, and major capital expenditures might impede technological adaption. The government should take necessary measures to replenish the groundwater reservoirs and also execute research projects that strengthen ground knowledge of AVIS.

Keywords: energy and water scarcity, groundwater exploitation, agrovoltaic, irrigation system

1. Introduction

Today's world faces several challenges, including population expansion, climate change, dwindling food supplies, energy crisis, and water shortages. The challenges are intertwined in a manner, and their cumulative impact is devastating, particularly for the agriculture sector [1]. According to the latest United Nations (UN) projection, the world's population will exceed 9.6 billion by 2050, implying significant food insecurity in near future [2]. In order to catch the pace of prevailing nutritional demands, agriculture communities ought to be enhanced by at least 1.5 times compared to

2012 [3]. However, recent statistics revealed that the production capacity needs to be increased by 70% for coping with the population hunger in 2050 [4]. On the other hand, the resources utilized for alleviating the production capacity are becoming increasingly scarce. For instance, the agriculture sector mainly relies on adequate supplies of water and energy. Agriculture consumes 70% of the global scale freshwater withdrawals for the growth of agriculture commodities [5, 6]. However, it has been identified that the freshwater reserves are shrinking due to climate change consequently impacting the agri-food supply chain thereby exploring alternative solutions such as desalination [7, 8]. On the other hand, according to Food and Agriculture Organization (FAO), agrarian activities mainly account for 30% of the global cumulative energy production [4, 9]. The Asian countries have a major footprint in the agriculture sector. In this spectrum, they consume 4,000,000 TJ of energy in 2018 to perform agriculture operations [4]. **Figure 1** presents the region-wise temporal variation of the energy utilized by the agriculture sector [4, 10, 11]. The origin of this energy is fossil fuels, which enable the modern farm mechanization, centering fertilizer application and enhancing the harvesting and post-harvesting efficiencies. However, the depletion of fossil fuel reserves limiting the farm efficiency [12, 13].

Pakistan stands in the lane of agricultural countries that produces its major food supplies from arable land. The agriculture sector employs 39–42.3% of the country's labor force and contributes 19.3% of gross domestic growth (GDP) [14, 15]. According to the Pakistan economic survey (PES), the agriculture growth in 2020–2021 was recorded at 2.67%, which is 1.33% less compared to 2018 (4.0%) [15]. Similarly, the share of the GDP obtained from the agriculture sector has gradually decreased due to minimization of surface water availability, high energy pricing, pest attacks, climate change, and other influencing factors that limit the agriculture productivity from its potential [15, 16]. Pakistan consumes 93.8% of its total freshwater withdrawals for agrarian activities. However, a dramatic shortfall in surface water supplies has been observed in the last few years. For instance, from 2018 to 2019, the freshwater shortfall was estimated at 18.5% [15]. According to Indus River System Authority (IRSA), an acute water shortage of 38% was recorded in 2022, which was aimed to irrigate the Kharif crops [17, 18]. Similarly, the pricing of on-farm energy sources is mounting.

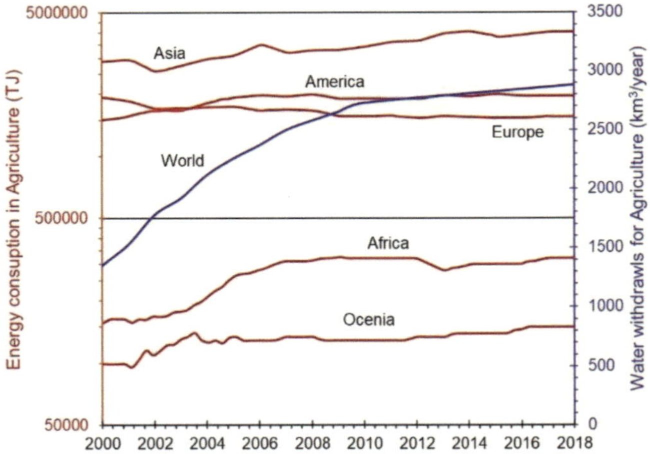

Figure 1.
Region-wise temporal variation of energy consumption and water withdrawal for agricultural activities reproduce here from [4, 10, 11].

The farmers are suffering from a shortage of fuel supplies. In this context, a sustainable alternative water and energy source are principally required in the country to improve crop productivity and for meeting the nutritional demands.

Groundwater is an alternative reliable resource of an adequate supply of freshwater. The dependency on the groundwater is increased in the last few years due to the depletion of surface water supplies [19]. Fortunately, Pakistan is blessed with the largest underground reservoir after China, India, and the USA, providing 60% and 90% of the irrigation and drinking water supplies, respectively [20]. Farmers are extensively pumping the groundwater in order to accomplish their water needs. For doing so, lifts pumps or tube wells are the ultimate sources to extract the underground water. In this context, a large number of tube wells are being installed across the country mainly driven with diesel as shown in **Figure 2**. Punjab mainly contains a high density of tube wells (995,456) followed by Sindh (71,454), Khyber Pakhtunkhwa (42,970), and Balochistan (39,567) with capacity varies between <10 and >25 horsepower (hp) [22]. Approximately 58% of tube wells in Pakistan have a capacity of 16–20 hp., necessitating massive fossil energy to operate [22]. The depletion and high pricing of the fossil fuels provoke food insecurity, which eventually destabilizes the country's economy furthermore, fossil fuel emissions are wreaking havoc on the ecosystem via global warming, shifting rainfall patterns, soil drying, and air pollution. The researchers concentrated their efforts in this area by constructing freestanding underground water pumping devices that could be fueled by renewable energy sources.

Solar photovoltaic (PV) technology has proven to be the most reliable on-farm energy source, capable to perform a variety of agro-operations [23]. Particularly for irrigation purposes a wide range of solar-driven pumps is being installed in order to accomplish the agricultural needs. However, due to the lack of groundwater governance in the country, massive groundwater extraction is committed by the farmers

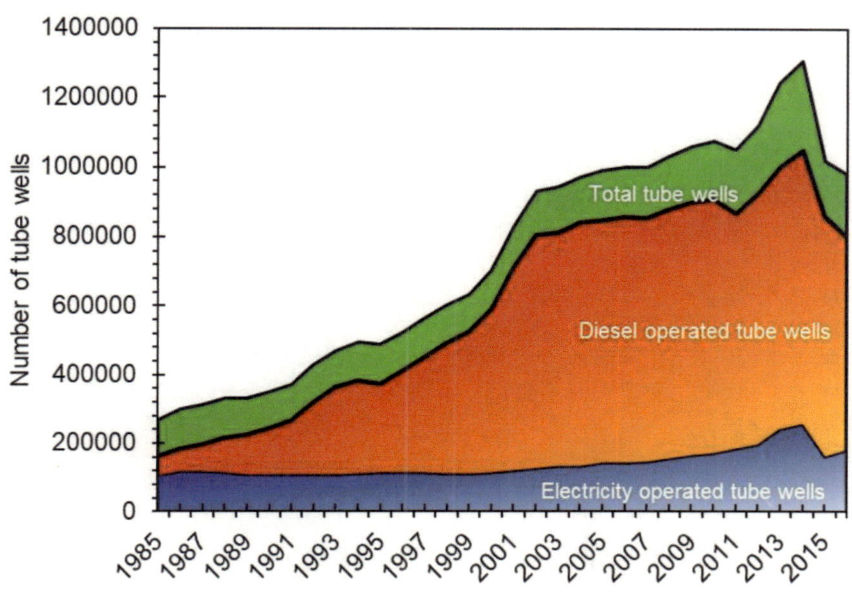

Figure 2.
Number of tube wells installed in Pakistan is segregated based on primary energy resources reproduce here from [21].

irrespective of their agriculture requirements. Consequently, stressing and exhausts the reservoir at a higher pace. In this context, alternative, viable policies, and smart irrigation technologies are principally required in order to mitigate the food insecurity arising from the water shortfall. Agrovoltaic irrigation system (AVIS) conception could be a remarkable and promising solution that not even metered the groundwater pumping but also utilized the culturable land for twin benefits (i.e., solar energy harvesting and agriculture production). The study presents the novel conception of AVIS and explores the prospects and challenges that need to be encountered for the potential adaptation of the technology at the farm level. In addition, the study highlights the consequences and possible remedies for the commercialization of the AVIS.

2. Agro-production scenario in Pakistan

Pakistan has two cropping seasons i.e., Kharif and Rabi for the production of crops. Major crops produced in the Kharif season are bajra, sugarcane, rice, moong, mash, maize, jowar, and cotton while wheat, masoor, and barley crops are produced in the Rabi season to accomplish the nutritional demands [24]. **Figure 3** shows temporal

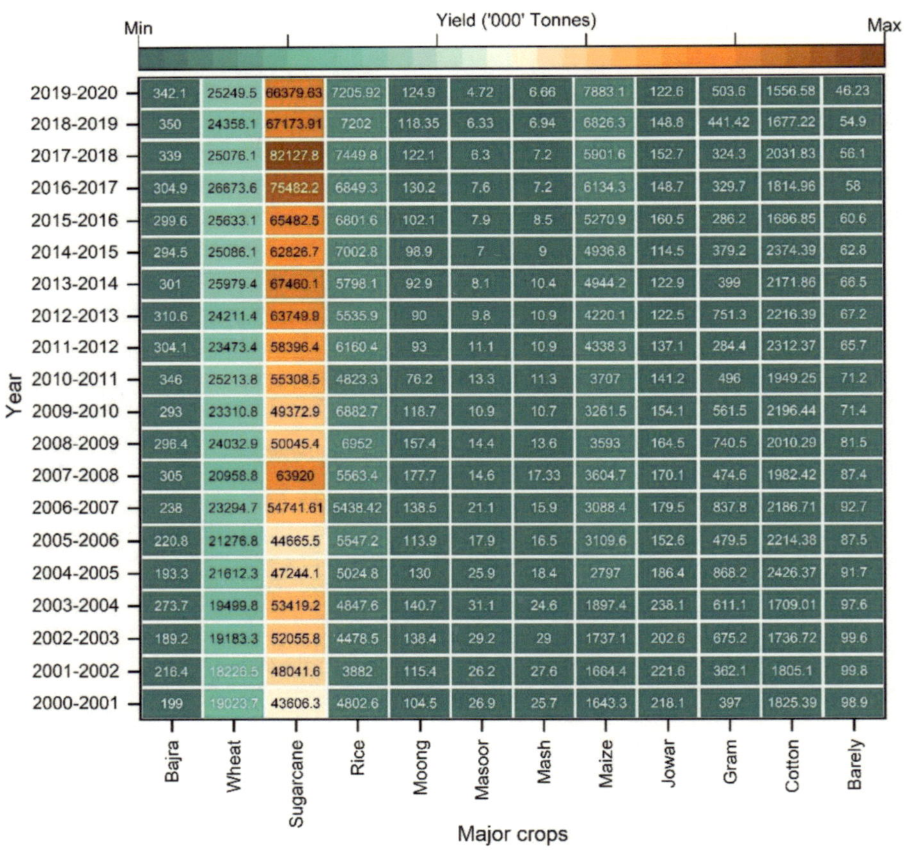

Figure 3.
Temporal variation of major crop yields for both Kharif and Rabi seasons in Pakistan from 2000 to 2020 reproduce here from Agriculture Marketing Information Service (AIMS) http://www.amis.pk/agristatistics/production.aspx.

variation in production capacity of major crops for both Kharif and Rabi seasons in Pakistan from 2000 to 2020. It has been realized that, in the Kharif season sugarcane, rice and maize have a major footprint thereby cultivating more than 70% of the agricultural land [25]. According to recent statistics revealed by the crop reporting service, in 2020, 1143.62 km^2, 2746.11 km^2, and 826.13 km^2 areas in Punjab have been utilized to grow sugarcane, rice, and maize, respectively [26]. On the other hand, in the Rabi season wheat has been extensively grown covering more than 70–75% of arable land. Similarly, in fruits and vegetable production, Pakistan has a major global market footprint, particularly in mango and potato production. **Figure 4** shows the temporal variation of the fruit and vegetable yields from 2000 to 2020. The matrix-shaped filled color gradient indicates that onion, potato, and mango production has periodically increased. If compared with 2000, the production of potato mango, and onion increases by 63.17%, 42.50%, and 25.55%, respectively in 2020. However, in 2022 it has been expected that mango production will fall by approximately 50% due to severe water shortages and heatwaves [27]. In addition, from **Figures 3** and **4** it has been realized that cropping pattern is more likely to lead toward the cultivation of high delta crops, which needs massive water supplies. However, the available water supplies are not likely to fulfill the potential needs of the crops.

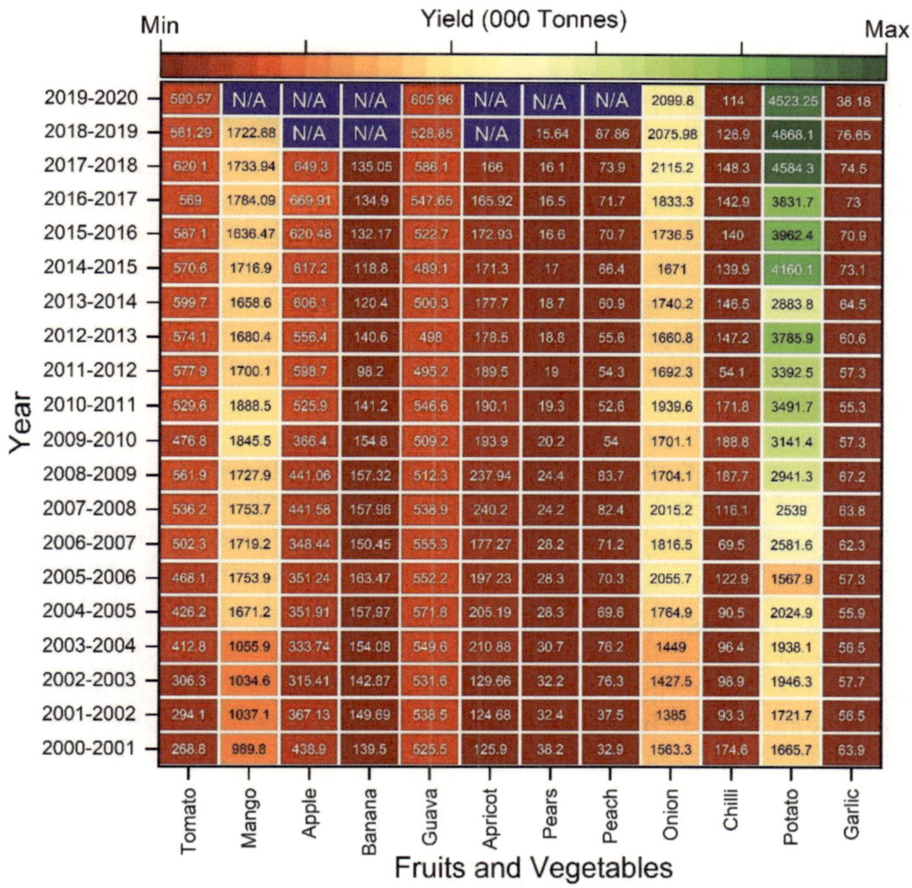

Figure 4.
Temporal variation of fruit and vegetable yields in Pakistan from 2000 to 2020 reproduce here from Agriculture Marketing Information Service (AIMS) http://www.amis.pk/agristatistics/production.aspx.

3. Sources of irrigation in Pakistan

Irrigation is the practice of watering crops or plants by digging pipes and ditches in the ground. The primary goal of irrigation is to supply water to crop fields making the land fertile. In Pakistan, irrigation water is obtained from three primary sources, including surface water, rainwater, and groundwater. Following is a brief description of these irrigation sources.

3.1 Surface water

Rainfall and melting snow form streams and storage reservoirs such as tanks, ponds, and damns, that are the primary source for the generation of surface water. Dams are erected along the river and water is channeled to agricultural areas through canals and irrigated the farmer's fields by means of gravity flow. However, the distribution of surface water from the canal head to the farmer's field triggers huge conveyance losses. In Pakistan, Indus Basin Irrigation System (IBIS) is the largest irrigation network that contributes to the conveyance of the surface water [28]. However, exponential growth in industrialization, urbanization, and population causes depletion of the surface water supplies in the IBIS [29, 30]. Additionally, the IBIS is highly vulnerable to adverse impacts of climate change owing to its geo-climatic situation. **Figure 5** represents the temporal variation of the surface water generated at the canal head and delivered at the farm end for both Kharif and Rabi crops. It can be seen that total surface water (both for Kharif and Rabi crops) generated at the canal head was 121.59 billion cubic meters (Bm^3) during 2019–2020 while 98.93 Bm^3 of the water delivered at the farm end resulted in huge conveyance losses. This amount of surface water could not fulfill the water requirements of both Kharif and Rabi crops. On the other hand, surface water irrigation is generally regarded to be free of energy cost because of gravity flow. However, an effective canal network requires operational energy for preservation e.g., removal of sediment, as well as vegetation and strengthening of canal banks [31].

3.2 Rainwater

Rainwater is another source of water that comes from rainfall to meet the irrigation requirements of crops. However, rainfall patterns and intensity are continuously decreasing because of the vulnerability of climate change. Climate change is impacting not just rainfall intensity but also the amount of annual rainfall [32]. The country's territory falls into arid to semi-arid regions where three-fourths of the country receives an annual rainfall of less than 250 mm, therefore rainfall alone is generally insufficient for growing crops, particularly in Baluchistan province having less water availability [33, 34]. During monsoon periods greater than 75% of the rainfall falls, which provides about 30 Bm^3 for irrigation but this is only enough to meet 15% of water requirements by the crops [35].

Figure 6 shows the temporal variation of rainfall patterns in Pakistan. For instance, in 2015 average annual rainfall was observed at about 546 mm while 407 mm in 2016, indicating considerable shifting of rainfall patterns throughout the country. Consequently, less rainwater availability necessitates further improvement of farming and irrigation infrastructure. On the other hand, the shifting of seasons triggers a relatively long monsoon season in which high rainfall may cause floods throughout different regions of the country. Nonavailability of storage facilities like dams and

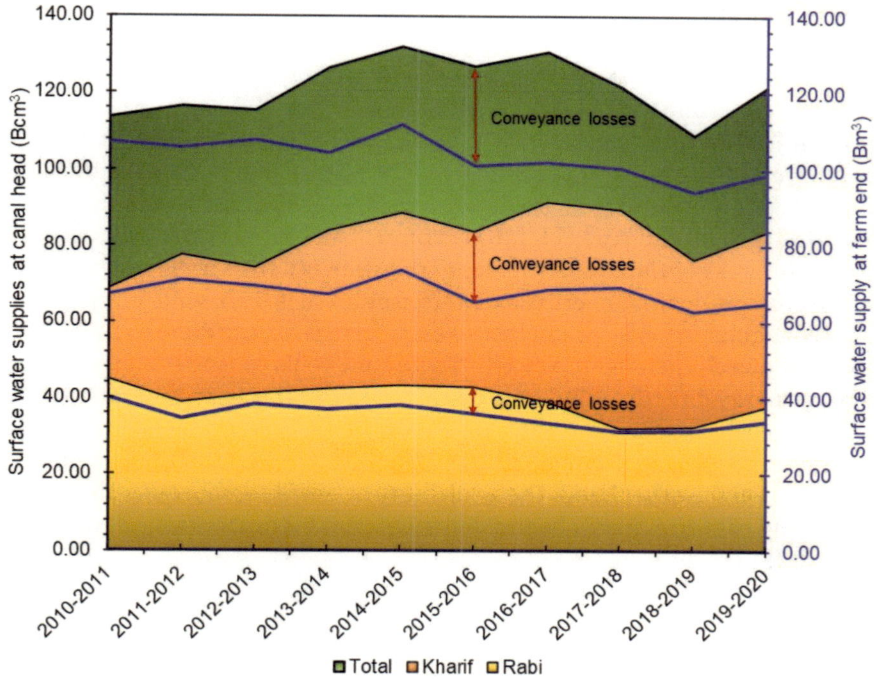

Figure 5.
*Temporal variation of the surface water generated in Pakistan at canal and farm ends reproduces here from the Pakistan Bureau of Statistics (**Figure 11**: Overall water availability).*

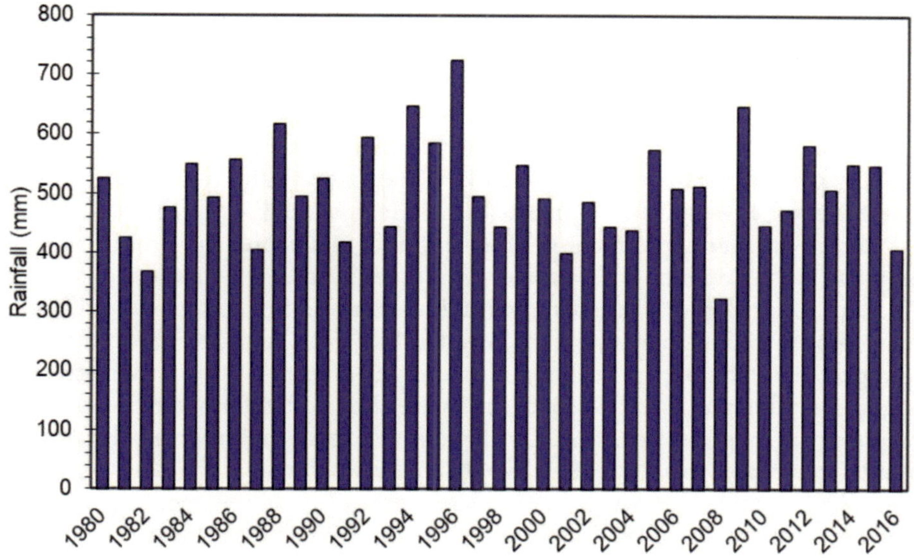

Figure 6.
Rainfall patterns in Pakistan reproduce here from [36].

other water management infrastructures directed a huge amount of water to the sea. In this regard, farmers are being shifted toward the utilization of groundwater in order to fulfill the crop water requirements.

3.3 Groundwater

Groundwater is one of the most essential sources of irrigation in Pakistan. Total groundwater availability in Pakistan is about 73 Bm^3 while 62 Bm^3 is being extracted annually. In Pakistan, agriculture has shifted from surface water to groundwater-fed irrigation over the previous 50 years. The groundwater share of the total irrigation water supply has increased from 8% in 1960 to 60% in 2010 due to the continuous expansion of agricultural lands [37]. Based on arid and semiarid climates that prevail in most of the areas of the country making irrigation is necessary for efficient and long-term crop production because evapotranspiration is high while rainfall is scarce and unpredictable [38]. The groundwater extraction is mostly done by means of electricity or diesel-operated tube wells. These tube wells consume a huge amount of primary energy directly or indirectly to lift groundwater from the water table. In this case, the cost of pumping 1000 m^3 of water from shallow and deep tube wells varies between 5$ and 15$ owing to varying energy prices in different regions of the country [39]. On the other hand, the availability of energy resources is challenging for developing countries like Pakistan. Furthermore, irrigation through groundwater is not possible in remote areas of the country where electricity is unavailable. In this perspective, a sustainable solution like Agrovoltaic is principally required to overcome the energy-nexus dilemma.

4. Energy status of Pakistan

4.1 Fossil fuel reserves

In developing countries, an increase in energy consumption is directly linked with population growth and industrial development. In recent decades, energy demand has risen faster than energy supply or energy production in the country. In Pakistan, total energy consumption exceeded from 139.56 TWh in 1986 to 606.74 TWh in 2020 with a 4.4% annual compound growth rate, and this consumption is predicted to reach 1162.07 TWh by the end of 2030 [40]. The country is highly dependent on non-renewable energy resources that primarily include fossil fuels e.g., coal, oil, natural gas, and peat for energy production. The rising energy demand and limited supply of energy due to the presence of fewer energy reserves have become a major challenge in Pakistan. On the other hand, being an oil importing country about 6.6 million tons of oil worth 3.4 billion dollars were imported in 2019 [41]. Another important aspect that impacts the economy of fuels and ultimately all sectors is the inflation rate. Energy generation and the agriculture sector are affected by the rise in fuel prices. In response to rising fuel prices and the depletion of fossil fuel reserves, the world is turning to renewable energy sources like solar thermal, wind, and biomass/biogas [42, 43]. Unfortunately, Pakistan has not been successful in implementing modern technologies for the use of renewable energy sources particularly for agrarian tasks. However, some renewable energy projects like solar, wind, and biomass have been initiated recently in some areas of the country [44–46].

4.2 Sever electricity shortfall

Pakistan is experiencing severe electricity shortages. Electricity consumption has increased throughout all economic sectors particularly the agriculture sector because

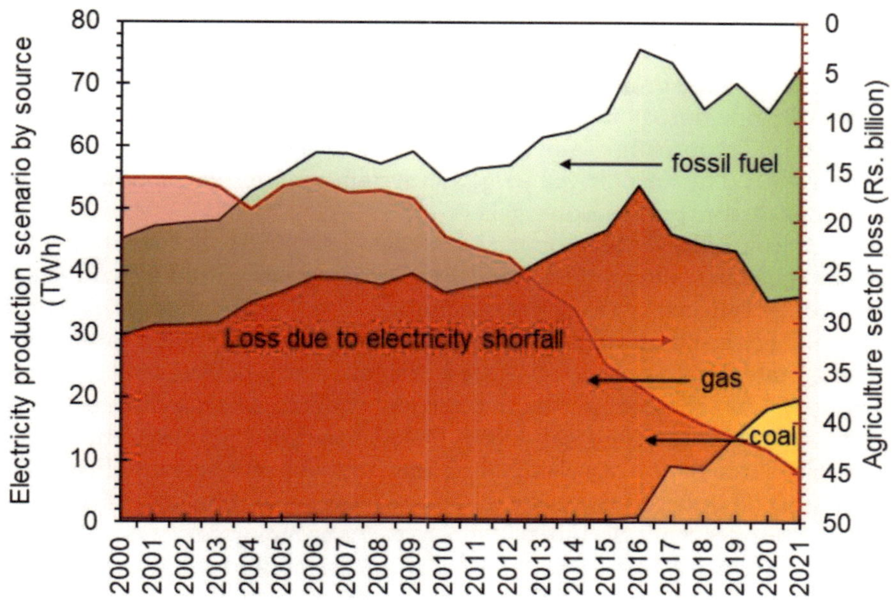

Figure 7.
Electricity production scenario by source and the cumulative loss to agriculture sector due to electricity shortfall reproduce here from [51, 52].

about 1 GW of energy is consumed annually by the irrigation practices only [47, 48]. In 2020, electricity generation was about 22 GW compared to a peak demand of about 25 GW, however, this peak demand is continuously increasing every year, therefore, the whole country is subjected to daily blackouts of about 3–4 h [49]. The electricity shortfall is mainly caused by multiple issues including inadequate power policies, exploitation of indigenous resources, high transmission and distribution systems, and persistent utilization of continuous antiquated thermal power plants that consume fossil fuels [50]. The severe electricity shortfall triggers a reduction in GDP growth of the country by 4%, because of closures of hundreds of manufacturing plants and less agricultural production [47]. **Figure 7** shows the electricity production scenario by different sources (coal, gas, and fossil fuels) that is not enough to meet the peak demand and the cumulative loss to the agriculture sector due to electricity shortfall and this shortfall is increasing every year. For instance, cumulative loss to the agriculture sector in 2020 was observed at Rs. 42.74 billion while in 2021 this loss was observed at about Rs. 45 billion (PKR).

5. Agrovoltaic concept for irrigation and potential in Pakistan

Recently, scientists and renowned engineers introduced a unique concept named "Agrovoltaic" for addressing food insecurity and on-farm energy constraint in tandem. The concept is to harness solar energy by putting photovoltaic (PV) modules into the same agricultural area that is already being cultivated to produce agrarian commodities. The approach promotes sustainable rural development, and the preservation of biodiversity and the ecosystem by forming synergies between renewable energy and agriculture. In addition, the PV modules shielded the crops from harsh weather conditions. The concept might be expanded to develop an intelligent vision-based irrigation system. **Figure 8** shows the generalized conception of AVIS. The AVIS concept mutually resolves the water and energy problems specifically for the countries that have groundwater reserves and solar energy harvesting potentials. The AVIS integrated Internet of Things (IoT) with the solar-driven irrigation system. The IoT comprising of soil sensors, weather sensors, crop sensors, and microcontrollers. The solar-driven irrigation system contains PV modules, AC or DC batteries, motors and pumps, sun trackers, etc.

The microcontroller is the computational hub of the AVIS which computes, planed, and regulates the components of the solar-driven irrigation system. The microcontroller collects data from the different sensors and executes the data processing activities. For instance, the soil sensor captures agricultural field information such as soil type, water holding capacity, and, most significantly, detects available moisture in the ground. The collected data aid in determining of irrigation volume needed in the field. Similarly, meteorological sensors included within the AVIS capture weather data such as solar radiation, daylight hours, ambient temperature, rainfall, etc. The obtained data will be used to evaluate the solar energy generation capacity and to cut off the irrigation system if the rain will forecast. The cropping data was utilized to calculate water demand based on crop water requirements and crop coefficients. The water data contains surface water availability as well as an estimate of the water shortage that can be met by pumping groundwater. Once the prerequisites were computed, the microcontroller equipped with AVIS actuated the solar pump to extract groundwater for a predetermined period of time.

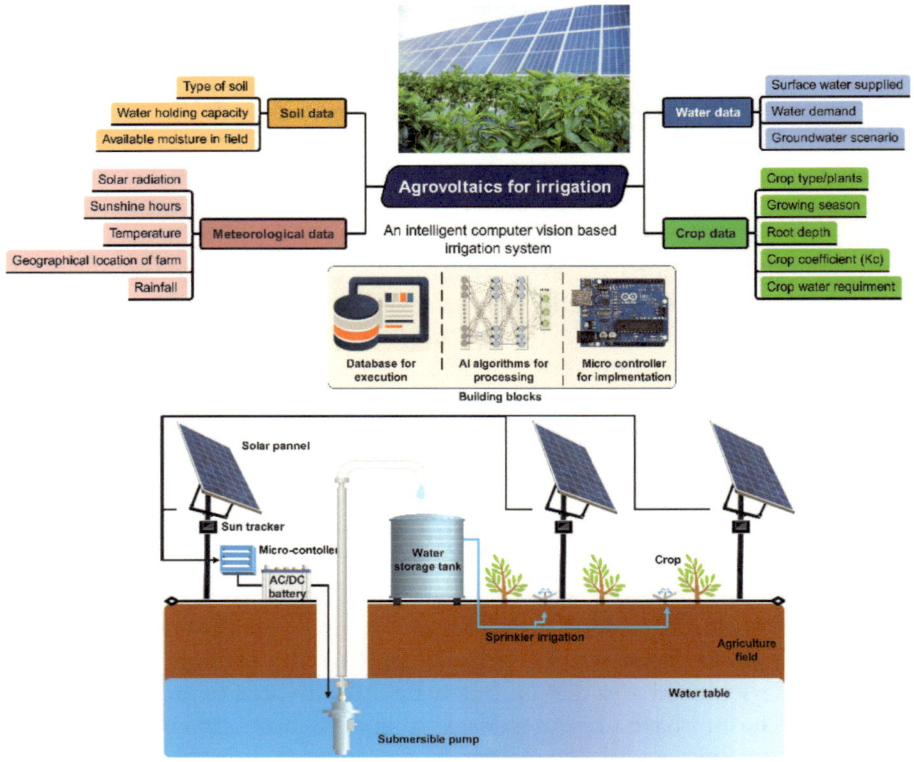

Figure 8.
Agrovoltaic irrigation system (AVIS) conception.

The working principle of the AVIS is based on the PV modules, which convert solar energy into electrical energy. The PV panels are interlinked with the solar motors either AC or DC motors to produce mechanical energy, which is then turned to hydraulic energy by the surface pumps or submersible pumps. The energy supply to the solar pumps will automatically be disconnected once the field requirements are accomplished. During off sunshine hours, AC or DC batteries are integrated into AVIS as a backup energy supply. The ability of a solar pumping system to pump water is determined by three major variables: pressure, flow, and pump power. For design reasons, pressure may be defined as the effort done by a pump to raise a specific amount of water to a storage tank, which is estimated by the elevation head (difference between water source and storage tank). The water pump will need a specific amount of electricity, which must be supplied by a PV array. The irrigation efficiency may be adjusted by using a high-efficiency irrigation system (HEIS) that is powered by PV modules. The benefits of the AVIS are as follows:

- Avoid permanent wilting of the crops due to the nonavailability of the surface water
- Precisely computes the amount of water need to supply from groundwater after subtracting the rainfall water and surface water
- Scavenge pumping energy freely from solar radiations
- Avoid the overexploitation of the groundwater reserves
- Cut off the pressure on the fossil fuels for irrigation activities
- Timely detect the water demand
- Reduction of evapotranspiration and evaporation from the soil
- Minimize the effect of temperature extremes on the crops
- Produce electricity that could be supplied to the national electricity grid
- Eco-friendly, reliable, and durable with longer operating life

Pakistan is geographically positioned in the domain of the sunny belt of the world having long sunshine hours and receiving high solar irradiation, which makes it an ideal locality for solar energy-driven technologies. The daily mean global radiation on the horizontal surface in Pakistan ranges between 1900 and 2200 kWh/m^2, which can potentially generate 1.9–2.3 MWh of energy [53, 54]. The sunshine hours vary between 2000 and 3000 h per year, which reflects massive solar energy harvesting potential. The average solar radiation intensity ranges between 36.05 and 287.36 W/m^2 in the country. Solar radiation intensities of more than 200 W/m^2 were recorded in Sindh from February to October; in practically all parts of Balochistan from March to October; in NWFP, Northern Areas, and Kashmir from April to September; and in Punjab from March to October [55]. During the course of the year, the average solar radiation intensity in Pakistan, namely in the southern parts of Punjab; Sindh; and Balochistan, ranges from 1500 to 2750 W/m^2 day^{-1} for a period of 10 hours every day. In the locations listed

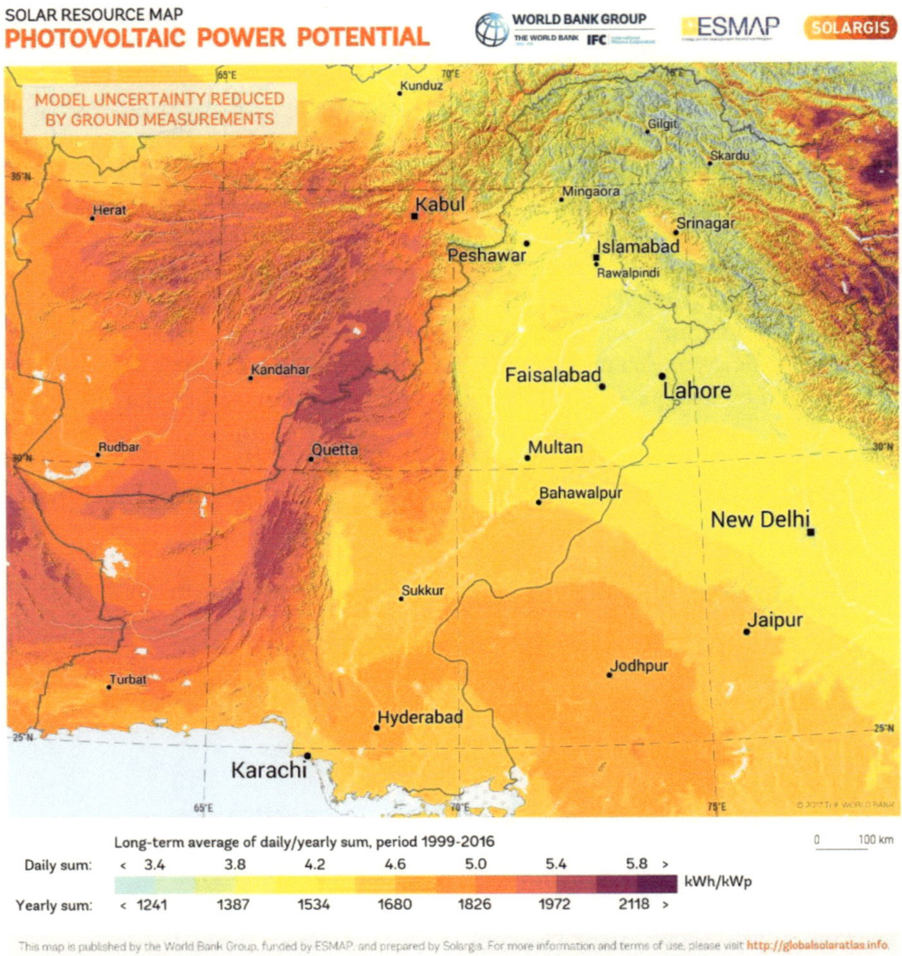

Figure 9.
Photovoltaic power potential in Pakistan is taken from the global solar atlas.

above, it is possible to generate between 45 and 83 megawatts (MW) of power every month through an area of 100 m^2 [55, 56]. **Figure 9** presents a solar resource map of PV power potential, which indicates generating electric power between 3.4 and 5.6 kWh day^{-1}. In this spectrum, the abundant supply of solar energy is promising to supply primary energy to AVIS.

6. Components of Agrovoltaic irrigation system

In this section, the authors of this chapter provide a detailed discussion of the components of the Agrovoltaic irrigation.

6.1 PV cell/generator

The term PV refers to electricity generators consisting of two semiconducting layers principally used in the construction of the PV cells. The negative layer of the

DOI: http://dx.doi.org/10.5772/intechopen.106973

PV cell releases electrons under sunlight. If an external circuit is present, the free electrons move to the positive layer and create an electrical current. So far, the PV cell material science has led toward the 4th generation (4G) PV modules as described in **Figure 10**. The 1G PV cells are thick monocrystalline and multi-crystalline silicon films, which not only leads to highly efficient but also expensive. The efficiency of the thick monocrystalline and multi-crystalline PV cell materials was reported at 26.3% and 21.3%, respectively [58]. The 2G PV cells are synthesized from thin amorphous silicon or polycrystalline Si (silicon), CIGS (copper-indium-gallium-selenium), and CdTe (cadmium telluride) aimed to minimize the cost by employing thin film materials. In comparison to 1G, the performance of the 2G filmed PV cells are inadequate thereby lowering the efficiency. For instance, the thin film chalcogenide such as CdTe has an efficiency range between 19.5% and 21% whereas, in the case of CIGS (minimodule) the efficiency further drops to 18% [59]. In the case of amorphous silicon, the efficiency declined to 10.2% [58]. In this context, a larger surface area will be required to produce electricity equivalent to 1G PV cells. However, the 2G PV cells dramatically reduced the unit cost of electricity generation, which was the key achievement. The 1G and 2G PV modules are highly penetrated in the solar markets having 85% of the market share [60, 61]. The 3G PV modules use nanocrystalline films staked with multilayers of inorganics based on III-V materials such as Gallium arsenide (GaAs), Germanium (Ge), and gallium indium phosphide (GaInP) aimed to improve the efficiency of the system with significantly low production cost [62]. The efficiency of GaAs and GaInP was reported at 37.9% [58]. Despite the acceptable success of 3G cells, major improvements in device performance are necessary if this technology is to compete in terms of cost per watt with prior PV generations. The 4G PV cells introduced the polymers aimed to improve the optoelectronic properties and

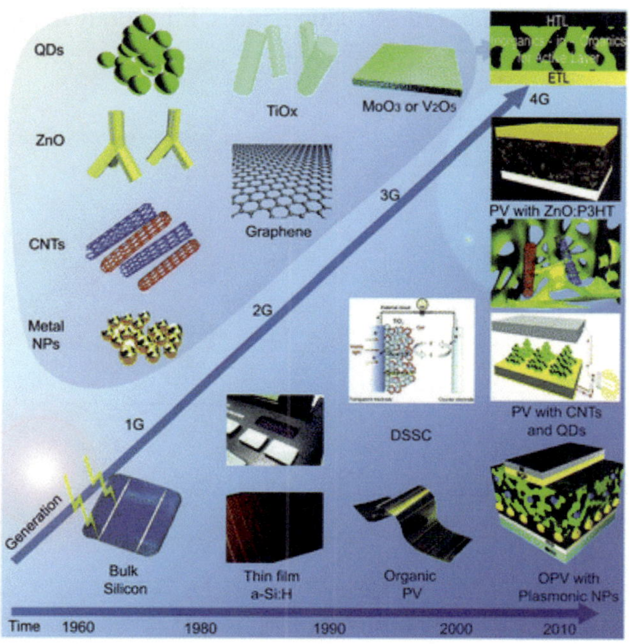

Figure 10.
Timeline of photovoltaic device generations, from 1G to 4G, with nanomaterial components that make up half of 4G devices [57].

low-cost thin PV panels/modules. Solar cells having several p-n junctions constructed of various semiconductor materials are known as multi-junction (MJ), which has been widely investigated in the literature. In reaction to different wavelengths of light, the p-n junction of each material will create an electric current. In this case, the cell efficiency was reported beyond 45%. **Figure 11** presents the temporal improvement of the cell efficiency developed by the National Renewable Energy Laboratory, Golden, Co.

6.2 Cooling and cleaning mechanism

The PV system's output dropped as the panels' temperature climbed over 25°C. Dirt on solar panels might have a negative impact on their performance. In order to reverse this declining trend and improve field performance, the solar panels' surfaces must be cleaned and cooled in some way. A tiny sprinkler mounted in front of the system is used as a cooling mechanism that could improve the system performance by 7–9%.

6.3 Batteries

Batteries are an important component of the AVIS in order to supply the power to the motor during off sunshine hours. Generally, lead-acid batteries and deep cycle batteries of types AC and DC are most frequently used in solar-powered irrigation systems [63]. The lead-acid battery is 80% efficient which reflects in storing 25% more energy in the battery. However, in most studies deep cycle batteries are recommended for solar energy storage due to their prominent attributes such as being discharged to a low energy level, rapid recharged, and no regular maintenance or topping-up required. The capacity of the battery should be sufficient enough to bear the load and smoothly run the connected appliances. If properly maintained, a deep

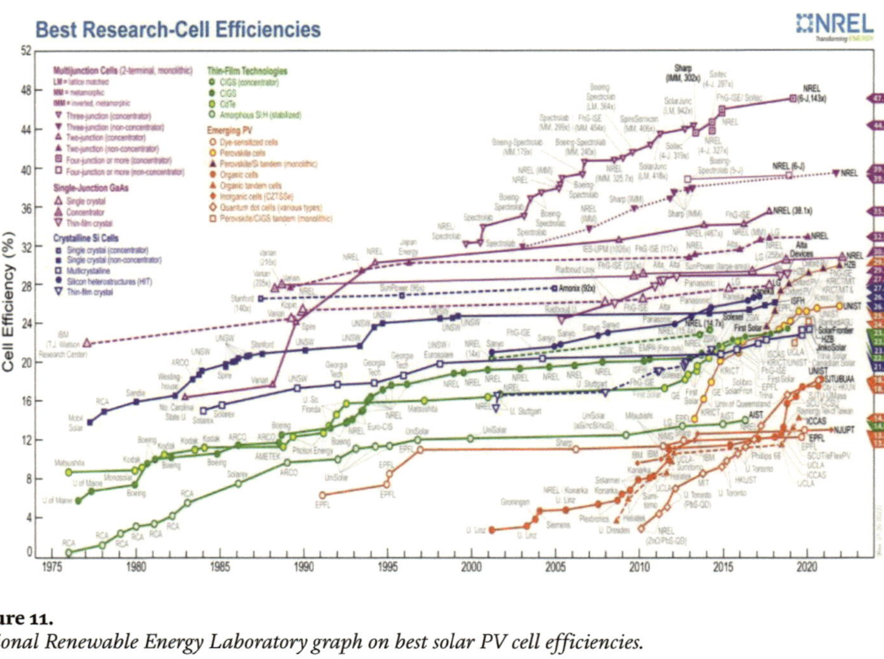

Figure 11.
National Renewable Energy Laboratory graph on best solar PV cell efficiencies.

cycle battery can prolong up to 3–4 years compared to the lead-acid battery, which is limited to 2–4 months.

6.4 Solar motors and pumps

Solar-driven pumps either surface or submersible are the heart of the AVIS. The harvested solar energy is directly being utilized to drive the AC or DC motors, which mainly transform the electrical energy into mechanical energy. The mechanical energy is then utilized to develop hydraulic energy for lifting the groundwater from the deep levels. If compared submersible pumps lift groundwater from more depths as compared to the surface pumps. However, surface pumps or centrifugal pumps are more often adopted due to their special attributes. The selection of the pumping unit is mainly dependent upon the groundwater depth. Higher groundwater depth needs high power pumps and vice versa. Different vendors are available in the market such as Shurflo, Grundfos, Lorentz, Dankoff, SolarJack, etc., [63] which produce different capacities of the solar pumps.

6.5 Sun tracker

The earth revolves around its orbit thereby, the solar radiation incident on the solar collector changes. The sun tracking device is vital in this case since it aims to direct solar energy perpendicularly to the sun during the entire sunshine hours. The device is capable to improvised the solar energy extraction 10−70% [19]. The cited study reveals that the sun tracker generates 57% more, however, increases the installation and maintenance cost of the irrigation system [64]. The sun trackers are classified based on the single axis and dual axis differentiated based on the degree of freedom **Figure 12**. The sun trackers are coupled with photo sensors, which create voltage difference when solar radiation incident on it and accordingly adjust its best orientation. Dual axis sun tracker entails two axes, primary axes and a secondary axis. The primary axis adjusts the solar panel with respect to ground rotation whereas the secondary axis provides tilt movement of the solar panel. However, it has been equipped with installation complexities and also mounts the project cost.

6.6 Micro controller

In order to ensure the smooth operation of the solar pump, it is essential to have both a maximum power point tracking system (MPPT) and a variable frequency inverter (VFI). Various configurations of variable frequency drive (VFD) are being investigated by coupling with and without MPPT. The VFD controller provides square wave output, which results in higher-order harmonics in the output, causing extra losses and pulsating torque in the motor. On the other hand, the maximum power point was not able to identify if the MPPT is not mounted, consequently, the system operates at a fixed DC voltage. Furthermore, the groundwater head found an influential entity for manipulating the average power tracking efficiency. Yadav et al. [67] investigated and compared the impact of sine wave MPPT and VFD on the tracking efficiency corresponding to varying the water head. It was reported that the tracking efficiency ranges between 99.30 and 99.60% when the water head fluctuates from 10 to 20 m. However, in the case of VFD tracking efficiency dropped to 72.20%. In addition, the sine wave MPPT coupled with a low pass filter eliminates the higher-order harmonics.

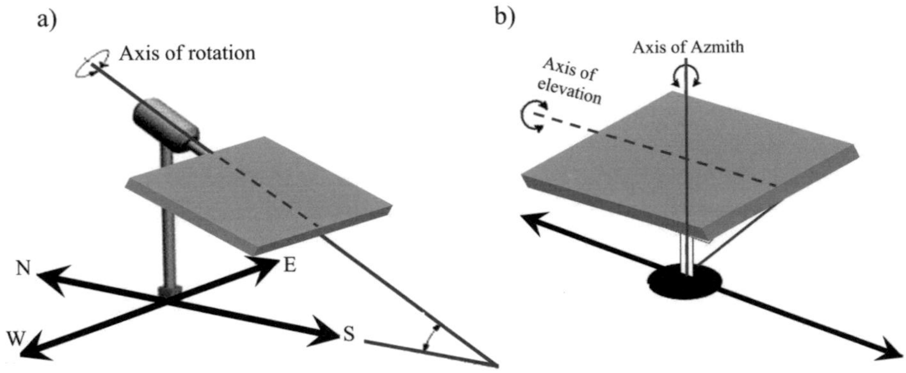

Figure 12.
Single axes (a) and dual axes (b) sun tracker [65, 66].

The MPPT was also found sensitive to environmental variables such as shading and temperature. In this context, various tracking algorithms are being investigated. The incremental conductance method (INC) works on the principle of comparing how voltage and current change. The INC technique, however, has a step size difficulty, particularly in rapidly changing weather conditions. Aside from that, the control system necessitates a costly and intricate circuit and is incapable of dealing with partially shadowed conditions, such as the shadow created by clouds and trees. Recently, it was realized that dual MPPT coupled with INC and dormant particle swarm optimization (DPSO) could handle the partial shading effect and mitigate the voltage fluctuation and spikes [68]. MPPT based on extremum-seeking control has a rapid convergence time and strong steady-state performance because the operating voltage or current of solar PV arrays can be dynamically tuned to maximize output power. Gray Wolf's optimization-based MPPT method was investigated and found efficient in terms of fewer operating parameters, higher efficiency, and outperformed under partially shaded conditions [69]. The artificial neural network (ANN) based MPPT method was also investigated for MPPT [70]. The ANN obtained records of environmental variables from sensors and give the output signal of maximum power point generating conditions. Ramaprabha et al. [71] also investigated the ANN algorithm and found it efficient for all insolation levels. Lin et al. [72] proposed a radial basis function network (RBFN) and an improved Elman neural network (ENN) for MPPT and found them effective. Similarly, another study concludes the ENN implementation due to its stable response and low fluctuation as compared to perturb and observe (P&O). One can find the details relevant to various types of ANN models from the cited articles [73, 74]. However, the variable step size-based ANN MPPT possesses high accuracy, with a quick convergence response time as compared to the step size-based ANN MPPT algorithm [73].

The microcontroller can also be used to determine the optimum irrigation requirements. Rajkumar et al. [75] investigated an automated switching mechanism of the water pump by collecting data from the temperature, humidity, and soil moisture sensors. It was reported that the microcontroller successfully regulates the pumping time. Gao et al. [76] designed a fuzzy irrigation control strategy for real-time estimation of real crop water requirements based on the data received from sensors. Based on experimental results it was realized that the system precisely regulates the solenoid valve, which turns to improve efficiency and proven food security. Normally, the sensor data contains noise that needs the preprocessing of the collected data. In this context, the recursive adaptive filter method was employed to remove the noise, handle the missing values, and normalize the data [77]. Cordeiro et al. [78] developed a deep learning model that was capable to anticipate the soil moisture availability in the agricultural land and addressing the sensor missing data and failure ambiguities. Among the studied algorithms, the k-nearest neighbors (KNN) algorithm bypasses the problems and accurately predicts the irrigation water need. Another study determined the soil moisture using gradient boosting with regression tree (GBRT) and found the best results for smart irrigation planning [79]. Abdulaziz et al. [80] utilized the Lagrange multiplier method in order to optimize the consumption of pumping power and water consumption. The model gives satisfactory results in optimizing the water consumption however, lacked in power utilization. The cited study [77] focused on the preprocessing of irrigation data and estimating plant growth using an adaptive neuro-fuzzy inference system (ANFIS) based on the soil, water level, temperature, and moisture conditions. The ANFIS proved to be optimum due to a precision value of 81% with an accuracy of 84.6% [77].

7. Challenges of agrovoltaic irrigation system

In Pakistan, the alternative energy development board (AEDB) has been actively involved in developing and promoting renewable energy technologies because of the massive resource potential across the country. For solar-driven technologies, the region is envisaged as highly feasible. In this spectrum, various megaprojects are being executed with collaborative efforts. One can find the details of these projects on the official website of the AEDB. In 2015, the federal government of Pakistan takes initiative to subsidize the farmers having cultivated areas greater than 5 hectares for the installation of solar-driven irrigation pumps under the project entitles solar irrigation for agricultural resilience (SoLAR). The project's worth was USD 93.2 million, providing an 80% subsidy for installing 30,000 solar irrigation pumps. However, there no such schemes/policies are being designed for AVIS despite its enormous potential. The challenges could be in the form of economic barriers, nonavailability of technical and skilled personnel, and technological barriers. In this section, the authors analyzed the potential challenges that need to ponder in order to develop successful AVIS.

7.1 Lack of awareness

Agrovoltaic itself is an emerging conception so far investigated in developed countries and possesses plenty of room for technical improvements. Lack of awareness regarding the new energy-water efficient AVIS is the potential bottleneck [9]. The farmers are not informed of relevant cost/economic, environmental benefits, and the potential barriers that need to be encountered for the concept implementation. In addition, the farmers hesitated to invest in emerging technologies due to their limited resources. The leading researchers need to execute the small projects on a university scale in order to develop local knowledge relevant to technology. After that, dissemination activities are performed to spread awareness among the farmers. It is imperative that there be a centralized information hub with a single point of contact for quick and simple access to AVIS data that is made available to all farmers in each and every district.

7.2 Lack of skilled workforce

The shortage of qualified and experienced installation workers, project managers, and engineers in both developing and developed countries is a significant barrier to the growth of AVIS. Likewise, In Pakistan, lack of qualified and semiskilled personnel has hampered the widespread use of the AVIS. This leads to inadequate support infrastructure, maintenance, and after-sales services, which significantly impacts farmers' opinions on AVIS adaptability. In addition, the farmers are not technically trained to operate and understand the complex operating mechanism of the AVIS. In this framework, training facilities that give technical expertise to farmers must be established in each area. Furthermore, short-term courses and vocational training courses must be established in order to develop a job-ready trained workforce.

7.3 Risk of declining groundwater

Recent studies reveal that groundwater is declining at a higher pace in the Indus plains of Pakistan due to climate change, low surface water supplies, installation of large-scale tube wells, over-exploitation, and lack of groundwater governance policies [29, 81, 82]. Although, farmers' access to groundwater benefited them to ensure food security and accomplished agriculture requirements. However, many farmers pump groundwater beyond their agriculture requirements thereby massive pressure, which leads to a declining groundwater table. For instance, Shakoor et al. [83] reveal that if the historical trend of groundwater pumping continues, the water table in Punjab might drop up to 18 m by 2030. Similarly in Balochistan, the groundwater is depleted at the rate of 2–3 m annually [82]. The lowering groundwater table is required to install high-capacity pumps due to depreciation of pumping efficiency beyond the depth of 500 cm [84]. In addition, if standalone solar-driven irrigation pumping systems are installed then there is fear of a more rapid drawdown of groundwater due to the assumption of being increasing pumping by the farmers. However, the Pakistan Council of Research in Water Resources (PCRWR) analyzed that actual solar energy is not capable to exploit the groundwater reserves through solar pumping, irrespective of considering the influential factors such as installation capacity and farmer behavior thereby not widely accepted [21]. The AVIS could be a potential solution to the stated problem in a manner of harvesting free solar energy and pumping the groundwater according to crop water requirement. The AVIS motor switches off the irrigation system once the field requirements are fulfilled.

8. Conclusions

Energizing the agro-food supply chain and access to sufficient water have been essential aspects that bring prosperity and stability to the agriculture sector. However, currently, Pakistan's agriculture sector is profoundly suffering from both energy and water crisis. The major culprits are the exponential growth of population, limited freshwater and primary energy reservoirs, electricity shortfalls, and dried agricultural lands, leading to food insecurity in the country. In 2021, the agriculture sector accounts for 45 billion PKR loss due to power shortages, which directly influence cultivation and irrigation activities. The groundwater and solar energy are alternative reliable solutions to accommodate water and energy shortages, respectively. In this framework, solar pumps are installed across the country however they exhausted groundwater resources in several regions of Pakistan due to inadequate irrigation technologies, irresponsive behavior of farmers, and lack of groundwater governance policies. Agrovoltaic irrigation system (AVIS) could be a remarkable solution that controls the overexploitation of groundwater by utilizing non-payable energy. The present study aimed to explore the prospects and challenges that need to encounter for local implementation of the AVIS. In addition, the study discusses and reviewed the components of the AVIS. Because of the massive solar energy harvesting potential in the country, a significant amount of solar radiation transforms into electrical energy via photovoltaic modules if installed on the land that is cultivated with the crop. The potential benefits of the AVIS include the utilization of the same land for twin benefits (i.e., energy and crop production), metering the groundwater extraction, saving crops from extreme weather conditions, and powering irrigation systems free of cost. However, the impediments such as complex operating mechanisms required skilled/technical personnel for installation, sophisticated maintenance, absence of sufficient ground knowledge, and high capital investment. In order to ensure food security and agriculture prosperity, the government takes the necessary initiatives and empowers the AVIS adaptation by involving leading researchers and engineers mutually involve in research projects that strengthen the AVIS knowledge and conduct dissemination activities and workshops to train and attract the farmers.

Acknowledgements

This work is carried out in the Department of Agricultural Engineering, Bahauddin Zakariya University, Multan-Pakistan. The work is financially supported by the Director Research/ORIC grants of Bahauddin Zakariya University (Multan), awarded to Principal Investigator Dr. Muhammad Sultan.

Conflict of interest

The authors declare no conflict of interest.

Author details

Hafiz M. Asfahan[1†], Muhammad Sultan[1*†], Fiaz Ahmad[1], Faizan Majeed[1,2], Md Shamim Ahamed[3], Marjan Aziz[4], Redmond R. Shamshiri[5], Uzair Sajjad[6], Muhammad Usman Khan[7] and Muhammad Farooq[8]

1 Department of Agricultural Engineering, Bahauddin Zakariya University, Multan, Pakistan

2 Department of Agricultural and Biosystems Engineering, University of Kassel, Witzenhausen, Germany

3 Department of Biological and Agricultural Engineering, University of California, Davis, USA

4 Department of Agricultural Engineering, Barani Agricultural Research Institute, Chakwal, Pakistan

5 Department of Engineering for Crop Production, Leibniz Institute for Agricultural Engineering and Bioeconomy, Potsdam, Germany

6 Department of Energy and Refrigerating Air-Conditioning Engineering, National Taipei University of Technology, Taipei, Taiwan

7 Department of Energy Systems Engineering, Faculty of Agricultural Engineering and Technology, University of Agriculture Faisalabad, Faisalabad, Pakistan

8 Department of Mechanical Engineering, University of Engineering and Technology, Lahore, Pakistan

*Address all correspondence to: muhammadsultan@bzu.edu.pk

† These authors contributed equally to this work.

References

[1] UNESCO. UN-Water 2020. The United Nations World Water Development Report 2020: Water and Climate Change. Paris: UNESCO; 2020

[2] United Nations. Department of Economic and Social Affairs. News 2019. Available from: https://www.un.org/development/desa/en/news/population/world-population-prospects-2019.html. [Accessed: June 01, 2022]

[3] Islam SMF, Karim Z. World's demand for food and water: The consequences of climate change. Desalination-challenges Opportunity. 2019:1-27

[4] Renewable I, Agency E, Fao AO. Renewable energy for agri-food systems. 2021

[5] Connor R, Koncagül E. The United Nations World Water Development Report 2014: Water and Energy. 2014. Epub ahead of print 2014. DOI: 10.46632/rmc/1

[6] Hussain G, Aleem M, Sultan M, et al. Evaluating evaporative cooling assisted solid desiccant dehumidification system for agricultural storage application. Sustainability;**14**:1479

[7] Abdel-Fatah MMA. Water Treatment and Desalination. Rijeka: IntechOpen

[8] Ali ES, Asfahan HM, Sultan M, et al. A novel ejectors integration with two-stages adsorption desalination: Away to scavenge the ambient energy. Sustain Energy Technol Assessments. 2021;**48**:101658

[9] Rahman MM, Khan I, Field DL, et al. Powering agriculture: Present status, future potential, and challenges of renewable energy applications. Renewable Energy. 2022;**188**:731-749

[10] Ritchie H, Roser M. Water Use and Stress, Our World in Data

[11] Status G, Needs AA. Progress on change in water-use efficiency. 2021. Epub ahead of print 2021. DOI: 10.4060/cb6413en

[12] Tariq GH, Ashraf M, Hasnain US. Solar Technology in Agriculture. United States: IntechOpen; 2021

[13] Patterson J. Exploitation of unconventional fossil fuels: Enhanced greenhouse gas emissions. Greenhouse Gases. 2012:147-168. DOI: 10.5772/31975

[14] Pakistan at a Glance. Unitied Nations: FAO; 2022. p. 1

[15] Ministry of Finance. Pakistan economic survey

[16] Ali A, Bahadur Rahut D, Behera B. Factors influencing farmers' adoption of energy-based water pumps and impacts on crop productivity and household income in Pakistan. Renewable and Sustainable Energy Reviews. 2016;**54**:48-57

[17] Khan I, Hasan M. Water Shortage Worse than Expected. Pakistan: IRSA; 2022

[18] Caldera U, Sadiqa A, Gulagi A, et al. Irrigation efficiency and renewable energy powered desalination as key components of Pakistan's water management strategy. Smart Energy. 2021;**4**:100052

[19] Islam MR, Sarker PC, Ghosh SK. Prospect and advancement of solar irrigation in Bangladesh: A review. Renewable and Sustainable Energy Reviews. 2017;**77**:406-422

[20] Qureshi RH, Ashraf M. Water Security Issues of Agriculture in Pakistan. Islamabad: Pakistan Academy of Sciences; 2019

[21] Shah MAA, Akbar MZ. Solar Irrigation in Pakistan a Situation Analysis. International Water Management Institute; 2021

[22] Ayaz J. Potential of solar irrigation water pumps in Pakistan. In: Intersolar Conference. Munich, Germany; 2015

[23] Meah K, Ula S, Barrett S. Solar photovoltaic water pumping—Opportunities and challenges. Renewable and Sustainable Energy Reviews. 2008;**12**:1162-1175

[24] Aziz M, Rizvi SA, Sultan M, et al. Simulating cotton growth and productivity using AquaCrop model under deficit irrigation in a semi-arid climate. Agriculture. 2022;**12**:242

[25] Mujtaba A, Nabi G, Masood M, et al. Impact of cropping pattern and climatic parameters in lower Chenab Canal system—Case study from Punjab Pakistan. Agriculture. 2022;**12**:708

[26] Crop Reporting Service. Directorate of Agriculture, Crop Reporting Service, Punjab. 2022. http://www.crs.agripunjab.gov.pk/reports [Accessed: June 01, 2022]

[27] Pakistan's mango production. Arab news. 2022

[28] Shah MAA, Anwar AA, Bell AR, et al. Equity in a tertiary canal of the Indus Basin irrigation system (IBIS). Agricultural Water Management. 2016;**178**:201-214

[29] Imran MA, Xu J, Sultan M, et al. Free discharge of subsurface drainage effluent: An alternate design of the surface drain system in Pakistan. Sustainability. 2021;**13**:4080

[30] Tariq MAUR, van de Giesen N, Janjua S, et al. An engineering perspective of water sharing issues in Pakistan. Water. 2020;**12**:477

[31] Singh H, Mishra D, Nahar NM. Energy use pattern in production agriculture of a typical village in arid zone, India—Part I. Energy Conversion and Management. 2002;**43**:2275-2286

[32] Syed A, Raza T, Bhatti TT, et al. Climate impacts on the agricultural sector of Pakistan: Risks and solutions. Environmental Challenges. 2022;**6**:100433

[33] Abubakar SM. Pakistan 7th most vulnerable country to climate change, says Germanwatch. Daw Novemb; 9

[34] Aziz M, Khan M, Anjum N, et al. Scientific irrigation scheduling for sustainable production in olive groves. Agriculture. 2022;**12**:564

[35] Bhatti MA, Akhtar MJU. Increasing irrigated agriculture productivity for poverty reduction in Pakistan. In: Proceedings of the Second South Asia Water Forum. Islamabad, Pakistan; 2002. pp. 14-16

[36] Safdar F, Khokhar MF, Arshad M, et al. Climate change indicators and spatiotemporal shift in monsoon patterns in Pakistan. Advances in Meteorology. 2019;**2019**:1-14

[37] Qureshi AS, Gill MA, Sarwar A. Sustainable groundwater management in Pakistan: Challenges and opportunities. Irrigation and Drainage. 2010;**59**:107-116

[38] Khan HF, Yang YCE, Ringler C, et al. Guiding groundwater policy in the Indus Basin of Pakistan using a physically based groundwater model. Journal of Water Resources Planning and Management. 2017;**143**:5016014

[39] Qureshi A. Groundwater governance in Pakistan: From colossal development to neglected management. Water (Switzerland). 2020;**12**:1-20

[40] Go P. Integrated energy planning for sustainable development. Pakistan Energy Demand Forecast (2021-2030). 2021;**1**:1-35

[41] Yaqoob H, Teoh YH, Goraya TS, et al. Energy evaluation and environmental impact assessment of transportation fuels in Pakistan. Case Studies in Chemical and Environmental Engineering. 2021;**3**:100081

[42] Al-Juboori O, Sher F, Khalid U, et al. Electrochemical production of sustainable hydrocarbon fuels from CO_2 Co-electrolysis in eutectic molten melts. ACS Sustainable Chemistry & Engineering. 2020;**8**:12877-12890

[43] Razzaq L, Farooq M, Mujtaba MA, et al. Modeling viscosity and density of ethanol-diesel-biodiesel ternary blends for sustainable environment. Sustainability. 2020;**12**

[44] Rafique MM, Bahaidarah HMS. Thermo-economic and environmental feasibility of a solar power plant as a renewable and green source of electrification. International Journal of Green Energy. 2019;**16**:1577-1590

[45] Yaqoob H, Heng Teoh Y, Ahmad Jamil M, et al. Feasibility study of a 50 MW wind farm project in Pakistan. Journal of Advanced Research in Fluid Mechanics and Thermal Sciences. 2021;**74**:27-42

[46] Khan NA, el Dessouky H. Prospect of biodiesel in Pakistan. Renewable and Sustainable Energy Reviews. 2009;**13**:1576-1583

[47] Tao J, Waqas M, Ali M, et al. Pakistan's electrical energy crises, a way forward towards 50% of sustain clean and green electricity generation. Energy Strategetic Review. 2022;**40**:100813

[48] Mirza SJ. Solar energy applications viable in Pakistan. Alternative Energy Development Board. Available from: www.aedb.org/ae-technologies/solarpower/applications

[49] National Electricity Power Regularity Authority. State of Industry Report 2020. 2020

[50] Das VG, Uqaili MA, Memon HR, et al. Overcoming electricity crisis in Pakistan: A review of sustainable electricity options. Renewable and Sustainable Energy Reviews. 2017;**72**:734-745

[51] Ritchie H, Roser M. Fossil fuels. Our world data; 2

[52] Shahbaz M. Costs of widening electricity gap. Dawn: The Business and Finance Weekly. 2019. Available from: www.dawn.com/news/1481860 [Accessed: May 2019]

[53] Ilyas SZ. Review of the renewable energy status and prospects in Pakistan. International Journal of Smart Grid;**5**:167-173

[54] Muhammad F, Raza MW, Khan S, et al. Different solar potential co-ordinates of Pakistan. Innovation Energy Research; **6**:173

[55] Adnan S, Hayat Khan A, Haider S, et al. Solar energy potential in Pakistan. Journal of Renewable and Sustainable Energy. 2012;**4**:32701

[56] Arif MN, Waqas A, Butt FA, et al. Techno-economic assessment of solar water heating systems for sustainable tourism in northern Pakistan. Alexandria Engineering Journal. 2022;**61**:5485-5499

[57] Jayawardena KDGI, Rozanski LJ, Mills CA, et al. 'Inorganics-in-organics': Recent developments and outlook for 4G polymer solar cells. Nanoscale. 2013;**5**:8411-8427

[58] Wazed SM, Hughes BR, O'Connor D, et al. A review of sustainable solar irrigation systems for sub-Saharan Africa. Renewable and Sustainable Energy Reviews. 2018;**81**:1206-1225

[59] Green MA. Radiative efficiency of state-of-the-art photovoltaic cells. Progress in Photovoltaics: Research and Applications. 2012;**20**:472-476

[60] Association SEI. Solar Market Insight Report 2012 Q3. SEIA org. 2012

[61] Crago CL, Koegler E. Drivers of growth in commercial-scale solar PV capacity. Energy Policy. 2018;**120**:481-491

[62] Green MA. Third generation photovoltaics

[63] Saleem M, Ashraf M, Fripp J. Solar-Powered Pumping Systems for High Efficiency Irrigation: Design Manual for Practitioners. A Jt Rep Publ by PARC, PCRWR ICARDA-Pakistan under USDA Proj "Watershed Rehabil Irrig Improv Pakistan; 42

[64] Mousazadeh H, Keyhani A, Javadi A, et al. A review of principle and sun-tracking methods for maximizing solar systems output. Renewable and Sustainable Energy Reviews. 2009;**13**:1800-1818

[65] Alata M, Al-Nimr MA, Qaroush Y. Developing a multipurpose sun tracking system using fuzzy control. Energy Conversion and Management. 2005;**46**:1229-1245

[66] Wu C-H, Wang H-C, Chang H-Y. Dual-axis solar tracker with satellite compass and inclinometer for automatic positioning and tracking. Energy for Sustainable Development. 2022;**66**:308-318

[67] Yadav K, Sastry OS, Wandhare R, et al. Performance comparison of controllers for solar PV water pumping applications. Solar Energy. 2015;**119**:195-202

[68] Başoğlu ME, Çakır B. Comparisons of MPPT performances of isolated and non-isolated DC–DC converters by using a new approach. Renewable and Sustainable Energy Reviews. 2016;**60**:1100-1113

[69] Brunton SL, Rowley CW, Kulkarni SR, et al. Maximum power point tracking for photovoltaic optimization using ripple-based extremum seeking control. IEEE Transactions on Power Electronics. 2010;**25**:2531-2540

[70] Hiyama T, Kitabayashi K. Neural network based estimation of maximum power generation from PV module using environmental information. IEEE Transactions on Energy Conversion. 1997;**12**:241-247

[71] Ramaprabha R, Mathur BL, Sharanya M. Solar array modeling and simulation of MPPT using neural network. In: 2009 International Conference on Control, Automation, Communication and Energy Conservation. IEEE; 2009. pp. 1-5

[72] Lin W-M, Hong C-M, Chen C-H. Neural-network-based MPPT control of a stand-alone hybrid power generation system. IEEE Transactions on Power Electronics. 2011;**26**:3571-3581

[73] Messalti S, Harrag A, Loukriz A. A new variable step size neural networks

MPPT controller: Review, simulation and hardware implementation. Renewable and Sustainable Energy Reviews. 2017;**68**:221-233

[74] Dangi P, Gawre SK, Ojha A. Dynamic performance analysis of neural network based MPPT under varying climatic condition. In: Control Applications in Modern Power Systems. Springer; 2022. pp. 163-176

[75] Rajkumar MN, Abinaya S, Kumar VV. Intelligent irrigation system—An IOT based approach. In: 2017 International Conference on Innovations in Green Energy and Healthcare Technologies (IGEHT). IEEE; 2017. pp. 1-5

[76] Gao L, Zhang M, Chen G. An intelligent irrigation system based on wireless sensor network and fuzzy control. Journal of Networks. 2013;**8**:1080

[77] Vivekanandhan V, Sakthivel S, Manikandan M. Adaptive neuro fuzzy inference system to enhance the classification performance in smart irrigation system. Computational Intelligence. 2022;**38**:308-322

[78] Cordeiro M, Markert C, Araújo SS, et al. Towards smart farming: Fog-enabled intelligent irrigation system using deep neural networks. Future Generation Computer Systems. 2022;**129**:115-124

[79] Campos GS, Rocha AR, Gondim R, et al. Smart & green: An internet-of-things framework for smart irrigation. Sensors. 2019;**20**:190

[80] Abdulaziz WB, Croock MS. Water and power optimized allocation for smart irrigation system. International Journal of Electrical Engineering Informatics. 2022;**14**:161-170

[81] Watto MA, Mugera AW. Groundwater depletion in the Indus Plains of Pakistan: Imperatives, repercussions and management issues. International Journal of River Basin Management. 2016;**14**:447-458

[82] van Steenbergen F, Kaisarani AB, Khan NU, et al. A case of groundwater depletion in Balochistan, Pakistan: Enter into the void. Journal of Hydrological Regional Studies. 2015;**4**:36-47

[83] Shakoor A, Arshad M, Ahmad R, et al. Development of groundwater flow model (MODFLOW) to simulate the escalating groundwater pumping in the Punjab, Pakistan. Pakistan Journal of Agricultural Sciences; **55**:629-638

[84] Qureshi AS. Groundwater governance in Pakistan: From colossal development to neglected management. Water. 2020;**12**:3017

Chapter 2

Sustainable Irrigation Management for Higher Yield

Fahd Rasul, Hassan Munir, Aftab Wajid, Muhammad Safdar, M. Salman Ayub, Sobia Shahzad, Rehan Mehmood, M. Adnan Shahid, Abid Sarwar, M. Danish Majeed, Umair Gull, Wajid Nasim Jatoi, Muhammad Mubeen, Summera Jahan and Shakeel Ahmed

Abstract

Sustainable irrigation is sensible application of watering to plants in agriculture, landscapes that aids in meeting current survival and welfare needs. Sustainable irrigation management can help with climate change adaptation, labor, energy savings, and the production of higher-value and yield of crops to achieve zero hunger in water-scarce world. To ensure equal access to water and environmental sustainability, investments in expanded and enhanced irrigation must be matched by improvements in water governance. Sustainable irrigation must be able to cope with water scarcity, and be resilient to other resource scarcities throughout time in context of energy and finance. The themes and SDGs related to clean water, water resources sustainability, sustainable water usage, agricultural and rural development are all intertwined in the concept of "sustainable irrigation for higher yield." Sustainable irrigation management refers to the capability of using water in optimum quantity and quality on a local, regional, national, and global scale to meet the needs of humans and agro-ecosystems at present and in the future to sustain life, protect humans and biodiversity from natural and human-caused disasters which threaten life to exist. Resultantly higher yields will ensure food security.

Keywords: sustainable irrigation, higher yield, sustainable agriculture, modern irrigation, crop water requirements

1. Introduction

Agriculture has a crucial part in human water resource use [1]. Approximately 70% out of the total available freshwater consumption is applied for water to support the agricultural output of the world [2]. About 18% of worldwide farmland yet produces almost 40% of food [3, 4]. Water resources have an impact on the productivity of a variety of anthropogenic activities that sustain livelihood [5, 6]. Increased demand for agricultural supplies has put a strain on the world's freshwater resources in recent decades, leading to their unsustainable use in many cases. With over one-fourth of the world's land area experiencing acute water shortage, [7] approximately 35% of people globally live in and around water-deficient places [8], Overexploitation of water resources usually occurs at the price of economic progress, resulting in environmental damage [9]. Thus, water is a necessity of almost all processes of production and means of production which represent the life-sustaining liquid as fuel for production systems. Crop plants cannot survive without water and limited supply prove havoc on the production levels. Thus, a sustained water supply is imperative to guarantee a higher yield of crops along with other climatic, edaphic, and genetic factors.

The water requirement for different crops, rainfall frequency, intensity, and effectiveness along with moisture regimes of soil is showing the irrigation requirement of crops.

$$\mathrm{IR} = \mathrm{WR} - (\mathrm{ER} + \mathrm{S})$$

Where IR stands for irrigation requirement, WR is water requirement, ER exhibits effective rainfall and S is soil moisture contribution. The factors affecting education requirement are given as single crop irrigation need, area of the crop, and farm level distribution of water losses, all the factors are expressed in cm/ha or cm, mm.

1.1 Net irrigation requirement

It is defined as the quantity of water required in the form of depth to bring the soil moisture to its field capacity level for the evapotranspiration demand of the crops. It is also defined as the differentiation between the field ability and moisture content of the soil before irrigation (**Table 1**).

Crops	Requirement	Production	Shortfall
Food-grains	50.0	31.5	18.5
Sugarcane	82.0	46.4	35.4
Cotton (lint)	3.5	2.7	0.8
Pulses	1.9	1.4	0.5
Oilseed	3.3	1.5	1.8
Vegetables	14.3	9.0	5.3
Fruits	16.1	9.0	7.1
Total	**171.0**	**102.8**	**69.4**

Source: Ref. [10].

Table 1.
Crops production, requirement and shortfall analysis for yield gap mitigation through sustainable water resource use in Pakistan.

1.2 Gross irrigation requirement

Gross irrigation need is the term used to describe the overall amount of water used for irrigation. Net irrigation requirements, water application losses, and other losses are included. The approximate losses at different phases of crop development can be considered to find the gross irrigation need for farms.

1.3 Irrigation frequency

The time interval between the two successive irrigations' during crop periods is known as irrigation frequency. It is showing the total number of dry days between irrigations during dry throughout the crop period. It is based on the pace at which plants absorb water, the field capacity of the soil, and the soil moisture present in the root zone. As a result, it depends on the crop, soil, and environment. In general, irrigation should be applied when the effective root zone, where most of the roots are concentrated, is about 50% and not more than 60% depleted of the available moisture. The interval (days) between two irrigations at the time of maximum crop growth, or peak crop consumption, is the irrigation frequency to be employed when constructing irrigation systems.

1.4 Irrigation period

The number of days that can be allowed for applying one irrigation to a specific design area during the crop's peak consumption time is known as the irrigation period.

1.5 Growth duration

The time it takes for various crops to grow varies greatly. Seasonal crops are like sorghum, maize, groundnuts, pulses, etc. that can only grow for one growing season. Crops like cotton, red gram, chilies, etc., whose growth duration spans two seasons, are called two seasonal crops.

1.6 Critical phenological stages of cereals sensitive to moisture variation

Germination stage—is the emergence of radicals from seed.

Tillering—the division for differentiation and development of tillers.

Shoot elongation—the phonological stage standing for internodal expansion.

Booting—swelling and development of grain holding structure or peduncle.

Heading/inflorescence initiation—ear head emergence from the leaf sheath.

Flowering—the appearance of flowers.

Grain development—grain formation from fertilization to maturity which is further classified.

into Milky stage—milk-type fluid development.

Dough stage—dough raw material development.

Ripe stage—fully mature embryo just before shattering.

2. Irrigation type for crop type

All type of irrigation does not suit all crops, the diversity of crops responds differently to applied irrigation and all crops have varied requirements of water which also varies with the soil types and weather. Cereals generally have fewer requirements for water, but it does not fit with Rice and maize which require a higher amount of water. The delta of water of the same stature crop may also vary as in the case of millet, sorghum, and maize which are comparable in terms of the duration of crop, but their delta of water is quite different. Moreover, the varieties of the crop also have varying levels of gene expression for drought and flooding tolerance in the two extreme conditions.

The pulses group mainly including the beans require comparatively less water and are mainly short duration in semi-arid regions with plenty of sunshine justified adaptation to drought is clear and the water productivity of beans is higher compared to other crops of similar duration, especially if compared with vegetables which require much more water compared to beans. Rice and sugarcane are highly hydrophilic and are major consumers of water among all field crops. Thus, the need arises to link the irrigation type with crop type and then through the study of crop type can fine-tune the water requirements for phonological crop stages which could be critical to be influenced severely by moderate drought even.

Irrigation types vary and are mostly dependent on technological development and resource availability and influenced by topography as well. High-efficiency irrigation systems are more technical and water-saving solutions that have been finding their way into modern agriculture and transforming the irrigation system of the world. Only the places with higher rainfall do not require such technical intervention a in case of south China s province here rainfall throughout the year I sufficient to manage the whole year's crop of sugarcane and double cropping of rice as well. Desertification on the other hand is putting pressure on arable land for low and marginal productivity of crops. Modern water-saving technologies like drip irrigation, sprinkler, and center pivot systems are proving essential for drylands around the world and bringing barren land into cultivation thus bringing a contributive role in the planet's food security [11].

3. Water scarcity, climate change, and crop yield uncertainties

Several uncertainties exist for sustainable crop yields in changing climate and in water-scarce regimes. In this study, the climate change impact on Pakistan's irrigation resources and food shortage are examined. According to the report, Pakistan's economy (21%) is built primarily on agriculture; nevertheless, the nation is struggling with crop food shortages, high inflation, and a lack of irrigation water. According to the study, even though Pakistan ranks 135th in the world in terms of per capita GHG emissions with 309 M tons of CO_2 equivalent emissions, which is only 0.8 percent of global emissions, the country is severely hit by changing climate compared to other nations because of trans-border emissions of greenhouse gases. As a result, there has been a continuous rise in temperature of 0.76°C in the nation as a whole and 1.5°C in Pakistan's mountainous regions, which are home to over 5000 glaciers in the KHH mountains. According to the study, Pakistan had a progressive decline in surface water availability per person from 5260 m^3 annually in 1951 to just 1000–1066 m^3 in 2008. Because the country, with its huge infrastructure and network of the largest integrated irrigation system globally, lacks adequate water reservoirs to hold extra water, the glaciers are melting and causing extreme events like floods as a result of rising temperatures, particularly in glacier-covered regions, and associated variations in precipitation pattern, Therefore, because glaciers tend to retreat quickly, a large amount of water is wasted into the ocean along with harm to crops, food reserves, billions of dollars worth of cattle, infrastructure, and land resources. In Pakistan, 84 out of 137 districts (primarily in Khyber Pakhtunkhwa, FATA, and Baluchistan) are said to be undersupplied with both crop- and animal-based food, or around 61 percent of the total. The main cause of the current food deficit is the low crop output brought on by the temperature increase and the lack of irrigation water. According to the study, farmers are switching from cultivating water-loving crops including rice, wheat, cotton, and sugarcane to low-water-requirement crops and vegetables. Moreover, the crops' production is lower because of heavy evaporation and the harsh summertime heat. According to the report, Pakistan's output of key basic foods like wheat, rice, and sugarcane has decreased over the first 10 years of the twenty-first century [12].

4. Soil water management

To actively take part in keeping soil water content at an ideal level for all specified goals, including environmental requirements, is to practice soil water management. An ideal state often involves striking a balance between conflicting demands and the need to take the soil water system's long-term viability into consideration.

To effectively manage irrigation systems, a full understanding of soil water is needed. Irregular water evaporation increased demand and crop water stress can occur from under-watering, which can reduce the amount of accessible water below essential levels. Two factors: 1) the Texture of the soil, which influences due to its water-holding capacity and vary soil moisture content, water potential, and infiltration; and 2) root depth, which finds the amount of water that is accessible for plants. Utilizing irrigation water resources more effectively may result in being able to control the soil water balance. For managing irrigation systems, the two most significant indicators of soil water are (1) Soil moisture content, which refers to the total quantity of water in the soil, and 2) Soil water potential, a measurement standing for ready availability of water to plants.

4.1 Soil structure constitution and composition

Soil includes the particles, air, and water the three major components (see **Figure 1**). Water and air are trapped in the soil particles in pore spaces a measure of which is the porosity of the soil.

4.2 Soil water potential

Irrigation management is not only about how much water is in the soil but also depends on the ability of plants to pump up or access water. The dynamic condition of soil water with moisture as a standard reference is known as its potential, which is often stated as energy per unit of volume (bars or centi bar unit). Gravitational potential, matric potential, and solute potential are the three main factors that make up total soil water potential. The gravitational potential results from gravity's force dragging the water in the soil downward. Matric potential also referred to as soil water tension, is the term used to describe the force that the soil matrix exerts on the water

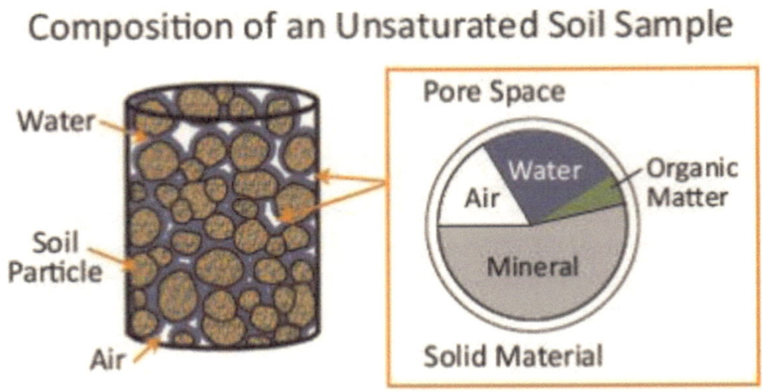

Figure 1.
Ref. [13].

through capillarity and adhesion. Salts that have dissolved cause the solute potential in the soil water.

4.3 Available soil water

The term "available soil water" or "available water capacity" (AWC), which refers to water that is available for plant use, refers to the water held between the field capacity and the permanent wilting point.

4.4 Infiltration

The rate at which water infiltrates the soil or, if water application is continued, the total volume of water that infiltrates the soil over time, can be used to define infiltration. The entire depth that has been penetrated following a certain amount of time for water application is known as cumulative infiltration.

4.5 Soil water balance

An understanding of the volume of water contained in the soil at any one time is necessary for planning irrigation. Growers who can control the soil water balance can prevent applying too much or too little irrigation water. The soil's texture and the plant's stage of development influence the root zone's ability to store water. Soil textureal classification is helpful to understand the ability of soil to hold water. For most crops, such as maize and soybeans, 50% of the water capacity can be used before plant stress sets in. To calculate the total amount of soil water that is accessible, multiply the rooting depth by the water storage ability:

$$\text{Total available water} = \text{rooting depth} \times \text{water storage ability.}$$

5. Water resources management

Water resources management is categorized into two types

DOI: http://dx.doi.org/10.5772/intechopen.107153

- On-farm water management
- Off-farm water management

5.1 On-farm water management

On-farm irrigation water management includes adjusting factors like irrigation timing and volumes, flow rates, and water control systems. These and many other factors can be adjusted to meet desired agricultural production targets while staying within the limits given by soils, crops, climate, water availability, and economics, as well as social and other factors.

There are several advantages to proper on-farm water management. In general, an effective on-farm water management scheme aids in the maximization or optimization of output. It can assist minimize water and energy usage, allowing more water and energy to be used to irrigate more area while also lowering the cost of an irrigation system. It can prevent fertilizer loss due to excessive water application, lowering the quantity of fertilizer required to meet targeted production targets. A competent management program ensures that root zone salinity is kept below acceptable limits, and that soil water logging and excessive deep percolation losses are reduced or avoided. It can aid in the elimination of issues like erosion and the management of crop diseases caused by insufficient or excessive water application. Water management can help you save time on the machine and the job.

The engineer, technician, or farmer must first have a thorough understanding of the irrigation system before implementing an efficient irrigation water management scheme. They must be knowledgeable of the many design and management options available.

For example, more land may be irrigated in that location if water is used more efficiently in a project upstream. Users downstream, on the other hand, may have to rely on upstream users' return flows to keep irrigating. Implementing a good on-farm water management program is a highly specific and planned procedure that includes several components. Certain requirements needed as below:

1. Evaluate the farming system, considering the soils, crops, and irrigation systems. This contributes to the identification of the main issues with the current management system. The assessment also aids in estimating the advantages of adopting the needed adjustments.

2. Find the system's design, uplift, and management options and choose the best option (s).

3. Confirm that the system has been adjusted following the chosen design and/or revision.

4. Create a complete system management plan that includes watering timings, safeguarding, and other management factors.

5. Technicians and farmers will be trained so that they can carry out the program that has been set up.

6. Throughout the season(s), check the system for any necessary adjustments. The phase of system monitoring is critical to successful water management.

Following the implementation of the program, both technical experts and the farmer handle monitoring.

6. Advanced technologies adoption

6.1 Drip fertigation

Water is the most vital natural resource on the planet. Without water on the globe, no life form can survive. It is required for all major activities such as food production, as well as sectors such as energy, production, and manufacturing. Although we are aware that soil water content and quality are dwindling, this could be due to inefficient water usage, drastic climatic change, or the use of more chemicals to boost productivity. With the rapid expansion of the area under micro-irrigation, fertigation is gaining traction in several countries. The notion of fertigation is new to the Indian subcontinent, but it is gaining popularity and making 'Fertigation' easy to adopt. Drip fertigation, which is defined as the injection of fertilizers, additives, or water-soluble compounds into the irrigation system, is a significant technology in this case. Furthermore, fertigation is linked to chemigation, which is injecting chemicals into an irrigation system. In a drip fertigation system, water is dispensed at a slower pace through the drippers, and nutrients from the fertilizers are carried along with the water.

6.2 Solar powered tube well

Renewable resources of energy such as solar power can be utilized for a solar-powered tubewell that can be operated during the day. Because of their higher efficiency and delivery head, submersible pumps are commonly employed as solar-powered tube wells. For calculating horsepower to run the pump depending on groundwater levels can be used to estimate the number of solar panels. The multistage submersible pump is powered by a DC motor linked. For the installation of solar tube wells, a three-stage submersible pump is required. The discharge of the solar tube well fluctuates throughout the day depending on the position of the sun. The tube well's peak discharge is expected between 12:00 and 2:00 p.m.

6.3 High-efficiency irrigation system

Pressurized irrigation systems, such as drip system irrigation and sprinkler system of irrigation, are commonly referred to as high-efficiency irrigation systems. Sprinkler irrigation systems have an irrigation efficiency of 70–80%, while drip irrigation systems have an irrigation efficiency of 80–95%. Both systems have different levels of applicability depending on the terrain, soil, water source, and crop. Both technologies are considered water conservation methods because they provide superior control over irrigation water application.

6.4 Raised bed planting technology

Growing crops on raised beds is one of the enhanced irrigation systems used all over the world, and it has several advantages. Crop yields on the bed, for example, are boosted by better nitrogen management, root aeration, efficient use of irrigation, and reduced lodging risk [14]. In comparison to flat sowing, this approach reduces seed rates without sacrificing crop yields. Better crops stand and yields are also ensured by improved root development in bed planting [15]. Above important, compared to traditional sowing, bed planting promises significant water savings of 35 to 45 percent, as well as the elimination of crust formation on the soil surface [16]. When compared to flat sowing, the next crops on permanent beds have shown higher yields and water savings ranging from 20 to 40%.

The following are the main benefits of bed planting for wheat crops:

Percent saving of water	Yield increase percent
30–50%	25%

Other benefits which raised bed technology can give are improved water and fertilizer use efficiencies along with the least weeds and no lodging of the standing crop.

6.5 Laser land leveling

Traditional land leveling techniques, including utilizing an engineer's level and staff rod, are time-consuming, exhausting, ineffective, and costly. For farmers to accurately level their fields, the Department of Agriculture, Government of Punjab, adopted Laser Land Leveling technology. The technique of laser land leveling involves employing drag buckets that are equipped with lasers to smooth the ground surface with a maximum allowable variation from its average height of no more than 2 cm. With the help of laser land leveling, it is possible to level the fields at a slope of zero, distributing irrigation water evenly across the head, middle, and tail of the fields. The GPS and laser-guided equipment used for laser land leveling are installed on high-horsepower tractors and soil movers. Therefore, laser land leveling of irrigated areas can also result in the following advantages, either directly or indirectly. The land receives an even distribution of water, which increases irrigation effectiveness. In laser-leveled fields, about 30% of the water is conserved; as a result, more land can be watered. The leveled field will yield 20 percent more because of more consistent germination. Due to the equal distribution of fertilizer, its effectiveness and efficiency will both increase. Erosion risks would be reduced. It increases the effectiveness of machine usage (**Figure 2**).

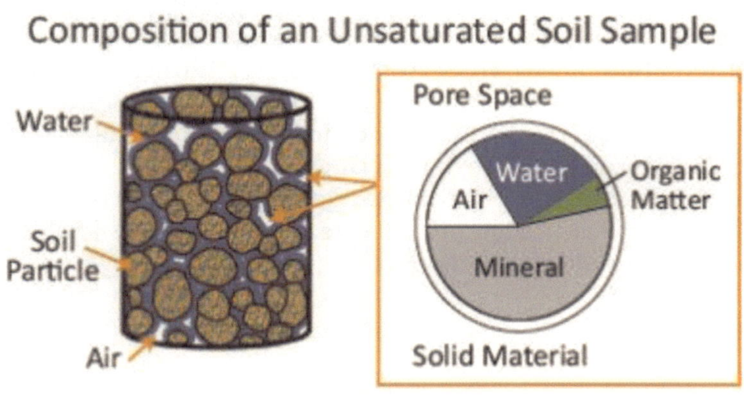

Figure 2.
Laser land leveling at WMRC, UAF, experimental area. Source: Water Management Research Center, University of Agriculture, Faisalabad Pakistan.

6.6 Zero tillage

No-tillage, also known as zero tillage, is a farming technique in which crop leftovers are left on the field after harvesting but the ground is not plowed. The following crop is directly sown using no-till planters (Zero Tillage Machines). For instance, in a rice-wheat rotation, wheat is planted using a Zero Tillage Drill right into a field that has previously produced rice without the use of any tillage equipment. Such methods, especially on sloppy terrains, are very effective in reducing soil erosion caused by wind or water. Due to the roots of the previous crop remaining in the soil and crop residue covering the ground surface to lessen the effects of heavy rainfall, zero-tillage reduces soil erosion.

7. Problems of irrigation system

The irrigation and drainage infrastructure of Pakistan, which is now experiencing serious issues, will determine the country's agricultural destiny. Some of these issues include rising salinity and water logging, overuse of fresh GW, low water use efficiency, and unequal and erratic supply. Rigid system design and inadequate drainage, low delivery efficiency and inequitable distribution of water, waterlogging and salinity, and over-exploitation of groundwater in fresh areas represent major problems in Pakistan‘s irrigation system.

7.1 Rigid system design

Despite the greater distribution control provided by the development of barrages, reservoirs, and link canals, the irrigation system is nevertheless managed following outdated canal diverting patterns that frequently do not match water demands. Due to inadequate reservoir capacity as well as the seasonal pattern in river flows, which can provide about 85% of the water throughout the summer, limited water supply occurs at the beginning and end of the summer and throughout the winter. The inconsistency between water supply and demand reduces agricultural productivity.

7.2 Inadequate drainage

Because of its flat terrain and lacking well-defined drainage channels, the Indus Plain has an urban drainage problem, which is being made worse by the construction of roads, railroads, flood bunds, and water systems that obstruct natural drainage flows. In irrigated areas, drainage has been a priority since the 1960s, and numerous sizable drainage programs are still in operation today. About 6.5-million-hectare acres of the 16.7-million-hectare acres of gross canal-controlled land need to be drained, of which 1.86 million-hectare acres are being worked on right now. It is a significant task to provide drainage to such a huge area. The predicted water table depth for a region of 2.38 million hectares is less than five feet.

7.3 Low delivery efficiency and inequitable distribution

Canal delivery is incredibly inefficient as a result of aging, excessive use, and poor maintenance. From the canal head to the root zone, delivery efficiency is between 35 and 40 percent on average, with the majority of losses happening in watercourses. There is less water available for agriculture because of the significant surface water loss, which enhances water logging and salinity. Excess water and water losses during irrigation are frequently returned to rivers and utilized once more downstream in irrigation systems with drainage. As a result, there is less efficiency damage to the river basin as compared to efficiency loss to individual systems. Unfair distribution is a key concern as well.

7.4 Waterlogging and salinity

According to the World Bank (1992), Pakistan may be unable to produce about 25% of the important crops that may be produced there because of soil salinity. In an area such as the Indus Basin, which has a flat topography, insufficient natural drainage, permeable soils, and just a semi-arid climate with significant evaporation, it is unavoidable that water tables will rise and salinity will grow. As a consequence of the greater diversion of stream flow for irrigated and seepage through canals, and water-courses, especially irrigated regions, the groundwater level has gradually risen. By the 1960s, many SCARPs had already been established. Nevertheless, despite these precautions, the gross protected area is wet in around 30% of instances, with about 13% of those instances being classified as extremely waterlogged.

7.5 Inadequate operation and maintenance (O&M)

Because of neglected maintenance and overuse, Pakistan's irrigation and drainage system has been deteriorating. Provinces agreed to keep spending on surface irrigation and subsurface saline drainage facilities at 1988 levels as part of Bank Projects.

7.6 Poor investment planning

In Pakistan, there are three stages to the investment planning process for drainage and irrigation. A sectorial plan provides a framework for intermediate- to long-term Sectorial development, five-year plans are used for short-term planning, and the Annual Development Programme allows monies annually (ADP). Sectorial planning has previously attracted a lot of attention. Plans like the Revised Action Programme (RAP) and the Water Sector Investment Planning Study (WSIPS), which were developed with assistance from abroad, thoroughly evaluate the needs and objectives of each sector.

8. Sustainable irrigation for best crop yields

Sustainable irrigation is the sensible use of all activities associated with the watering of plants, whether in agriculture, landscape, or ornaments, in such a way that it aids in meeting current survival and welfare needs without jeopardizing future generations. The term "sustainability" is frequently used to refer to the management, use, and protection of natural resources in such a way that future generations will have access to them [17–19]. In anthropocentric words, "sustainability" refers to a state in which current generation demands are met "without jeopardizing future generations' ability to fulfill their wants." [20]. As a result, sustainable irrigation encompasses the need to examine a variety of factors, particularly those relating to the deterioration, loss, or depletion of resources such as soil, water, and energy, as well as biodiversity and environmental preservation. Because water is a renewable resource, it is possible to aspire for great sustainability when it comes to irrigation. Of course, water resources may be non-renewable on a local level, like in arid regions with low rainfall inputs and large non-renewable groundwater reserves. The exploitation of groundwater (commonly referred to as "groundwater mining") is a classic example of unsustainable water usage in such areas [21]. The aspect of sustainable irrigation encompasses not only the geographical location where irrigation is used but also the production and transportation of necessary equipment and supplies, as well as discharges and waste consequences. The consequences of the construction, operation, and maintenance of the works, which are either directly or indirectly required for irrigation and are frequently located over great distances, should not be overlooked.

Irrigation must be able to adapt to changing climates to remain sustainable. As a result, it must combat droughts as well as the impacts of global climate change on a broader scale. As a response, relying on other disciplines like crop selection and development, automation and telecommunications, institutional governance, and others might be critical.

Irrigation can help with climate change adaptation, labor and energy savings, and the production of higher-value crops. On the other hand, irrigation agriculture must be made more egalitarian, efficient, and sustainable to achieve zero hunger in an increasingly water-scarce world. Irrigated agriculture is contributing to, and being affected by, increased strains on freshwater resources, with more than 60% of global irrigated cropland under significant water stress. To ensure equal access to water and environmental sustainability, investments in expanded and enhanced irrigation must be matched by improvements in water governance. Improved data and knowledge on water resources and their use and well-defined water rights are cornerstones of better water governance [4].

Finally, just as sustainable irrigation must be able to cope with water scarcity, it must also be resilient to other resource scarcities throughout time, such as energy and finance. The themes of "water resources sustainability, "sustainable water usage," and "agricultural and rural development" are all intertwined in the concept of "sustainable irrigation."

Water resource sustainability refers to the ability to use water in sufficient quantities and quality on a local, regional, national, and global scale to meet the needs of humans and ecosystems now and in the future to sustain life and protect humans from natural and human-caused disasters that threaten life.

"Water usage that supports human society's potential to persist and thrive indefinitely without jeopardizing the integrity of the hydrological cycle or the biological systems that rely on it" is what sustainable water use means.

9. Sustainability concerns in irrigated agriculture

Irrigation is a considerable change in the physical, environmental, and social aspects of the area. Existing equilibriums are disrupted in the process, and new ones emerge over time. Fact, the essential principle of irrigation development is that new conditions meet mankind's goals better than old ones. This premise has proven to be correct in many ways. Irrigation was a major driving force in the growth of many ancient civilizations, and it continues to be so today. While just approximately a 6th of the world's agricultural land is irrigated, this part produces about a third of the world's food. The irrigated area of the world is not uniform. Different locations are afflicted by various sustainability issues. Many of the sustainability challenges raised by irrigated agriculture in affluent countries are like those raised by modern high-input agriculture, which places a heavy demand on natural resources and often exceeds the capacity for environmental assimilation. Many of the concerns about irrigated agriculture's sustainability in emerging countries, on the other hand, stem from the general development problems that these countries face, such as a lack of public capital, macroeconomic reliance on agricultural commodity exports, widespread poverty, population pressure, insufficient management, and human resource development, institutional and regulatory shortcomings, and so on [22].

9.1 Water resources

Many countries, particularly those in the dry climatic zone with high rates of population growth, urbanization, and industry, are finding water to be a precious resource. Increased water competition in these nations will have a considerable influence on irrigated agricultural water supply [23]. The competing industries' water demand will always be a modest proportion of the naturally accessible supply in most countries. In most nations, the need to use water more efficiently will steadily rise. Most occurrences of groundwater mining are caused by a lack of sufficient legislation and enforcement, as well as a lack of awareness of the environmental implications.

9.2 Land resources

The deterioration of land resources because of agricultural usage is a major cause of worry across the world. Irrigation development has worsened this problem by generating conditions that have led to deforestation and soil erosion accidentally. This is especially true of land degradation in river diversion schemes upstream catchment regions [24].

9.3 Waterlogging and salinity

Irrigation has resulted in waterlogging and salinization of irrigated land on a huge scale. The incidence of this dual issue is usually limited to dry regions. About half of the world's already irrigated territory, or 270 million hectares, is thought to be desert.

10. A long-term irrigation system for small landholdings in rain-fed agriculture Punjab Pakistan

The drip irrigation system plays a vital role for fruits and vegetables in Pakistan, but the primary hindrance to the widespread acceptance is for small landowners. Because the drip system was obtained from local merchants, it was also less expensive. In 2015 and 2016, field trials were conducted on vegetables (potato, onion, and chilies) and fruits to analyze the productive and economic effects of low-cost drip irrigation (olive, peach, and citrus). While comparing with other systems this system saved 50% on water expenditures and created 27–54% net revenue. Drip and furrow irrigation systems have obtained water use efficiency (WUE) of 3.91–13.30 kg/m^3 and 1.28 4.89 kg/m^3, respectively. According to the current study, low-cost drip irrigation increased yield by more than 20% [25].

11. Sustainability enhancement and management

There are no obvious flaws in this technology that make irrigation development unsustainable in the long run. Salt accumulation/mobilization and accompanying downstream water deterioration, as well as the development of waterborne infections, are the only two sustainability issues organically linked to irrigation technology. Careful planning and the implementation of mitigation measures can help solve these issues.

The most typical causes of water logging and salinity are poor planning, inadequate irrigation efficiency, and development issues. The consequences of planned interventions can be better predicted with the development of comprehensive model-based planning and design approaches. This will help in the battle against flooded and saline regions throughout the world.

Improving irrigation efficiency is a requirement that will be pushed upon the irrigation industry, gladly or unwillingly, for the welfare of the sector. Many of the above-highlighted sustainability issues are directly or indirectly connected to the already permitted irrigation water waste.

- Government policy changes aimed at improving cost recovery, as well as regulatory and legal frameworks.
- Irrigation department institutional reforms to improve performance and water conservation, as well as make them more responsible to end-users.
- Water pricing incentives to discourage over-irrigation
- Improved water supply reliability and equity through more decentralized water management and expanded engagement of (groups of) end users in system design and real water management.
- Improving the abilities of operational people and providing an extension to farmers to help them understand and make better decisions.

12. Smart irrigation system

Smart irrigation technology uses weather data or soil moisture data to determine the irrigation need of the landscape. Smart irrigation technology includes: These products maximize irrigation efficiency by reducing water waste while maintaining plant health and quality. Using intelligent irrigation devices, outdoor water savings are possible. In contrast to conventional effective automated timers, which irrigate according to a user-determined set schedule, smart irrigation control systems and sensors have been created to reduce external water usage by irrigation depending on plant water needs. This technology can be used to construct a smart controller by attaching a sensor to an existing water distribution timer or as a whole controller. Utilizing smart irrigation systems in the landscape may help to cut back on water usage outside. Both tiny, private landscapes and huge, professionally managed landscapes can increase productivity. The functions of each product, as well as their benefits and drawbacks, are covered in the sections that follow.

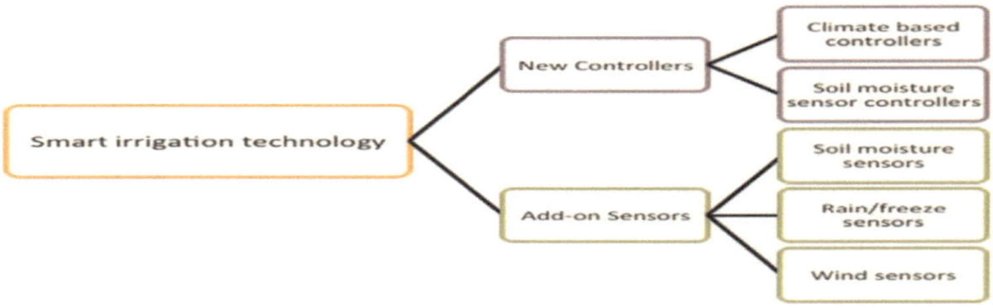

13. Irrigation system

Irrigation is the technique of artificially delivering water to agricultural fields to produce crops. If the amount of water available to the plants from rainfall is insufficient, irrigation water is used to supplement it. To accomplish this goal, an irrigation system must be created, which includes the planning, design, building, operation, and maintenance of numerous irrigation works:

- Source: River, Reservoirs, Alternate sources (groundwater, treated wastewater)
- Control structures: Barrages, Head Regulators
- Distribution system: Irrigation Canals and Tertiary Irrigation System

Each irrigation system has unique advantages and disadvantages depending on different factors such as:

1. Initial installment cost
2. Filed architecture
3. Soil texture and structure
4. Nature and availability of the water supply
5. Climate
6. Cropping patterns
7. Social preferences and structures
8. Historical experiences

Let us have a look at different types of irrigation and the methods used for irrigation.
Surface Irrigation.
Sprinkler Irrigation.
Localized Irrigation.
Drip Irrigation.
Centre Pivot Irrigation.
Sub Irrigation.
Manual Irrigation.

Irrigation may be used in a variety of ways to increase agricultural productivity. Irrigation systems are used in different ways depending on the soils, climates, crops, and resources available. Farmers use a variety of irrigation methods, including:

13.1 Surface irrigation

There is no irrigation pump in this system. Gravity distributes water throughout the terrain here. Water is ponded on an enclosed level field and allowed to penetrate basins, borders, or furrows in surface irrigation.

Water is applied to the field in either a controlled or uncontrolled manner.

- Uncontrolled: Wild flooding and Flood Irrigation
- Controlled:

Water is applied from the water channel outlet and controlled by borders and/or furrows. Surface irrigation is only used when there is plenty of water. The cheap initial design expense is countered subsequently by the high labor cost for water application.

- Water is delivered directly to the soil surface from a channel or open ditch at the field's upper reach.
- Water moves due to gravity force in surface irrigation
- It is the oldest method of irrigation (from 4000 years)
- Highly adopted method in the world
- In most countries, about 90% area is under surface irrigation
- In the USA, about 60% of the area is under surface irrigation

Following are some advantages and disadvantages of surface irrigation methods.

Advantages	**Disadvantages**
It is really simple to manage. Easily adapts to flat topography. It is necessary to keep costs down. It's possible to use it even if there aren't any drainage outlets. Allows for quick salt leaching. Allows for full use of rainfall. It is possible to attain high application efficiency. Adapts effectively to infiltration rates ranging from moderate to low. The short-term water supply works well. Small landholdings do not bother it.	To obtain high efficiency, the land must be flat (the greatest land elevation fluctuation should not exceed half the administered irrigation depth). Small field sizes are required for soils with high infiltration rates, which problems with automation. It's tough to get rid of extra water, especially when there's a lot of it. Plants that are partially submerged in water can occasionally provide longer periods (in low infiltration rate soils). Small irrigations are difficult to apply.

13.2 Sprinkler irrigation

Sprinklers from the movable platform or overhead high-pressure sprinklers deliver water from a central point. Water is sprayed over the area to be watered by pressurized water flowing via pipes to outlets.

Advantages	**Disadvantages**
By this method, we can increase efficiencies. Land leveling is not a mandatory method and can easily use on uneven land Can be used on all types of infiltration rates soils require less labor to operate this system and reduce labor costs.	The high initial cost to install the system consumes more energy as compared to other systems. Requires moderately high technology.

13.3 Localized irrigation

Water is delivered to each facility through a network of pipes at low pressure in this system. Water is continually delivered at very low rates to points or small areas in the field through small holes in plastic tubing or from emitters in this system. Only a portion of the field has been soaked.

Advantages	Disadvantages
It has many of the same benefits as sprinklers, but it can attain extremely high-water efficiency and can be used successfully in highly salty conditions.	It needs a large initial investment. It needs a high level of technological expertise. Saline soils may cause problems

13.4 Drip irrigation

Drops of water are provided at the roots of the plants in this style. Because it takes more upkeep, this method of irrigation is only utilized in orchards and high-value crops. Fertigation can also be done through this system.

13.5 Centre pivot irrigation

A sprinkler system that moves in a circular pattern distributes the water. This approach is most employed in commercial cooperative farms with a lot of acreages to grow. It aids in the attainment of precision agricultural aims. This technology can also be used for fertilization. Sub Irrigation Water is dispersed by raising the water table through a network of pumping unit valves, ditches, and canals.

13.6 Manual irrigation

This is a laborious and time-consuming irrigation technique. Manual labor is used to supply water using watering cans at this location. We can characterize irrigation in other ways also. Irrigation can be done by two different techniques:

Conventional Methods.

Modern Methods.

13.6.1 Conventional methods of irrigation

This method involves hand irrigation. A farmer brings water to farming areas by hand, with the aid of livestock, or from wells or canals. Depending on the localization strategy. The main advantage of this technique is its affordability. Its effectiveness is nonetheless minimal due to the uneven dispersion of water. Additionally, there is a substantial chance of water loss. Examples of typical systems are the chain pump, the lever system, and the pulley system. The most popular and commonly used of these is the pump system.

13.6.2 Modern methods of irrigation

The current technique compensates for the limitations of previous methods and so aids in proper water consumption. The significance of irrigation may be described as follows:

Agriculture suffers from insufficient and unpredictable rainfall. Droughts and famines are caused by insufficient rainfall. Irrigation improves efficiency even in areas with minimal rainfall. The productivity of irrigated land is higher than unirrigated land. Multiple cropping is not feasible in Pakistan because the rainy season varies by area. However, the climate allows for agriculture all year. In most areas of the region, irrigation infrastructure allows for the cultivation of more than one crop. Irrigation has contributed to the cultivation of the majority of the fallow land. Irrigation has helped to maintain productivity and yield levels. Irrigation enhances the availability of water supply for crops, which boosts farmer income.

GRAVITY OR SURFACE IRRIGATION.

Suitability of Surface Irrigation Methods

- Soils with a low to moderate rate of infiltration
- Leveled lands
- Lands with a slope less than 2–3%
- Accommodate all types of crops
- Labor intensive but nowadays equipment is available for automation

14. Irrigation resource constraints and climate change

Freshwater availability has an impact on almost all social and environmental elements of climate and demographic change, and also their implications for sustainability. Water scarcity is already a significant problem in many parts of the globe and is predicted to get worse as the population increases, the nutrition quantity is demanded, temperatures are rising, and rainfall pattern change. It also affects energy projects, anthropogenic water usage, and ecological use.

The world's freshwater resources are under tremendous pressure due to population growth, changing land uses brought on by agricultural development, and deforestation [26]. The future availability of freshwater for industrial use, agricultural production, and human consumption become more uncertain as global climate change intensifies. Depending on greenhouse emissions, the predicted range of global temperatures is comparable to 1980–1999 even by end of the 21st century anywhere between 1.1 and 6.4°C [27].

The size and sign of potential impacts are still up for debate, and the extent of predicted precipitation changes varies substantially relying on geographical region and spatial extent [28]. Even in certain locations where mean precipitation is expected to decrease, daily heavy precipitation occurrences are likely to increase [27]. Even though atmospheric CO_2 seems to have the potential to increase photosynthesis by close to 30%, such changes are anticipated to have a greater negative impact on

	Feel or appearance of soil and moister deficiency			
Available soil moisture remaining	**Loamy Sand**	**Sandy Loam**	**Loam and Silt Loam**	**Clay Loam or Silty**
				Clay Loam
	Course Quality	Reasonably Course Texture	Medium Texture	Fine and Very Fine Texture
0 to 25 out of a hundred	Waterless, moveable, solo-grained, flows through fingers	Waterless, moveable, flows through fingers	Powdered dry, sometimes a little covered but easily broken down into powdered condition.	Unbreakable baked, split, and sometimes has loose crumbs on the surface.
25 to 50 out of a hundred	Seems to be dehydrated and will not form a ball with density. [1*]	Seems to be dry and will not form a ball. [1*]	Rather crumbly but holds together from pressure.	Somewhat flexible will ball under pressure. [1*]
50 to 75 out of a hundred	Seems to be waterless and will not form a ball with gravity.	Tends to the ball under pressure but seldom holds it together.	Forms a ball somewhat plastic and will sometimes slick slights with pressure.	Forms a ball, ribbons out between thumb and forefinger.
75 percent to field capacity (100 out of a hundred).	Tends to switch together somewhat, and sometimes forms a very weak ball under pressure.	Forms a weak ball, breaks easily, and will not slick.	Forms a ball, is very pliable, slicks readily, and is relatively high in clay.	Effortlessly ribbons out between fingers, have a slick feeling.
At field capacity (100 out of a hundred).	Upon pressing, no free water seems on the soil but the wet framework of the ball is left on hand.	Upon pressing, no free water seems on the soil but the wet framework of the ball is left on hand.	Upon pressing, no free water seems on the soil but the wet framework of the ball is left on hand.	Upon pressing, no free water appears on the soil but a wet framework of the ball is left on hand.

[1*] *Ball is formed by squeezing a handful of soil very firmly.*
Source: https://sanangelo.tamu.edu/extension/agronomy/agronomy-publications/how-to-estimate-soil-mois ture-by-feel/

Table 2.
Guide for judging how much moisture is available for crops.

agricultural output [29]. The impact of climate change on agriculture water usage especially changes in net irrigation demands, demand, and agricultural water consumption has been the subject of several studies (**Table 2**).

15. Irrigation scheduling

It is the ability with which the farmer decides the amount and time of irrigation. It includes the Time of irrigation when irrigation is needed, Amount of irrigation (how much water should be applied?). What should be the response of the crop (seed yield or forage yield) to the irrigation applied? There are two alternative aims for scheduling irrigation. 1 For achieving maximum output per unit of land area 2.For Maximizing land area utilized in crop production.

15.1 Methods of irrigation scheduling

Fixed interval application: In the warabandi system there is a fixed interval for irrigation application.

Apply irrigation when your neighbor is doing so. A visual sign of crop-based upon their experience farmers apply irrigation when they observe the sign of starvation of water, it is only descriptive. The determination of moisture content of the soil: Gravimetric method, Gypsum block method, Tensiometer method and Neutron probe method.

15.2 Irrigation scheduling benefits

It enables the farmer to plan a water rotation schedule between fields to lessen agricultural water stress and boost yields. For the farmer, it cuts the cost of labor and water. Because it collects surface runoff, fertilizer costs are reduced. It raises net profits by enhancing agricultural quality and productivity. Water logging risk is decreased. It helps to control problems with root zone salinity. Using the "saved" water to irrigate non-cash crops, creates additional revenue.

15.3 Strategies for irrigation scheduling

15.3.1 Certain

Climate variables that have been measured; Average soil parameters were measured. Irrigation water quality. The amount of water used and when it was used Irrigation technique and a few crop parameters (crop height, development stage, DAP, LAI, root depth).

15.3.2 Uncertain

Reference evapotranspiration and Kc values are estimated. Estimation of crop water needs. Crop water uptake pattern, and crop response function to shortfall irrigation and/or excessive salt accumulation.

Irrigation Scheduling Approaches, Fixed Scheduling, Flexible Scheduling, and Flexible Scheduling Incorporating Rainfall are the three types of irrigation scheduling. Each category is described in detail below.

1. By taking the soil samples from the field and estimating the required depth of irrigation and net depth of water needed.

2. Estimate ET by multiplying it with Kc by the pan evaporation method.

3. Calculate the total crop water need (CWR) for the whole crop period, considering the duration of the growing season or the number of growing days.

4. Calculate the irrigation interval by the ratio between total CWR and net irrigation depth. 5. Adjust irrigation depth according to the irrigation interval set up in step.

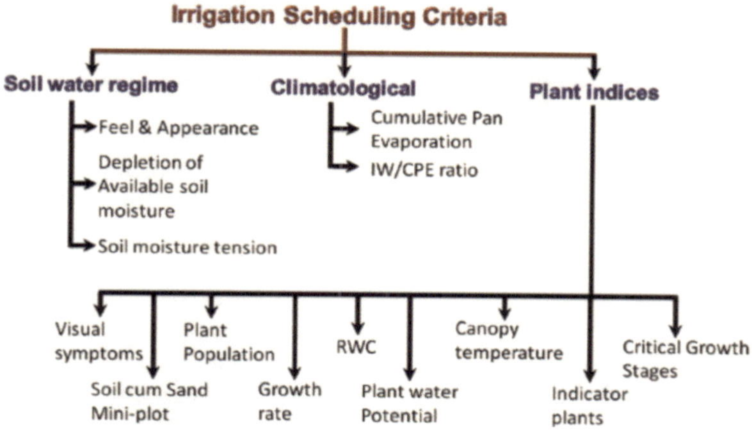

Author details

Fahd Rasul[1*], Hassan Munir[1], Aftab Wajid[1], Muhammad Safdar[2,3], M. Salman Ayub[1,4], Sobia Shahzad[5], Rehan Mehmood[2,3], M. Adnan Shahid[2,3], Abid Sarwar[3], M. Danish Majeed[2,3], Umair Gull[1], Wajid Nasim Jatoi[5], Muhammad Mubeen[6], Summera Jahan[7] and Shakeel Ahmed[8]

1 Department of Agronomy, University of Agriculture Faisalabad, Pakistan

2 Agriculture Remote Sensing Lab (ARSL), University of Agriculture Faisalabad, Pakistan

3 Department of Irrigation and Drainage, University of Agriculture Faisalabad, Pakistan

4 Valley Irrigation Systems Pakistan (Valley – California, USA)

5 Department of Botany, Islamia University of Bahawalpur, Pakistan

6 Department of Environmental Sciences, Comsats University, Vehari Campus, Pakistan

7 Department of Botany, University of Gujrat, Pakistan

8 Department of Agronomy, Bahauddin Zakariya University Multan, Pakistan

*Address all correspondence to: drfahdrasul@uaf.edu.pk; drfahdrasul@gmail.com

References

[1] Green PA, Vörösmarty CJ, Harrison I, Farrell T, Sáenz L, Fekete BM. Freshwater ecosystem services supporting humans: Pivoting from water crisis to water solutions. Global Environmental Change. 2015;**34**:108-118. DOI: 10.1016/j.gloenvcha.2015.06.007

[2] Rockström J, Falkenmark M, Allan T, Folke C, Gordon L, Jägerskog A, et al. The unfolding water drama in the anthropocene: Towards a resilience-based perspective on water for global sustainability. Ecohydrology. 2014;**7**:1249-1261. DOI: 10.1002/eco.1562

[3] Chartzoulakis K, Bertaki M. Sustainable water management in agriculture under climate change. Agriculture and Agricultural Science Procedia. 2015;**4**:88-98. DOI: 10.1016/j.aaspro.2015.03.011

[4] Food and Agriculture Organization. Food and Agriculture Data. USA: FAOSTAT; 2019

[5] Sullivan C. Calculating a water poverty index. World Development. 2002;**30**:1195-1210. DOI: 10.1016/S0305-750X(02)00035-9

[6] D'Odorico P, Davis KF, Rosa L, Carr JA, Chiarelli D, Dell'Angelo J, et al. The global food-energy-water nexus. Reviews of Geophysics. 2018;**56**:456-531. DOI: 10.1029/2017RG000591

[7] Alcamo J, Döll P, Henrichs T, Kaspar F, Lehner B, Rösch T, et al. Global estimates of water withdrawals and availability under current and future "business-as-usual" conditions. Hydrological Sciences Journal. 2003;**48**: 339-348. DOI: 10.1623/hysj.48.3.339.45278

[8] Rock Strom J, Williams J, Daily G, Noble A, Matthews N, Gordon L, et al. Sustainable intensification of agriculture for human prosperity and global sustainability. Ambio. 2017;**46**:4-17. DOI: 10.1007/s13280-016-0793-6

[9] Savenije HHG, van der Zaag P. Water as an economic good and demand management: Paradigms with pitfalls. Water International. 2002;**27**:98-104. DOI: 10.1080/02508060208686982

[10] ADB (Asian Development Bank). Water Resources Strategy Study. Draft Report Vol. 1. Islamabad, Pakistan: ADB; 2002

[11] Reinders FB. Micro irrigation: A world overview. In: Proceedings of the 6th International Micro Irrigation Congress, South Africa. Vol. 2000. Florida, USA: Apple Academic Press & CRC Press Taylor & Francis Group; 2000

[12] Keller J. Evolution of drip/micro irrigation: Traditional and non-traditional uses. In: Keynote Address at the International Meeting on Advances in Drip/Micro Irrigation, December 2002. Spain: Apple Academic Press & CRC Press Taylor & Francis Group; 2002

[13] Kiselev MV, Voropay NN, Dyukarev EA, Preis YI. Temperature regimes of drained and natural peatlands in arid and water-logged years. In: IOP Conference Series: Earth and Environmental Science. Vol. 386, No. 1. IOP Publishing; 2019. p. 012029

[14] Syre KD, Moreno Ramos OH. Application of Raised Bed Planting Systems to Wheat. Wheat Special Report No. 31. Mexico, D.F: CIMMYT; 1997

[15] Peries R, Chris B, Bruce W. Raised bed cropping leading to improved root

proliferation in heavy duplex soils prone to waterlogging. In: Proceedings of the 6th Symposium of the International Society of Root Research. Nagouya, Japan: Southern Victoria Australia; 2001, November 2001

[16] Fahong W, Xuquing W, Sayre KD. Comparison of conventional flood irrigated, flat planting with furrow irrigated, raised bed planting for winter whear in China. Field Crops Res. 2003; **87**:35-42

[17] Sands GR, Podmore TH. A generalized environmental sustainability index for agricultural systems. Agriculture, Ecosystems and Environment. 2000;**79**:29-41. DOI: 10.1016/S0167-8809(99)00147-4

[18] FAO. Sustainability Assessment of Food and Agriculture Systems. Guidelines Version 3.0. Rome, Italy: FAO; 2013

[19] Borsato E, Galindo A, Tarolli P, Sartori L, Marinello F. Evaluation of the grey water footprint comparing the indirect effects of different agricultural practices. Sustainability. 2018a;**10**:3992. DOI: 10.3390/su10113992

[20] FAO. Sustainable development and natural nesources management. In: Twenty-Fifth Conference, Paper C 89/2 —Sup. Vol. 2. Rome, Italy; 1989

[21] Konikov E, Likhodedova O. Global climate change and sea-level fluctuations in the black and Caspian seas over the past 200 years, in: Geology and geoarcheology of the black sea region: Beyond the flood hyphotesis. Geological Society of American. 2011;**473**:59-69. DOI: 10.1130/2011.2473(05)

[22] DGIS. Sustainable Land Use. the Hague, the Netherlands: Development Cooperation Information Department, Ministry of Foreign Affairs; 1992

[23] World Bank. Water Resources Management, a Policy Paper. Washington DC, USA: The World Bank; 1993

[24] ISRIC. The Extent of Human-Induced Land Degradation. Wageningen, the Netherlands: International Soil Reference and Information Centre; 1991

[25] Aziz M, Rizvi SA, Iqbal MA, Syed S, Ashraf M, Anwer S, et al. A sustainable irrigation system for small landholdings of Rainfed Punjab, Pakistan. Sustainability. 2021;**13**(20):11178

[26] Nejadhashemi AP, Wardynski BJ, Munoz JD. Large-scale hydrologic modeling of the Michigan and Wisconsin agricultural regions to study impacts of land use changes. Transactions of the ASABE. 2012;**55**(3):821-838

[27] Solomon S, Qin D, Manning M, Averyt K, Marquis M, (Eds.). Climate Change 2007-the Physical Science Basis: Working Group I Contribution to the Fourth Assessment Report of the IPCC. Vol. 4. UK: Cambridge University Press; 2007

[28] Bates B, Kundzewicz Z, Shaohong W. Climate Change and Water. Geneva Switzerland: Intergovernmental Panel on Climate Change Secretariat; 2008

[29] Long SP, Ort DR. More than taking the heat: Crops and global change. Current Opinion in Plant Biology. 2010; **13**(3):240-247

Chapter 3

An Overview on Techniques Involved in Recharging Ground Water and Its Impacts

Muthuminal R. and Mohana Priya R.

Abstract

The movement of water from the surface of the earth into the sub surface areas are said to be as a hydrologic process which helps in improving the water table at the ground level. This process of water movement towards downward direction is said to be as Groundwater recharge or deep drainage or deep percolation. Ground water recharge could be achieved either by natural method or by artificial methods which involves anthropogenic processes. The ground water recharge has a superior impact while in consideration of various complexities such as climate, land surface and biosphere processes, and characteristics in the unsaturated and saturated subsurface. So in this chapter, we would like to briefly explain about the techniques involved in recharging the ground water table and also about the methods involved in estimating the ground water available along with the importance of agriculture globally along with the policies incorporated to maintain the ecological demand and also about the impacts caused to the environment due to the insufficient water at the ground water table.

Keywords: techniques, impacts, methods, benefits, trends and policies

1. Introduction

We all know that, in order to lead a peaceful life, we are in need of three basic things such as food, water and shelter. The main source of water for all living beings is through the fresh water which has reliable pH and also for the growth of nutritious food through agriculture. Ground water is one among the largest fresh water supply at the world for potable water, irrigation water etc. [1, 2]. At the other hand, Ground water is in need to recharge periodically through recharging methods for the replenishment of resources, since they are available as a limited resource i.e. not in an infinite manner. Hence, for the purpose of avoiding scarcity and to cope up with the need of ground water, it is really essential to recharge the ground water table as it is linked to the water resource sustainability and resilience amidst the changing climatic conditions and demand. For the purpose of planning and managing the sustainability of recharging the water table, it is necessary to consider the conditions

Figure 1.
Sprinkler irrigation with groundwater.

such as total volume, location and timing according to the demand of water. Though the recharge of ground water table is one among the complex process which needs to quantify the requirement in an appropriate manner which is a tedious challenge to meet out at the hydrology. As the quantification of water cannot be done through the already exiting direct measurement methods, hence we need to adopt indirect observations for the purpose of monitoring the processes involved, properties and its quantities.

The water flow that takes place from downhill from surface water to aquifer is known as ground water recharge. This otherwise known as deep percolation or deep drainage. The most common method of recharging the ground water table is through entering the aquifer. The water recharge happens at the vadose zone which is located below the plant roots and is always denoted as flux to water table. At this process, the movement of water from water table into the saturated zone is said to be as ground water recharge. Recharging of water can be done through two methods that is through natural method and through manmade process. Some of the man made process are artificial recharge methods which are carried out through rain water harvesting where the rain water is channeled into the subsurface zone by bore holes.

The ground water recharge has a greater impact while in consideration of various complexities such as climate, land surface and biosphere processes, and characteristics in the unsaturated and saturated subsurface. The components such as Groundwater abstraction, artificial groundwater recharge and irrigation, management of hydraulic structures, infrastructure, and flood management are all manmade variables plays a vital role in the ground water recharge [1, 3]. Hence, for measuring the recharge of water table has been devised through several estimating methods (**Figure 1**).

2. Why ground water recharge is necessary?

For the purpose of utililization of water depending over the demand, the ground water is pumped out from the aquifer which leads to the depletion of it. For instance,

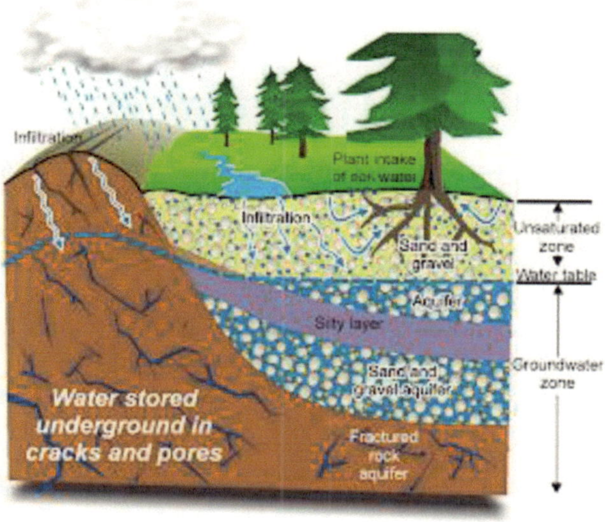

Figure 2.
Groundwater recharge.

due to the usage of water for the purpose of irrigating plant in agriculture sector, approximately 50 billion gallons per day is extracted from ground water table. Hence, due to the extraction of ground water as an enormous amount may lead to dry wells, poor water quality, soil subsidence and even deviation in plate boundaries. This over pumping of ground water from the soil causes the ground to sink as the soil compacts and also the soil is not capable to support the land above it. As the water table has been over pumped, the level of water at the ground decreases and necessitates deeper well drilling for the source.

The agriculture sector is dependent over the ground water for the purpose of irrigating at all over the world, as surface water availability is limited for irrigation and due to the depletion of ground water table it has become a serious threat to the farming. Though the surface water availability is low, for the places far from the availability is dependent over the water conveyance system which is a cheaper and quicker mode to transport.

Simply put, climate change disrupts the water cycle, causing less precipitation than normal, while population growth and increased food demand push agricultural businesses to pump more water out of the ground. This is known as water stress and warrants artificial groundwater recharge [4].

Ground water depletion is also caused due to the availability of surface water at lakes and streams in a reduced manner. As we all know that, the source of ground water is also achieved through surface water in an intricate way, less groundwater may imply less water flowing into wetlands and other ecosystems that rely on it. Hence the result will be less vegetation and less water quality which has high impact on Farmers and ranchers who often rely heavily on groundwater. **Figure 2** shows the process of ground water recharge [5].

3. Irrigation trends, potential and risks

The crop yield is usually dependent over the ground water as it is responsible for strengthening and radiating the cropping system which has an upper hand over household food security and income of Farmers In larger parts of Asia, North Africa and Middle East locations depends over Groundwater as it has boosted agricultural production in an enhanced manner. Due to the vast usage of land and private wells in an unregistered condition, the ground water usage has become unsustainable leading to a serious threat in rural and urban areas in the forms of pollution and over exploitation of resources that has an impact over socio economic benefits.

Due to the risk caused under the exploitation of ground water has lead to the need of developing Groundwater development research centers and policies that quantify both the potential and the risks, pooled with innovative management and policy solutions which can help guide future reliable and flexible groundwater use.

4. Global

The world is surrounded by 99% of sea water, at which only one percentage covers the water which could be adopted as potable water. This 1% of potable water covers the fresh water that is available at the streams and the ground water which is stored under the aquifers. Among which the 40% of ground water is currently utilized for the world's irrigation purpose alone, as it provides about 13% of total food production and also 44% of irrigated food production globally [6, 7].

Though the food production and agricultural crop yield has significantly increased, at certain parts of the world the depletion of ground water has been occurred leading to replenishment levels.

An estimated amount of 14 to 17 percent of food produced with groundwater relies on unsustainable mining of groundwater resources globally. Regionally, the reliance on depleting groundwater sources for food production is greatest in South Asia, the OECD countries, East Asia, and the Near East and North Africa (MENA), where 15 percent, 16 percent, 21 percent, and 25 percent of total crop production from groundwater, respectively, is unsustainable [6].

Overexploitation causes groundwater tables to descend, water quality to depreciate, environmental dilapidation, pumping costs to ascend, and crop yields to plummet. Additionally, overreliance on depleting groundwater resources poses momentous risks, jeopardizing prospect food production and global food security as agricultural land becomes less prolific, if not nonproductive. These issues are progressively more being discussed at the national and international levels, such as at the World Food and Agriculture Organization's Global Forum for Food and Agriculture.

5. Optimizing groundwater policies to boost sustainable groundwater use

Effective groundwater use has been constrained due to incorrect policies. The ground water is underutilized in many parts of India since there is abundant seasonal rainfall in the Indian states. The electric tube well pumps method were allowed to take by farmers in order to preserve the state ground water underground water legislation which was enacted in mid 2000s.the agriculture growth estimated to be 5% per in the year 1980s and 1990s, it fell to 2% in 2000s due to legislation where farmers has limited access to electric pump which consumes high expensive diesel pumps and electricity. State's actual abundance of groundwater was reflected by scientist and state officials, who reviewed the changing policies which was done few years later after amendment of ground water legislation. The smallholder farmers are encouraged to use to use available water to boost agriculture productivity by relaxing the electric pump processing and flat electric connection were introduced to farmers. This policy change was implemented in the year 2011 under the recommendation of farmers. This policy results in higher net returns, better water quality and higher value outputs as this policies brought modification where over 140,000 new electric connections for tube well were established. This improves the irrigation on 250,000 ha for approximately 1.3 million water users [8]. Researchers and officials are now reviewing the policy to see if any additional changes are required to protect the state's interests [6, 9].

6. Managing groundwater via incentives from the energy sector

Apart from water source, for sustainable groundwater irrigation incentives can be helpful to the poor farmers. In India, Ground water is considered to be the critical factor for crop growth, since the implementation farm power subsidies in 1970s, most farmers have electric connections. This leads to depletion of ground water. Solar power is considered to be climate change mitigation strategy which was introduced recently. This solar power is considered to be green alternative to electric pumps and versatile power. If carefully designed programmes are not implemented, it may endanger groundwater sustainability due to the drastic reduction in pumping cost. Effectively treated solar power is considered to be a 'cash crop'. The intervention of solar power results in production of excess power which urges the state's local electricity company to buys back the excess power from farmers. These profits of farmers can be used as an economic incentive to irrigate their crops effectively where groundwater and energy can conserved.

In India carbon emissions has been cut down by 4–5 percent per year by the use of solar power. The world's first solar cooperative was formed in Gujarat where researchers recently testing the 'smart solar pump" model. At the same time excess solar power can be purchased back at transaction costs than purchasing power from individual farmers. Sustainable groundwater use can be ensured if the model is scaled up. This model also helps in increasing the agricultural productivity and profitability, which reduces the demand for fossil – fuel energy and reduces the carbon emission into the atmosphere. **Figure 3** shows the application of technology and incentives achieved through them [10].

Figure 3.
Incentives via energy sector.

7. Supporting urban aquifers via inter- sectoral water transfers

Researchers are studying the business of rural–urban water transfers and aquifer recharge with wastewater to address the significant challenges that urbanization poses to aquifers and their management. In Bangalore, India, groundwater recharge is done by directing the urban wastewater into reservoir called peri-urban tank. The revitalization reservior had been dry for more than 18 years, now it is making groundwater accessible to peri-urban and urban users once again. The Llobregat delta near Barcelona in Spain and the Mashhad plain in Iraq follows the same trend, where the freshwater is exchanged with treated wastewater by farmers and city people, while wastewater also replenishes the city's aquifer. Researchers are investigating various models to increases the efficiency in exchanging the freshwater with treated water.

8. Enhancing community management through experimental games

To preserve the surface irrigation, forests, and protected areas, there are numerous successful models available but there is lack of models to validate the ground water. Many experimental games has been introduced by researchers in Andhra Pradesh, India to understand the social and biophysical contextual variables to solve groundwater challenges which influences the crop production and daily needs of people. The farmers refused to follow the legislation of groundwater use as a result the aquifers are used excessively which leads to reduction of water in water table. Researchers found out this due to lack of information among farmers, who believes the depletion of ground water is due to no proper rainfall, but they had not realized that crop choosing also plays major role in reduction of water in aquifer, since few crops need lot of water contribution. The experimental games plays major role in allowing the farmers to understand the connection more clear. Water management is improved by implementing by this games which proves to effective starting point [6, 11]. **Figure 4** shows about the management of ground water [12].

Figure 4.
Management of ground water.

9. Estimation methods

In order to maintain the sustainability, the process of ground water recharge is adopted at which the recharge rates are usually difficult to measure. The difficulty in quantifying the rate of ground water recharge has become a tedious process because [13], it is in need to measure the related processes such as transpiration, evaporation and infiltration for achieving balance in the environment. **Figure 5** shows the cycle of water flow [14].

9.1 Physical

For the purpose of estimating the recharge at the ground water table, it is necessary to employ physical methods for determining soil physical principles. The direct physical methods are also incorporated to quantify the amount of water that passes through the soil pores below the root zone. While the indirect methods depend on determining or identifying soil parameters, which is used for measuring the potential of actual ground water recharge. At this method, the quantified ground water denotes that amount of drained ground water during no rain season and if the area of catchment is already known, then the recharge rate could be identified through the base flow itself.

9.2 Chemical

This kind of method could used while the ground water is found at a deeper zone, where the chemicals like chloride, isotropic tracer [15] are utilized in presence of relatively inert water soluble substance poignant through the soil.

9.3 Numerical models

The ground water recharge is identified using numerical codes such as Hydrologic Evaluation of Landfill Performance, UNSAT-H, SHAW, WEAP, and MIKE SHE. HYDRUS1D, a 1D programme that is available online. These kinds of codes usually use datas obtained by monitoring climate and soil to determine the amount of

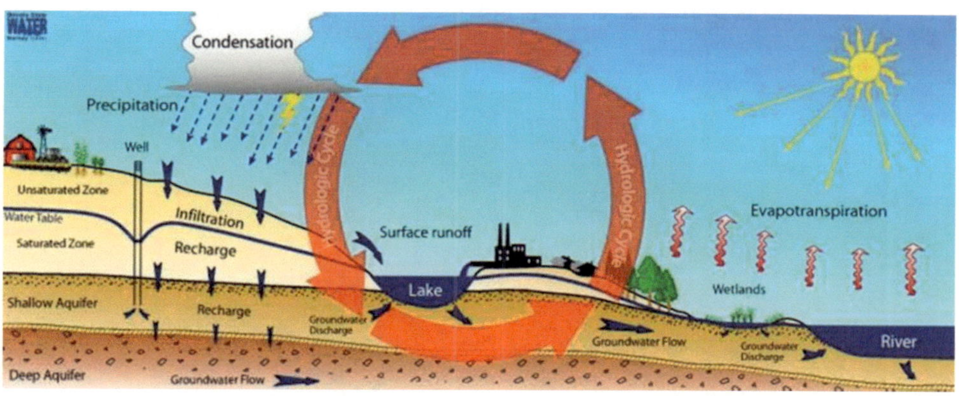

Figure 5.
Cycle of water flow.

recharge to be done. At these numerical models, the flow of water at the vadose zone in the aquifer could be modeled with the consideration of Richards's equation.

10. Techniques involved ground water recharge

There are two methods to recharge water table or ground water (**Figure 6**).

10.1 Natural method

The Natural methods involve precipitation on soil, which enters into the soil. The other form of natural recharge occurs when there is leakage of water from lakes, ponds, streams etc.

10.2 Artificial method

This is efficient method in comparison with natural method, the amount of water consumed per capita is decreasing day by day and to increase the ground water recharge we need artificial method. This artificial methods involves various techniques (**Figure 7**).

10.2.1 Basins or percolation tank

A percolation tank is a anthropogenic method which consist of highly permeable land in its reservoir, which allows the excess runoff water to percolate and it will increase the storage of ground water. The percolation tank should be constructed using second to third order steams, particularly on extensively fractured and worn rocks with lateral continuity downstream; and the recharge area downstream should contain enough wells and cultivable land to benefit from the increased ground water. The percolation capacity in the tank bed determines the tank's size. Percolation tanks are designed in a way to store 0.1 to 0.5 MCM capacities. There is also a column of 3 and 4.5 cm provided for ponded water.

The objective of percolation tank is ground water recharge; it allows leakage under bed's seat. The majority of percolation tanks are earthen dams, with masonry structures serving only as spillways. For dams up to 4.5 m in height, keying and benching

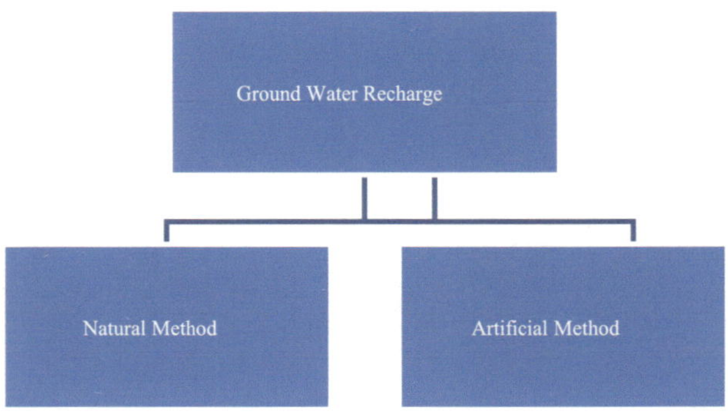

Figure 6.
Methods recharge.

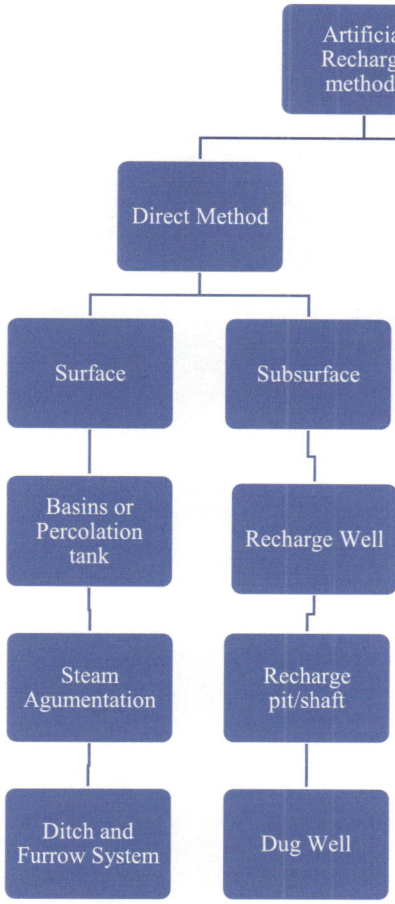

Figure 7.
Artificial recharge.

between the dam seat and natural ground is sufficient, and cut off ditches are not necessary (**Figure 8**).

10.2.2 Ditch and furrow system

Shallow, flat-bottomed, and closely spaced ditches or furrows provide maximal water contact area for recharge water from the source stream or canal in locations with variable topography. This technique is less sensitive to silting and preparation of soil is very less [16].

10.2.2.1 Lateral ditch pattern

The surface runoff water is streamed into a feeder canal/ditch, from which smaller ditches at right angles are formed. The flow rate from feeder cannal to ditiches is monitored by Gate valves .the uniform. The factors such uniform velocity, maximum wetted surface and furrow depth are determined by topography. A return canal transports the excess water, as well as any remaining sediment, back to the main stream.

Figure 8.
Percolation tank or basins.

10.2.2.2 Dendritic (tree-like) pattern

The water is streamed into the main canal. This pattern is continued until all the water are infiltrated into the ground through smaller ditches which are connected with main canal in tree like pattern.

10.2.2.3 Contour pattern

Ditches are excavated in accord with the area's ground surface outline. The ditches are designed in such way to move front and back to navigate the spread area multiple times and switchback if ditch get close to the stream. The ditch joins the main stream to downstream at its lowest point, returning extra water to it.

10.2.2.4 Steam augmentation

The most important source of ground water reservoir is leakage that occurs from natural streams or rivers. When the total available water supply in the stream/river exceeds the rate of infiltration, the excess is lost as runoff. The infiltration can be increased by installing check bunds where runoff water can be stopped when the total available water exceeds the infiltration range.

To recharge the stored water in short span of time, there must be check dam with sufficient thickness of porous bed or weathered formation. The Height of the structure should be less than 2 m and the stored water in the structures are restrained to stream course. Series of check dams are constructed at various places to control the maximum runoff.

10.2.3 Recharge well

The main important components of groundwater recharge are:

- Catchment
- Conveyance

- Filtration
- Recharge well

10.2.3.1 Catchment

The area used for collection of water is called catchment the area can be smooth or unsmooth surface, such as rooftop, plot (garden, driveway etc). the water flowing through catchment need not to be clean, once they percolates through soil the dirt particles in water can be removed since soil acts a natural filter. The only care should be taken when using chemical residues, even small amount would have impact on whole ground water. Utmost care must be given while considering gardens and farmlands since there are high chances of using fertilizers and pesticides (**Figure 9**).

10.2.3.2 Conveyance

There are different conveyances for different surfaces. Gutters or downpipe acts as conveyor for rooftops. Channels/pipes are used to carry water for plots. Storm water drains in layouts, campuses, and large residences convey runoff from various land use catchments (Eg: parks, roads, other paved areas, gardens and excess rooftop runoffs). As a result, storm water drains are an important conveyance system for recharge (**Figure 10**).

10.2.3.3 Filtration

Filtration is a process to remove the dirt or residual. In this method, soil act as filtration medium.

10.2.3.4 Recharge well

There are multiple ways to incorporate Recharge wells in the real estate development. The run off rain water, the catchment area decides the design and number of

Figure 9.
Catchment.

Figure 10.
Conveyance.

recharge well. Large numbers of recharge wells has to built in valley points, since they get highest rainfall. The valley point can be identified from topography study.

10.2.4 Recharge pit

Recharge pits are constructed for recharging a trivial aquifer. They can be circular rectangular or square. They are generally 1–2 m wide and 2–3 m deep. The Water that is diverted to ground water table should be residual free. In order to achieve that condition the pits after excavated are filled with stones, sands and boulders which act as a filter medium. Cleaning of the pit area should be done from time to time. Small buildings can have roof top area up to 100 m^2. If the pit is trapezoidal, the side slopes should be steep enough to avoid silt deposition [17].

10.2.5 Recharge shaft

A recharge shaft may be dug or drilled. The recharge shaft are filled with pebbles, sand and boulders, where cleaning can be done by removing the top layers alone and refilling it. This type's ground water recharge is most commonly seen in area where the shallow aquifer is located. They are constructed where there is low permeability and ends in more permeable layer. Usually the depth of the recharge shaft varies from 10 to 15 m below the ground level [17].

10.2.6 Dug wells

Dug wells are the most common method of extracting water from ground water table. Dug well are bored manually since the labour charges are minimal. in this methods the dug well are bored deep down until it reaches the ground water level from where water can be obtained with the help of pumps and by other means. The diameter at initial phase is 1.5 meters and at final stage after lined with stones, bricks, or tiles to prevent disintegration, it is 1.2 meters but in some cases the diameter at initial phase is 15 meters. Since the excavation cannot do beyond the ground water table it will be only 10 to 30 feet deep. There are also high chances of contamination, since the well are shallow. In order to prevent the contamination there must be certain features (**Figure 11**).

10.2.6.1 Dug well construction features

In order to prevent the contamination, the well needs to be covered with cap or concrete curb that stand above the ground level and also the land surface around the well should be slope in nature so that water runs from the well flows into land and will not form pool near the well. The well needs to be cased with concrete and a cement grout or bentonite clay to avoid disintegration of well. The pump of the well should be separately placed rather than placing it near the well.

10.2.7 Induced recharge

It is an indirect method of artificial recharge involving pumping from aquifer hydraulically connected with surface water, to induce recharge to the ground water reservoir. When the cone of depression intercepts river recharge boundary a hydraulic connection gets established with surface source which starts providing part of the pumpage yield. In such methods there is actually no artificial build up of ground water storage but only passage of surface water to the pump through an aquifer. In this sense, it is more a pumpage augmentation rather than artificial recharge measure.

Figure 11.
Dug well.

Usually the ground water recharge is possible in a proper way as the abandoned channels in the hard rock regions act as a good site for the induced process of recharging. For this process, the check weir at stream channel helps in infiltration of water from surface reservoir to the abandoned channels which are then directed towards the aquifers.

During the unfavorable hydro geological conditions, induced recharge process has a greater advantage and also it helps in improving the quality of surface water which is generally improved due to its path through the aquifer material before it is discharged from the surface. Also, for the purpose of obtaining very high water supplies from the river bed deposits or waterlogged areas, collector wells are constructed. In India, these kinds of collector wells are constructed in the river bed such as Yamuna bed at Delhi and also in other places in Gujarat, Tamil Nadu, and Orissa as these wells are economical due to large discharges and lower lift even though initial capital cost is higher as compared to tube well. If the phreatic aquifer is situated near the river of limited thickness, then instead of horizontal wells, the vertical wells could be adopted which is rather effective than the other [17].

11. Creation of resilience to the environment

For agricultural production, the groundwater has the capability to improve the potentially of resilience in food-insecure areas around the world. Due to this the rural income of the farmers can increase and also to withstand climate shocks and water variability. However, in order for groundwater to contribute to the sustainable intensification of agriculture, it is critical to understand where to invest in groundwater development and how to manage groundwater resources in a sustainable manner. WLE has identified potentially usable groundwater resources in Africa, supported important policy changes to improve the sustainable use of groundwater in eastern India, and created maps and new tools that can be used to implement new groundwater policies [6].

12. Conclusion

Hereby, we conclude that the ground water recharge through artificial method is effective while compared to that of natural method. Though the installations of artificial methods are cost consuming rather than natural methods, the rate of ground water recharge level is in an increased manner along with the consideration of water quality. Since, the water consumption rate per capita area is increasing day by day; the need for creating the source of ground water recharge is mandated in order to satisfy the demand.

Author details

Muthuminal R.[1*] and Mohana Priya R.[2]

1 Department of Agriculture Engineering, SNS College of Technology, Coimbatore, India

2 Department of Food Technology, SNS College of Technology, Coimbatore, India

*Address all correspondence to: muthu.minal151@gmail.com

References

[1] Barthel R, Stangefelt M, Giese M, Nygren M, Seftigen K, Chen D. Current understanding of groundwater recharge and groundwater drought in Sweden compared to countries with similar geology and climate. Geografiska Annaler: Series A, Physical Geography. 2021;**103**(4):323-345

[2] Shiklomanov IA. Appraisal and assessment of world water resources. Water International. 2000;**25**:11-32

[3] Healy RW. Estimating Groundwater Recharge. Cambridge: Cambridge University Press; 2010

[4] Groundwater Recharge—Its Importance and Scalability. 2021. Available from: https://aquaoso.com/blog/groundwater-recharge

[5] Groundwater Recharge 2012. Available from: http://kayjayr-akshay.blogspot.com/2012/05/how-to-recharge-ground-water-and.html?m=1

[6] Building Resilience through Sustainable Ground. 2014. Available from: https://wle.cgiar.org/solutions/groundwater

[7] Villholth KG, Sood A, Liyanage N, Zhu T. Forthcoming. The role of groundwater and depleting aquifers in global irrigated food production. Nature Communications. Jun 2017;**1**:1-12

[8] Irrigation System Cost. 2021. Available from: https://rethority.com/irrigation-system-cost

[9] Buisson MC. What does pump sets electrification change? Impacts on cropping patterns, productivity and incomes in West Bengal. Paper presented at the ICID 26th Euro-Mediterranean Regional Conference and Workshops on Innovate to Improve Irrigation Performances. Workshop: Irrigation and Energy, Montpellier, France, October 12-15, 2015. p. 5

[10] New Approach to Farming is Transforming Agriculture. 2021. Available from: https://www.edf.org/ecosystems/resilient-agriculture/precision-agriculture

[11] Meinzen-Dick R, Chaturvedi R, Domènech L, Ghate R, Janssen MA, Rollins ND, et al. Games for groundwater governance: Field experiments in Andhra Pradesh, India. Ecology and Society. 2016;**21**(3):38

[12] Management of Ground Water. 2021. Available from: https://wle.cgiar.org/thrive/2013/02/12/strategies-managing-indias-groundwater

[13] Reilly TE, LaBaugh JW, Healy RW, Alley WM. Flow and storage in groundwater systems. Science. Jun 2002;**296**(5575):1985-1990

[14] Groundwater 2021. Available from: https://www.esri.com/news/arcuser/0408/groundwater.html

[15] Allison GB, Gee GW, Tyler SW. Vadose-zone techniques for estimating groundwater recharge in arid and semiarid regions. Soil Science Society of America Journal. 1994;**58**(1):6-14

[16] Guide on Artificial Recharge to Ground Water. 2010. Available from: http://cgwb.gov.in/documents/ArtificialRecharge-Guide.pdf

[17] Artifical Recharge by Sinha Ray with a Contribution on Sand Dams by Milu. Available from: https://www.samsamwater.com/library/TP40_6_Artificial_recharge.pdf

Chapter 4

An IoT-based Immersive Approach to Sustainable Farming

Pratik Ghutke and Rahul Agrawal

Abstract

Despite the fact that the agricultural process is more data-driven, exact, and intelligent than ever before, the reality is that today's agriculture industry is more data-driven, precise, and intelligent than ever before, regardless of public perception. Virtually every industry has been altered by the rapid expansion of Internet-of-Things (IoT)-based technologies, including "smart agriculture," which has transitioned from statistical to quantitative methodologies. Such large advancements are upending conventional farming practises and offering new chances in the middle of numerous issues. A new paper looks at the promise of wireless sensors and the Internet of Things in agriculture, as well as the challenges that may occur when these technologies are integrated with traditional farming methods. Using Internet of Things (IoT) devices and communication protocols, wireless sensors utilized in agriculture applications are fully investigated. Sensors for soil preparation, crop status, irrigation, insect and pest detection, and other agricultural applications are on the list. From sowing to harvesting, packing, and transportation, this technique is explained. This article also discusses the use of unmanned aerial vehicles for agricultural monitoring and other useful purposes, such as crop yield optimization. When feasible, cutting-edge IoT-based agricultural ideas and systems are presented. Finally, we highlight present and future IoT trends in agriculture, as well as possible research challenges, based on this comprehensive analysis.

Keywords: internet-of-things (IoTs), smart agriculture, advanced agriculture practices, sensors

1. Introduction

Agriculture is the economic backbone of India. Since agriculture is our major source of sustenance, life would be impossible without it. The farmer needs to labour every day of the week to produce the harvest, which lowers his revenue, so he must look for alternative sources of food, especially since horticulture is becoming less popular. Mechanization is so critical in the rural cycle. Accordingly, this work proposed a framework so that livestock farmers could productively carry out their horticultural practices from remote locations while providing fewer ideal opportunities for farmland. In this framework, all the equipment works alone with the help of sources of information from sensors that are continuously checking the rural land, and ranchers can screen whether everything is going well or some activity should have been done. The entire cycle is controlled and monitored by a programmable regulator. Solar primarily based power is the maximum considerable source of electricity inside the whole plants. Solar based electricity is not most effective a reaction to the contemporary electricity emergency, but additionally a sort of energy that is related to weather. The solar era is the era of efficient use of solar energy. Solar chargers (solar powers) are momently used to power street lights and water radiators.

1.1 Problem statement

The unexpected drop in the price of solar chargers is increasing their use in various fields. This innovation is used in horticultural water supply systems. Inside the cutting-edge nation of energy emergency in India, a solar powered water system infrastructure can be an inexpensive option for farm animals' farmers. it's far a conductive to imparting green energy that releases strength when the underlying is estimated. Water device design is a dangerous water detection method for the vicinity or soil that bureaucracy the primary basis of our yield shape. Water for the most element should be made available within the fields or through pits. This framework will lessen the responsibility of the rancher and hold enough soil pleasant for better development. From that factor on, the development was viable in that it introduced a ranch water system into the field, which quickly killed units. This mechanical fin is the entire structure of an electric water system washed out of the field. In term of GSM, there are two important developments in the water supply system. "GSM" is an arbitrary and basic moderator or handler. Global System for Mobile Communications (GSM) is a standard used to address infrastructure for mechanized data collection. Nowadays the agriculturist is focused on increasing output while keeping expenses down. This strategy necessitates an innovative approach to handle the problem and boost output and profitability while minimizing food production's environmental impact.

1.2 Research gap

Precision agriculture in the cultivation field necessitates several ecological criteria that indirectly analyze the aforementioned issue. However, because the intra-field variability in sugar beet production and quality is unknown, it must be assessed in terms of soil qualities and microclimate conditions. The crop environment's observed variability can therefore be managed by tailoring inputs to the places where they are needed. As a result, the precision agriculture model has been used and advanced in this study to track crop growth while taking into consideration field changes air temperature, soil moisture, soil type and other factors. Precision agriculture improves agricultural earnings and resource usage by reducing the use of traditional management practices.

1.3 Research objective

Precision farming has an impact on yield-based crop development depending on the soil type. Precision agriculture techniques have the potential to improve the Indian agricultural economy. The suggested research focuses on identifying limiting factors that have an impact on crop output. Providing precise water-based soil conditions predictions.

1.3.1 Irrigation system in India

In India, rural areas hold 18% of the US debt (GDP) and attract 49% of the workforce. According to records, 2022 of the military will be deployed in agricultural areas. 0205,000 siphons are continuously delivered to rural areas. Production of waste water devices, including siphon sets, uncontrolled water intake, unplanned water supply systems, water and energy savings, water loss and lack of expe.

Table 1 shows the effectiveness of different yields in India and the world. Unlike the general use of the world, India has a low yield potential. This is due to a decrease in the use of water in the water system as a result of the development of various misfortunes,

As shown in **Figure 1**. The repetition of dry season is increasing steadily, adversely affecting the usefulness of the crop. The dry season after that may be a repetition of the dry season, it is no shaggy dog story. The recurrence of drought in exceptional regions of India is proven in determine three. Streamlining water system productivity and electricity use is a vital take a look at inside the current water device state of

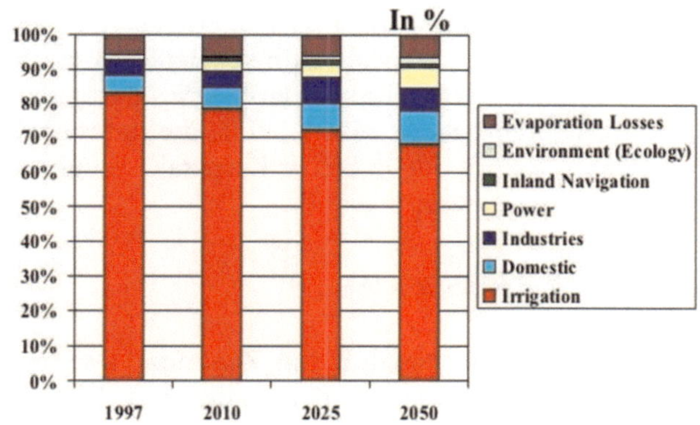

Figure 1.
Availability of water for irrigation [1].

Plant	India	World	Highest
Cotton	4.69	7.36	11.4 (China)
Maize	1.9	4.6	14.6 (Israel)
Oilseeds	0.87	1.87	4.09 (Germany)
Onions	10.7	17.8	46.8 (USA)
Paddy	2.97	3.9	8.9 (Egypt)
Potato	19.6	15.9	48.6 (Belgium)
Pulses	0.59	0.9	4.6 (Netherland)
Soya	0.95	2.34	3.58 (Turkey)
Sugarcane	69	63	1.17 (Zimbabwe)
Tomato	16.5	26.8	61.9 (USA)
Wheat	2.76	2.88	7.9 (Germany)

Table 1.
Crop productivity (MT/ha).

affairs. Controlling using water sources is extremely fundamental to reap maximum core yield in higher energy use. Maintaining water device efficiency is extremely challenging. The effectiveness of a water device is the ratio of water utilized by the dust to the filth that is distributed through the water system. The effectiveness of the water device has to be cohesive, i.e., the water utilized by the filth and the contribution of water to the effluent via the water device ought to be same. If the performance of the water system is more prominent than the consent, the structure of the water device cannot satisfy the dirty water hobby. If the efficiency of the water unit is not as high as the clutch, this redundancy is supplied by using the water machine structure, which induces wastage of water.

The water system's structure is prone to large amounts of water catastrophe, reducing the effectiveness of the water system. What's more, the water system productivity of the different structures is analyzed, given in **Table 2**.

Here in **Figure 2**, Experts are baffled by the frequency of droughts shown in official data. According to the agriculture ministry's drought management section, Assam would experience a drought-like condition once every 15 years. However, the trend shows that the state has had three droughts in the last 9 years (including this year). Though Bihar and Uttar Pradesh are only vulnerable once every 5 years, the state has seen three drought-like circumstances in the last 5 years, while Uttar Pradesh has experienced two. The statistics for the southern states, particularly Karnataka and Andhra Pradesh, are considerably more mixed. While the states have had numerous droughts in the last 5 years, a study suggests that Karnataka is vulnerable once every 5 years and Andhra Pradesh twice every 5 years.

1.3.2 Factors responsible for low productivity

Existence of Big Farmers: Even though India's Zamindari system has been abolished, rural large farmers continue to play a shadow role. These large landowners control rent, tenure, tenancy rights, and other aspects of renters' lives. As a result, the situation of tenants is deteriorating day by day. It is quite difficult to increase productivity using solely modern technologies in this type of tenure structure.

High Land-Man Ratio: Huge demographic pressures characterize Indian agriculture. According to the 2001 Census, over 72.2 percent of the entire population lived in rural areas, with agriculture employing nearly three-quarters of the total rural working population, or nearly 228 million employees (out of 310.7 million workers).

Sources of losses	Leakages	Evaporation	Total
Canals & Branch Canals	13.6	3.4	17.0
Distributaries	6.4	1.6	8.0
Water Courses in the Field	16.0	4.0	20.0
Field application Losses	13.2	3.3	16.5
Total	49.2	12.3	61.5

Comparative Efficiency of Irrigation System
Surface 30–40%
Sprinkler 60–70%
Drip Irrigation 80–90%

Table 2.
Water losses in percentage in India.

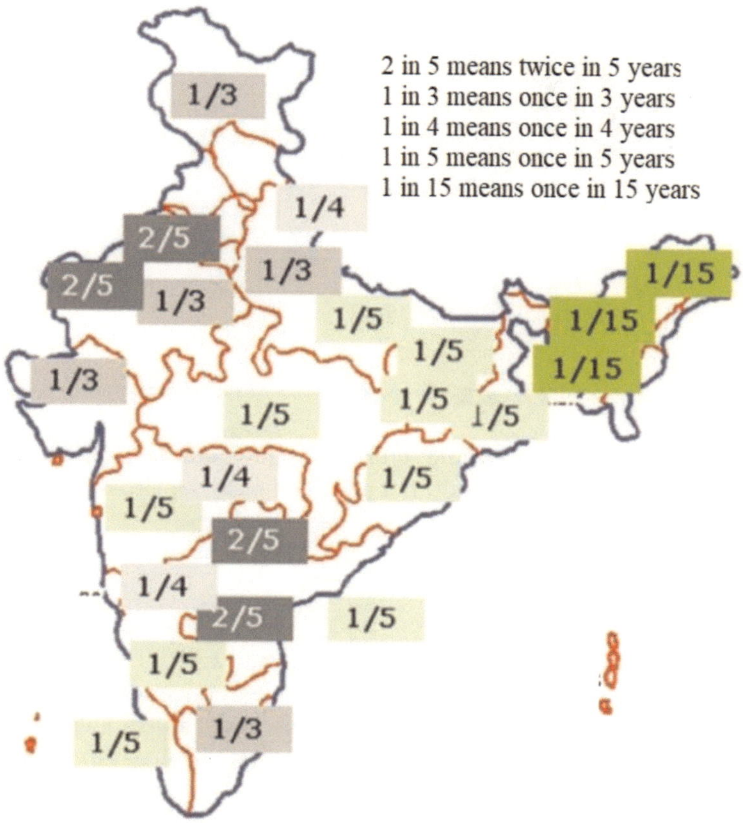

Figure 2.
Frequency of occurrence of drought in various parts of country.

Uneconomic land subdivisions occur as a result of population growth. All of these factors contribute to low production.

Rural Environment: In India, the rural social milieu is a major contributor to low productivity. Farmers in India are lethargic, illiterate, superstitious, have a primitive outlook, are conservative, unfit, and resistive to modern farming methods. Farmers' marginal productivity in agriculture is zero, due to the family-based farming method. Credit and marketing facilities that are irregular and insufficient: According to Raj Krishna's research, poor farmers are unable to effectively spend money in the land during the peak season of agriculture due to a lack of and insufficient availability of agricultural loans at a low rate of interest. Furthermore, crop marketing is regulated by intermediaries or touts. As a result of all of this, agricultural productivity was low. Modern technologies are lacking: In India, over 60% of cultivable land lacks irrigation facilities. In 2000–2001, only 75.14 million hectares (out of 87.94 million hectares) were irrigated. As a result, the green revolution's 'Package Program' is ineffectual across the vast majority of India's gross cultivated areas. Degradation of the Ecosystem: According to the Indian government, 329 million hectares (almost half of the country's land) have already been degraded. This leads to a yield loss of 33 to 67%. Furthermore, 5% of the land has been ruined to the point where it can no longer be utilized.

Figure 3 shows that interest in this issue has grown over time as measured by the number of papers published every year. The smaller number of papers for 2019 is

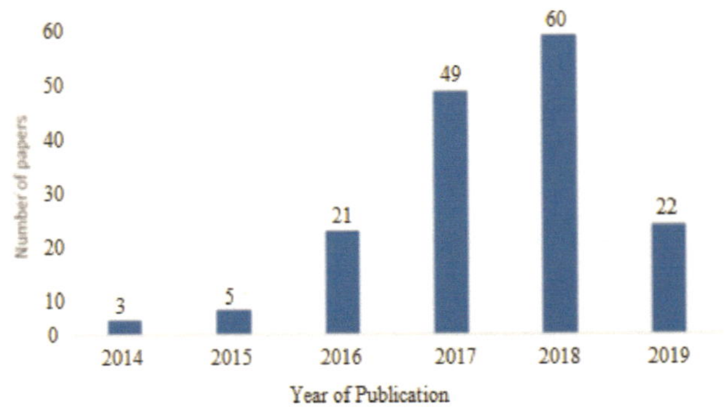

Figure 3.
Annual number of articles presenting IoT irrigation systems that have been published [2].

owing to the fact that the year was not quite through when the paper selection procedure was concluded. As a result, not all of the papers written in 2019 have been published.

1.3.3 Water management

There are various ways to distribute water irrigated agriculture is a type of agriculture that uses water as a source of input. The effectiveness of the various possibilities varies, in some circumstances, a specific technique for a given crop should be adopted. Irrigation can be done in a variety of ways, but they can all be categorized into the following groups: When it comes to how water is spread, we can examine the flood irrigation, (ii) spray irrigation, (iii) drip irrigation, and (iv) nebulizer irrigation. On the subject of sensing systems, we can discuss I unplanned irrigation, in which the amount of water is not calculated or estimated; (ii) planned irrigation, in which the water is supplied according to estimated demands over a year; and (iii) adhoc irrigation, in which the amount of water is estimated based on sensor readings. The great majority of the papers in this section propose to distribute water using pumps and valves in conjunction with sensors that assess ambient conditions in order to determine water needs. In this part, 83 of the 89 reviewed articles include detailed information about the planned irrigation system, while the remaining six just state that irrigation actuators are present (see **Figure 4**). There are 49 articles that just indicate that their system has motors/pumps (40 papers) or valves (nine papers) without providing any additional information. 19 of the studies that provide additional information use sprinklers (the most common irrigation technique) [3–21], eight utilize drip irrigation [22–30], two propose sprayers [22, 31], and the rest use highly specialized irrigation systems or it can be used on multiple systems [32]. Three papers [16, 18, 27] suggest using a fogging system in conjunction with the main irrigation system, whereas two papers [17, 28] suggest using fertigation in their systems.

1.3.4 Major applications

Every aspect of traditional agricultural methods can be significantly transformed by incorporating cutting-edge sensor and Internet-of-Things (IoT) technologies into farming practises. Smart agriculture has the potential to reach previously unimagined heights thanks to the current seamless integration of wireless sensors and the Internet

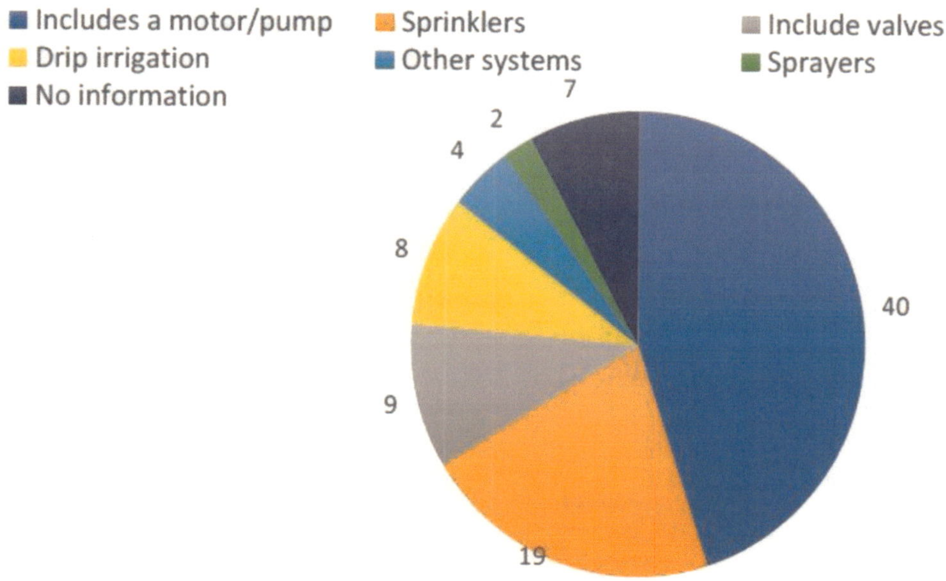

Figure 4.
Number of papers that propose different irrigation systems [2].

of Things. By utilizing smart agriculture approaches, IoT can help improve answers to many traditional farming difficulties, such as drought response, yield optimization, land suitability, irrigation, and insect management. **Figure 5** depicts a hierarchy of critical applications, services, and wireless sensors in smart agriculture applications. While key examples of how modern technology might aid in improving overall efficiency at various levels have previously been covered.

The Internet of Things (IoT) is beginning to affect a wide range of sectors and companies, spanning from manufacturing to health, communications, energy, and agriculture, in order to reduce inefficiencies and improve performance across all markets. When you think of the word, two words come to mind: [34, 35]. Current applications appear to be simply scraping the surface of IoT's potential, with the full extent of its influence and applications still to be seen. Given the recent increase, we may expect IoT technology to play a vital role in a number of agricultural applications. This is due to the capabilities of the Internet of Things, which include basic communication infrastructure (used to connect smart objects to sensors, vehicles to user mobile devices via the Internet), as well as a variety of services such as local and remote data collection, cloud-based intelligent information analysis and decision making, user interfacing, and agriculture operation automation. Such people have the power to change agriculture, which is currently one of our economy's least efficient sectors. **Figure 6** displays the important technology drivers in smart agriculture, whereas **Figure 7** depicts the major technological implementation roadblocks.

1.3.5 Demand for water

The water demand of the irrigation system is determined by estimating the amount of water required for best crop output. The estimated crop evapotranspiration (ETc) is used to determine the water demand; however, estimating the ETc requires knowledge of the reference evapotranspiration (ET0). ET0 was defined by Dorenbos

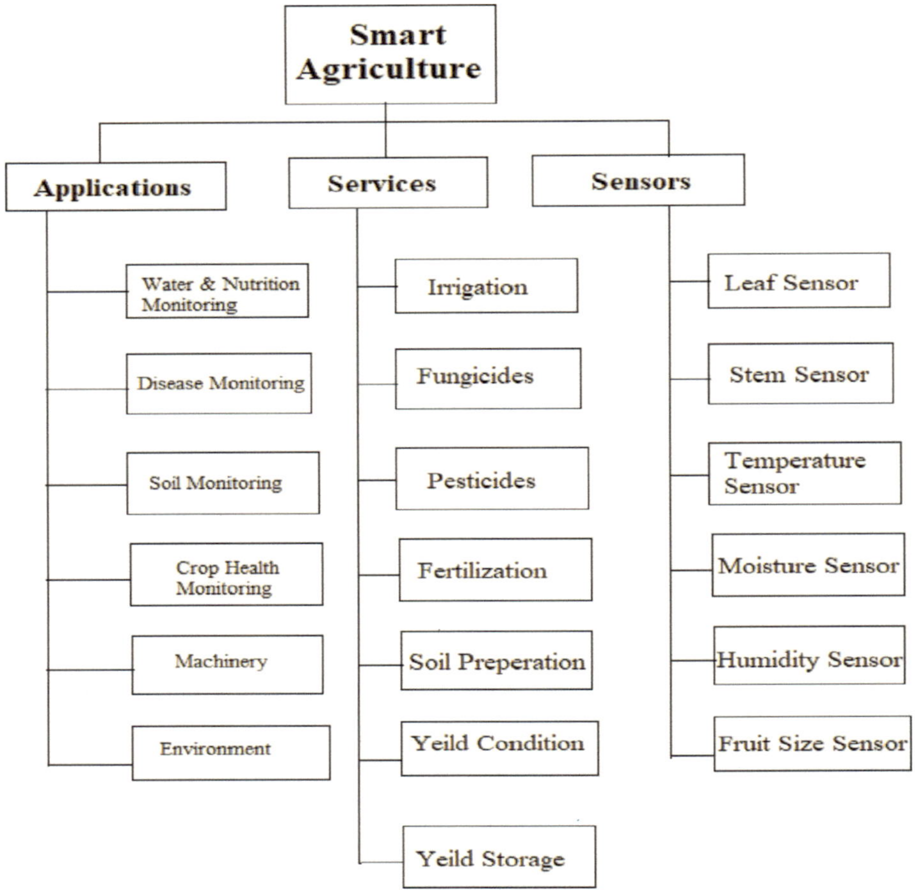

Figure 5.
A hierarchy of applications, services, and sensors exists for smart agriculture [33].

Figure 6.
Major challenges in technology implementation for smart agriculture [33].

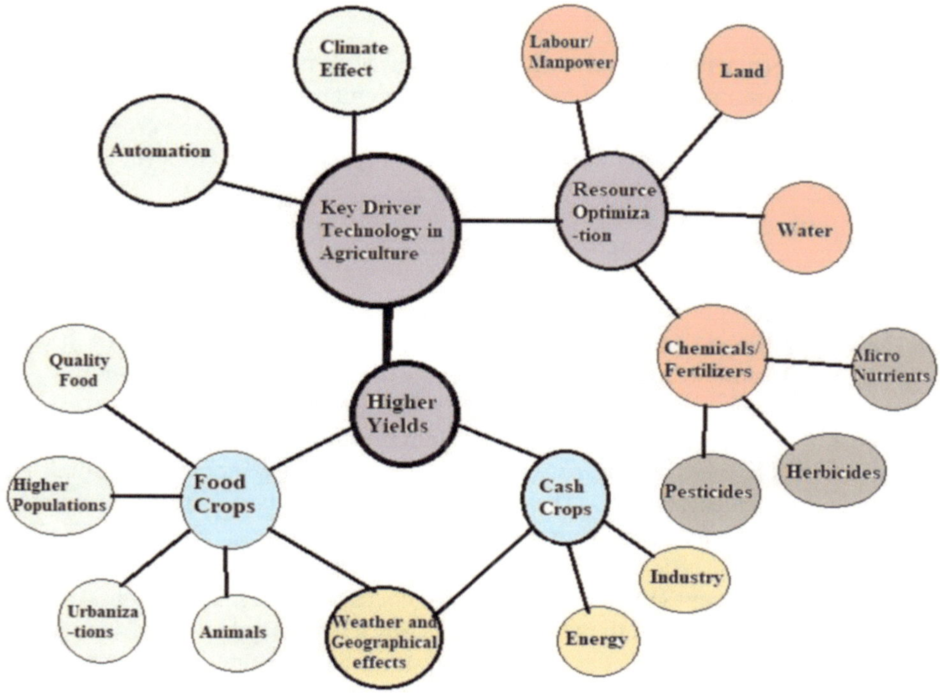

Figure 7.
The agricultural industry's key technological drivers [33].

and Pruitt [36] as a result of the total amount of under ideal conditions, water evaporated from the soil and a large area of grass-covered ground transpired a large amount of water (vigorous development and unrestricted access to water). The Penman–Monteith equation is the most extensively utilized approach [37], was used to calculate ET0, as illustrated below [2].

$$ET0 = \frac{0.408\,\Delta\,(Rn - G)\frac{900\;\gamma\;u2\;(es-ea)}{T}}{\Delta + \gamma\,(1 + 0.34u2)} \tag{1}$$

The ET0 value is measured in millimeters per day, Rn is the net solar radiation incident on the crop surface, G is the soil heat rate ($MJ/(m^2 day)$), is the psychrometric constant (kPa/C), u2 is is the steam pressure slope, expressed in kPa/C. is the wind velocity recorded at a height of two metres, es is the saturated steam pressure, and ea. is the actual steam pressure, all in kPa. Agrometeorological stations measured all climate factors to determine ET0, which is dependent on wind speed, sun radiation, air temperature, and relative humidity.

ETc was calculated using ET0 and the crop coefficient as inputs (Kc). The type of crop, the climatic circumstances, the soil's distinctive features, and the vegetative phase are all taken into account by Kc. The CNR (Chilean Irrigation Commission) bulletins [38, 39], as well as the paper headed "Reference Evapotranspiration, for the Determination of Water Demands for Agriculture in Chile" [40], tabulate values for each species and growth phase. Eq. (2) was used to compute crop evapotranspiration, ETc, where ET0 was changed based on the crop coefficient:

$$ET_c = K_c ET_0 \tag{2}$$

The crop coefficient is Kc, while ET0 is measured in millimeters per month (millimeters per month). The monthly net water demand was calculated using the Eq. (ND). ND is calculated using the difference between ETc and the crop's effective rainfall (Pe). ND refers to the water required by the crop's roots from the irrigation system.

$$ND = ET_c – P_e \tag{3}$$

The United States Department of Agriculture's Natural Resources Conservation Service (NRCS) developed a method for calculating Pe based on real rainfall [36]. It was estimated in this study using the monthly average of actual rainfall data from the national agroclimatic network (Agromet) [41]. Because any irrigation system water losses must be compensated, the irrigation system must provide more water than the net water demand (ND). Certain security elements were also implemented to ensure that the crop received at least the ND. The effects of deep percolation and surface runoff are factored into the drip irrigation system's application effectiveness (Ea), which was determined to be 90% efficient. Two further elements that influence water demand are the washing requirement (RL) and the coverage coefficient (Kr). The minimal amount of percolation water required to maintain a constant soil salinity and avoid an increase in salinity that could stymie crop development is known as RL. Water does not require to be applied to the entire anticipated surface of the crop when Kr is used. The value of Kr is determined to be less than or equal to unity. Equation shows the water requirement [2].

$$D = \frac{ND\ (1 + Rl)\ Kr}{Ea} \tag{4}$$

The irrigation schedule specifies how frequently and for how long water must be provided to the crops. The irrigation frequency interval and volume of water provided are determined by the amount of water kept in the root zone of the crops and how quickly it is consumed. The soil texture, soil structure (water percolation), effective root zone depth, crop type, and crop growth stage all influence irrigation frequency [3]. For high-frequency watering requirements, a short interval is defined (one, two, or more days). The goal is to maintain a consistent soil humidity [4]. The annual irrigation schedule, which detailed the frequency of irrigation for each month, was presented by an irrigation consultant. The daily irrigation demand (RID) in liters was calculated using Eq. (5) once the irrigation calendar was defined.

$$\text{RID} = \frac{DAc}{Di} \tag{5}$$

where Di is the number of irrigation days each month and Ac is the number of hectares of land covered by the crops. Ac was calculated using Eq. (6), which took into consideration the surface of each plant frame (PF) as well as the quantity of plants (Nplants)

$$A_c = PFN_{plants} \tag{6}$$

The length of time (ti) for which an irrigation system may run in order to provide enough water to meet the needs of the crops was determined using Eq. (7).

$$ti = \frac{RID}{qeNe} \tag{7}$$

where q_e is the volume flow rate supplied by the emitters in liters per hour, and N_e is the number of emitters [2].

1.3.6 Irrigation System's electricity demand

Drip irrigation, a water-saving irrigation technique that distributes water to crops through a pressurized network of valves, pipelines, and emitters, is one of the most widely used irrigation systems. The irrigation system pump is chosen based on the irrigation system head (necessary pumping pressure) and the amount of water that the crops demand. The irrigation system head takes into account the elevation head, the pressure drop due to friction in the pipes and singularities (i.e., valves), and the required working pressure by the emitters. The pump's electrical demand remains constant since the pumps in this study deliver a constant volume flow rate. Other research [5, 6, 42] suggested using a variable speed pump to modify water flow in response to changes in solar radiation, allowing for an enhanced irrigation regime. A control system that aligns water supply with solar radiation could aid energy optimization [7], particularly in off-grid environments; however, this option is not explored in this study. The optimal design achieves the lowest overall cost, which includes the operational cost (electricity cost) of pumping, which lowers as the pressure drop decreases, as well as the capital cost of the irrigation system. Pipe friction and singularity losses in valves and fittings are used to compute the pressure drop. The pressure reduces as the pipe diameter increases, cutting the operational cost; nevertheless, the capital cost climbs in lockstep. Reduced pressure loss can also be helped by selecting the right emitters and filters.

The pumping system was designed to manage the worst-case scenario, which occurred during the peak water demand month. The operational characteristics of the pump were determined using the pump characteristic curve, which was produced during an experimental standard test. The pump characteristic curve provided information on the system head (H), pump efficiency (), and electrical power required by the pump (Wp) as a function of the pumped volume flow rate of water (Q). Eq. (8) was used to calculate Wp:

$$Wp = \frac{Q\rho gH}{\eta p\eta m} \tag{8}$$

where ηp denotes the efficiency of a mechanical pump and m denotes the efficiency of an electric motor. In general, ηp was between 90 and 95 percent, and m was between 45 and 65 percent. Some high-efficiency pumps can attain up to 85 percent total pump efficiency (ηp ηm). Eq. (9) was used to compute the daily electricity requirement (Ed) once Wp was estimated (from Eq. (8):

$$E_d = W_p t_i. \tag{9}$$

1.3.7 Solar PV system design

Irrigation water demand is frequently seasonal, throughout the year, the PV system generates electricity. The solar PV system should be able to deliver the electricity required by the irrigation system in order to ensure a uniform distribution of the volume flow rate of water required by all crops. Reliable data on solar resources is required for proper PV system design. In this investigation, the Solar Explorer, an online tool developed by Chile's Ministry of Energy [10], was used. The Solar Explorer's solar PV model is used in this study. It's based on a Sandia National Laboratories model, which is outlined in the reference. Solar radiation data for each unique location was also collected using the Solar Explorer, which was then incorporated in the solar PV model. The reference goes into great detail on the radiation database and its accuracy. Internet of Things Research: Key Technology and Applications In this paper, the author discusses the importance of IoT and RFID. With proper administration and dependable transmission, IOT can connect all items anywhere, at any time. There are several strata to be found.

1. Layer of Access: Data is transferred from the sensing layer to the network layer via this layer.

2. Network Layer: To pool the knowledge resources of the network.

3. Layer of Middleware: To deal with real-time data processing.

4. Layer of Application: To combine the functions of the bottom system. In this work, RFID technology is utilized to distinguish enemy aircraft by machine. RFID devices can also be used for inventory control, transportation, high security, and high irresponsibility. One of RFID's most essential elements is antenna technology.

2. Common architecture designs for IoT irrigation systems in agriculture

This section will provide an overview of the most popular architectures for these systems. In IoT irrigation management solutions, multi-agent architectures are widespread [43, 44]. These structures provide a distinction between the numerous components that make up its structure. The difference is typically made depending on the architectural strata in which the elements are housed. For example, nodes higher in the hierarchy may act as a broker for nodes lower in the hierarchy [43]. The most of of designs are divided into layers or functional blocks that represent the main tasks that must be done [45]. These blocks or layers are considered generic and are found in the majority of IoT irrigation management system architectural designs. The essential components of these architectures are devices, connectivity, services, administration, applications, and security. IoT systems are made up of devices that are put in a specific location and may perform activities including detection, monitoring, control, and action. In order to convey the essential data, the devices must have interfaces that allow them to communicate with other devices. The information gathered by various sensors will be treated as a whole, and the results will be applied to various actuators. The data collected and the actions taken must then be sent between the devices. The use of communication protocols is required for this task. In the majority of circumstances, different communication protocols are used on the same IoT system in order for it to work together. Services may be required to complete tasks such as device discovery, device control, and data analysis. The user can interact with the system using the programmes. The user will be able to see data acquired through monitoring as well as data extracted once it has been processed utilizing the applications. On numerous occasions, the user can execute actions that he considers relevant to the scenario presented by the data, and the actions can also be performed automatically.

Finally, the security of the system may be considered. The three layers of IoT architecture have typically been believed to be perception, network, and application. After several research, an intermediary layer was built between the network and application levels. In cloud and fog computing environments, this layer, also known as the service layer, is used to store and process data. For the past few years, authors such as Ferrández-Pastor [46] have proposed a new architecture based on four layers: objects, edge, communication, and cloud. In their current architectural proposals, the authors use the edge layer to locate critical apps and perform basic control activities. According to [46], cloud (internet/intranet) can also include Web services, data storage, HMI interfaces, or analytic applications. An illustration of the architecture models is shown in **Figure 9**. These designs in Sensors 2020, 20, x 34 of 48, include devices, communications, services, administration, applications, and security. IoT systems are made up of devices that are put in a specific location and may perform activities including detection, monitoring, control, and action. To transfer the essential data, the devices must have interfaces that allow them to communicate with other devices. The information gathered by numerous sensors will be processed in general, and the results will be applied to various actuators. The observed data as well as the response actions must then be sent between the devices. Communication protocols are required for this task. In the majority of circumstances, different communication protocols are used on the same IoT system in order for it to work together. Services may be required to complete tasks like device discovery, device control, or data analysis. The programs enable the user to interact with the system. The user will be able to visualize information collected through monitoring as well as information taken from data after it has been processed using the applications. On numerous occasions, the user can take actions that he considers important to the scenario presented by the data, and these actions can also be taken automatically. Finally, assess the system's security. Traditionally, the three layers of IoT architecture have been thought to be perception, network, and application. Following several research, an intermediary layer between the network and application layers was built. In cloud and fog computing environments, this layer, also called the service layer, is used to store and process data. For the past several years, authors like Ferrández-Pastor [46] have proposed a new architecture that is built on four layers: objects, edge, communication, and cloud. In these current architectural methods, the authors employ the edge layer to locate critical apps and perform basic control operations. According to [46], cloud (internet/intranet) can also include Web services, data storage, HMI interfaces, or analytic applications. **Figure 8** shows a representation of the architecture.

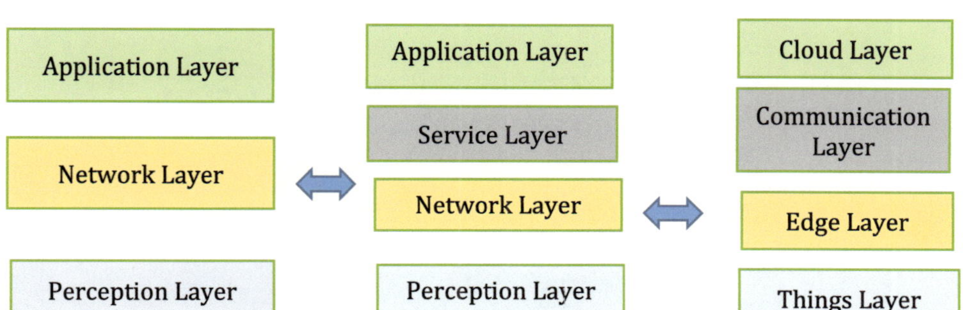

Figure 8.
Evolution of the layered model in IoT architecture [2].

Both 3-layered [43, 47] and 5-layered [48] designs are accessible in the assessed IoT systems for irrigation. The sensor nodes and actuators are usually found in the lowest layer. The middle layer has a gateway and is concerned with data transport. Finally, the third layer is often responsible for data storage and analysis. Cloud services, databases, and applications are common examples of third layers. The Internet of Underground Things [33] is considering an innovative approach to IoT deployments for precision agriculture. In-situ sensing, wireless communication in underground environments, and the interaction between architectural features like sensors, machinery, and the cloud are all identified as functions by the authors. In the case of IoUT, sensors are implanted underground. Wireless communication between above-ground and beneath devices was examined by the researchers. The route loss link between above ground and subterranean devices achieved −80 dBm over a distance of 50 metres. The distance between underground devices for −80 dBm was roughly 10 m. The authors also explore the impact of soil moisture on route loss.

2.1 Recommendations for putting a smart agriculture irrigation system in place

In this section, the researcher has presented an architecture suggestion for an IoT irrigation system. To ensure the optimal functioning of the IoT irrigation system for precision agriculture, the architecture should provide interoperability, scalability, security, availability, and robustness. Following a thorough analysis of other researchers' work, we have divided our architecture concept into four tiers, as shown in **Figure 9**, which we refer to as devices, communication, services, and applications. Furthermore, the communication and services levels should solve management and security concerns at the same time.

The first layer is the Device layer, which includes all of the devices that will perform detection, monitoring, control, and action functions. There would be four types of nodes in total. The water quality would be checked at the water monitoring node to verify if it was suitable for crop irrigation. The soil monitoring node would

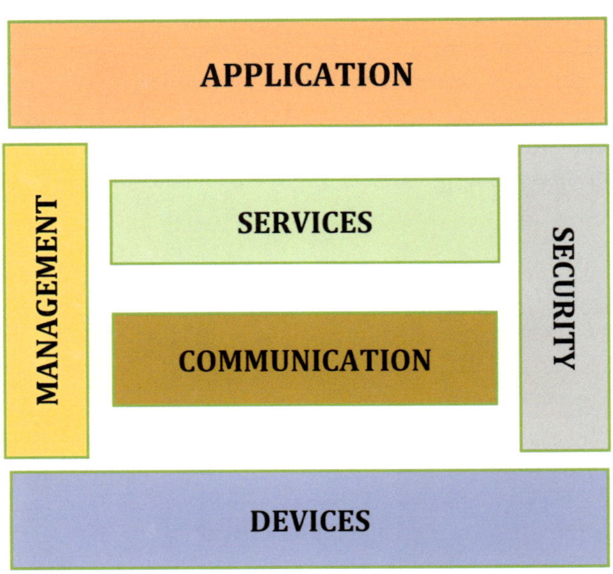

Figure 9.
Architecture proposal for an IoT irrigation system for agriculture [2].

monitor soil moisture, temperature, and other parameters, which would contribute in the irrigation schedule decision-making process. The weather monitoring node would measure air temperature and humidity, precipitation, luminosity, radiation, and wind parameters to facilitate decision-making. Finally, the decision-making process's operations would be carried out by the actuator nodes. The second layer is the communication layer, which has three blocks. The Hop-to-Hop communication block allows for the design of data link layer technologies as well as frame transmission with device layer data. In order to reach far-flung sites, frames will be transmitted from this block to the network communication block. The routing function may be assumed in this block in mesh networks, such as 802.15.4 networks. The end-to-end communication block is responsible for delivering the capabilities of the TCP/IP model's transport and application layers when communication spans various network contexts. Finally, the network communication block is responsible for network communication (routing), hop-to-hop communication at end-to-end blocks using IPv4 and IPv6 addresses, as well as ID resolution. It will also be in charge of overseeing service quality. The following layer is the services layer, which consists of three blocks. The services section includes IoT services as well as the ability to discover and search for them. Users are assigned services by the organization block based on their needs or available resources. Finally, in IoT-related business environments, service block modeling and execution will be triggered by application execution. Management and security are two elements that work on both the communication and service tiers. The management block is built using the fault, configuration, accounting, performance, and security (FCAPS) idea and architecture. This model represents the ISO Telecommunications Management Network [33]. The security block, which consists of four blocks, ensures the security and privacy of the systems. User and service authentication are handled by the authentication block. The authorization block is in charge of access control policies. Furthermore, access control decisions will be made based on access control regulations. To provide secure peer-to-peer communication, the key exchange & management block is used. Finally, the trust & reputation block is responsible for scoring the user and evaluating the level of trust in the service. The final layer is the application layer. It allows customers to interact with IoT technologies. This layer allows users to receive alarms, see acquired data in real time, and trigger actuators or actions that have not been configured automatically.

2.2 India's IoT farming challenges

- Inadequate knowledge about the local climate.
- There aren't enough sales of distribution data sources to go around.
- Inadequate ICT infrastructure and illiteracy in the use of technology.
- Farmers are under-informed on the advantages of smart farming.
- Machinery for the workplace is expensive. There is a need for more manual labour. Keep a written record of all you have done.
- a scarcity of market research competence and a research centre.

- Changes brought on by the weather.
- Agriculture is attracting the attention of young and educated individuals who have no desire to work in the industry.

2.3 Limitations

- Agriculture is a phenomenon that is completely dependent on nature, and man can forecast or regulate nature, such as rain, drought, daylight convenience, and pest management, among other things. As a result, IOT systems are used in agriculture on a regular basis.
- Smart agriculture is constantly looking for ease on the internet. The rural areas of developing countries were unable to meet these needs. Furthermore, the internet is sluggish.
- Fault detectors or data processing engines can lead to erroneous decisions, resulting in waste of water, fertilizers, and other resources.
- Smart farming, which is mostly based on instrumentation, necessitates that farmer understand and learn how to use technology. This could be the most difficult obstacle to overcome in implementing smart agriculture frameworks on a large scale across countries.
- It conjointly has some problems that ought to be half tracked properly to achieve the total good thing about it.

3. Conclusion

Water management is crucial in locations where water is scarce. This has an influence on agriculture, as agriculture consumes a substantial amount of water. Water management approaches are being studied in light of growing concerns about global warming in order it is necessary to ensure that water is available for agricultural production and consumption. As a result, the number of studies on irrigation water saving has grown over time. In this paper, we present a summary of the current state of the art in IoT irrigation systems for agricultural. When it comes to deciding irrigation, soil, and weather water quality, we have discovered out what parameters are most strictly observed. The most widely utilized IoT and WSN crop irrigation nodes, as well as the most widely used wireless technologies, were also identified. The most current breakthroughs in the use of IoT technologies for crop management and irrigation were also presented. In addition, a four-layer crop irrigation management system has been developed. Based on the proposed architecture, we are creating a smart irrigation system that analyses water quality before irrigation.

As a result, we are expanding the system that can monitor crops in fields where humans cannot provide protection. We're setting up a system in the field to keep track of valuable crops and ensure that all climatic requirements are met. In this place, we provide this type of system. As a result, this effective and dependable technology aids in agricultural monitoring. Aside from its core objective, the system makes a substantial contribution to global warming reduction. In a roundabout approach, plants' normal instincts are impeded. Plants can also be protected from fire using this method. Crop destruction is reduced as a result. As a result, the ecological balance is maintained. The research develops both an automatic watering system and a field monitoring system. The results of this study would catapult farming to another developmental level.

Author details

Pratik Ghutke[1]* and Rahul Agrawal[2]

1 Tulsiramji Gaikwad Patil College of Engineering and Technology, Nagpur, India

2 Guru Gobind Singh College of Engineering and Research Center, Nashik, India

*Address all correspondence to: pcghutke2214@gmail.com

References

[1] Carvin D, Philippe O, Pascal B. Managing the upcoming ubiquitous computing. In: 2012 8th International Conference on Network and Service Management (CNSM) and 2012 Workshop on Systems Virtualiztion Management (SVM). International Federation for Information Processing; 2012. pp. 276-280

[2] Fernández D, Sánchez P, Álvarez B, López JA, Iborra A. TRIoT: A proposal for deploying Teleo-reactive nodes for IoT systems. In: Proceedings of the International Conference on Practical Applications of Agents and Multi-Agent Systems, Porto, Portugal. MDPI; 21–23 June 2017

[3] Fathima N, Ahammed A, Banu R, Parameshachari BD, Naik NM. Optimized neighbor discovery in internet of things (IoT). In: Proceedings of International Conference on Electrical, Electronics, Communication, Computer, and Optimization Techniques (ICEECCOT). IGI Global Publisher; 2017. pp. 1-5

[4] Doorenbos J, Pruitt WO. Guidelines for Predicting Crop. Water Requirements; FAO Irrigation and Drainage Paper. Vol. 24. Rome, Italy: FAO; 1977

[5] Allen RG, Pereira LS, Raes D, Smith M. Evapotranspiración del Cultivo Guías Para la Determinación de los Requerimientos de Agua de los Cultivos. Rome, Italy: FAO; 2006. pp. 1-322

[6] CNR. Boletín N_6: Coeficiente de Cultivo. Santiago, Chile: CNR; 2010. pp. 1-7

[7] CNR. Boletín N_2: Programación de Riego Usando Estaciones Meteorológicas Automáticas. Santiago, Chile: CNR; 2010. pp. 1-14

[8] Melillán, C. Evapotranspiración de Referencia. Para la Determinación de las Demandas de Riego en Chile. 2018. Available from: http://agrimed.wixsite.com/riego/evapotranspiracin (Accessed: May 27, 2018)

[9] Agromet Red Agroclimática Nacional. Available from: http://agromet.inia.cl/ Accessed: July 1, 2017

[10] Gonzalez-Andujar JL et al. SIMCE: An expert system for seedling weed identification in cereals. Computers and Electronics in Agriculture. 2006;**54**(2): 115-123

[11] Andri P, Siti MW, Azhari. Expert system model for identification pests and diseases of Forest tree plantations. International Journal of Advance Soft Computing and its Applications. 2017; **9**(2):204-216

[12] Evangelos A et al. Integrating RFIDs and smart objects into unified internet of things architecture. Advances in Internet of Things. 2011;**1**:8

[13] Chen XY, Zhi-Gang J. Research on key technology and applications for the internet of things. Physics Procedia. 2011;**33**:561-566

[14] Rubeena MM, Denny J, Gokilavani M. Recent survey on IOT application: Smart agriculture. International Journal of Innovative Research in Advanced Engineering. 2019;**6**: MYAE10081. DOI: 10.26562/IJIRAE.2019.MYAE10081

[15] Anand N, Er VP. Smart farming: IOT based smart sensors agriculture stick for live temperature and moisture monitoring using Arduino, Cloud Computing & Solar Technology.

International Journal of Engineering Research & Technology. 2016

[16] Maras V, Popovic T, Gajinov S, Mugosa M. Optimal irrigation as a tool of precision agriculture. In: 2019 8th Mediterranean Conference on Embedded Computing (MECO): IJERT. 2019. pp. 1-4

[17] Algeeb A, Albagul A, Asseni A, Khalifa O, Jomah OS. Design and fabrication of an intelligent irrigation control system. Latest Trends on Systems.

[18] Balakrishna K, Rao M, Kumar YHS. A WSN application to optimize the irrigation for horticulture crops in real-time using climatic parameters. Journal of Advance Research in Dynamical and Control Systems. 2018; **10**:199-207

[19] Baoanh TN, Tan S-L. Real-time operating Systems for Small Microcontrollers. IEEE Micro. 2009;**29**: 30-45

[20] Arumugam SS, Ganeshmurthi M, Annadurai R, Ravichandran V. Internet of things based smart agriculture. International Journal of Advnaces in Computer and Electronics Engineering. 2018;**33**:8-17

[21] Boonchieng E, Chieochan O, Saokaew A. Smart farm: Applying the use of node MCU, IOT, NETPIE and LINE API for a lingzhi mushroom farm in Thailand. IEICE Transactions on Communications. 2018;**101**:16-23

[22] Talpur MSH, Shaikh MH, Talpur HS, et al. Relevance of internet of things in animal stocks chain Management in Pakistan's perspectives. International Journal of Information and Education Technology. 2012;**2**(1):29-32

[23] Rawal S. IoT based smart irrigation system. International Journal of Computers and Applications. 2017; **159**:7-11

[24] Guo T, Zhong W. Design and implementation of the span greenhouse agriculture internet of things system. In: Proceedings of the 2015 International Conference on Fluid Power and Mechatronics, Harbin, China. Research Gate; 5–7 August 2015

[25] Khattab A, Abdelgawad A, Yelmarthi K. Design and implementation of a cloud-based IoT scheme for precision agriculture. In: Proceedings of the 28th International Conference on Microelectronics, Giza, Egypt. Research Gate; 17–20 December 2016

[26] Nawandar NK, Satpute VR. IoT based low cost and intelligent module for smart irrigation system. Computers and Electronics in Agriculture. 2019;**162**: 979-990

[27] Barkunan SR, Bhanumathi V, Sethuram J. Smart sensor for automatic drip irrigation system for paddy cultivation. Computers and Electrical Engineering. 2019;**73**:180-193

[28] Parameswaram G, Sivaprasath K. Arduino based smart drip irrigation system using internet of things. International Journal of Engineering and Computer Science. 2016;**6**:5518-5521

[29] Kumar A, Surendra A, Mohan H, Valliappan KM, Kirthika N. Internet of things based smart irrigation using regression algorithm. In: Proceedings of the 2017 International Conference on Intelligent Computing, Instrumentation and Control Technologies (ICICICT), Kannur, India. IJESC; 6–7 July 2017

[30] Kodali RK, Jain V, Karagwal S. IoT based smart greenhouse. In: Proceedings

of the 2016 IEEE Region 10 Humanitarian Technology Conference (R10-HTC), Agra, India. IEEE; 21–23 December 2016

[31] Abidin SAHZ, Inrahim SN. Web-based monitoring of an automated fertigation system: An IoT application. In: 2015 Proceedings of the IEEE 12th Malaysia International Conference on Communications, Kuching, Malaysia. 23–25 November, 2015. pp. 1-5

[32] García L, Parra L, Jimenez JM, Lloret J, Lorenz P. IoT-based smart irrigation systems: An overview on the recent trends on sensors and IoT Systems for Irrigation in precision agriculture. Surveys of Sensor Networks and Sensor Systems Deployments Sensors. 2020;**20** (4):1042. DOI: 10.3390/s20041042

[33] Babu MSP, Murty NVR, Narayana SVNL. A web-based tomato crop expert information system based on artificial intelligence and machine learning algorithms. International Journal of Computer Science and Information Technologies. 2010;**1**(1):6-15

[34] Daskalakis SN, Goussetis G, Assimonis SD, Tenzeris MM, Georgiadis A. A uW backscatter-morse_leaf sensor for low-power agricultural wireless sensor networks. IEEE Sensors Journal. 2018;**18**:7889-7898

[35] Jain S, Vani KS. A survey of the automated irrigation systems and the proposal to make the irrigation system intelligent. International Journal of Computational Science and Engineering. 2018;**6**:357-360

[36] Karimi H, Navid H, Besharati B, Behfar H, Eskandari I. A practical approach to the comparative design of non-contact sensing techniques for seed sow rate detection. Computers and Electronics in Agriculture. 2017;**142**:165-172

[37] Mohanraj I, Ashokumar K, Naren J. Field monitoring and automation using IOT in agriculture domain. In: 6th International Conference on Advances in Computing & Communications, ICACC 2016, 6–8 September 2016. Vol. 93. Cochin, India: Procedia Computer Science; 2016. pp. 931-939

[38] Mizunuma M, Katoh T, Hata SI. Applying IT to farm fields a wireless LAN. Special Feature. 2003;**1**:56-60

[39] Balakrishna K. Fusion approach-based horticulture plant diseases identification using image processing. In: Chakraborty S, Mali K, editors. Applications of Advanced Machine Intelligence in Computer Vision and Object Recognition: Emerging Research and Opportunities. Hershey, PA: IGI Global; 2020. pp. 119-132

[40] Balakrishna K, Rao M. Tomato plant leaves disease classification using KNN and PNN. International Journal of Computer Vision and Image Processing (IJCVIP). 2019;**9**(1):51-63

[41] Balakrishna K. WSN-based information dissemination for optimizing irrigation through prescriptive farming. International Journal of Agricultural and Environmental Information Systems (IJAEIS). 2020;**11**(3):14

[42] Zampieri M, Carmona García G, Dentener F, Krishna Gumma M, Salamon P, Seguini L, et al. Surface freshwater limitation explains worst rice production anomaly in India in 2002. Remote Sensing. 2018;**10**:244

[43] González-Briones A, Castellanos-Garzón JA, Mezquita Martín Y, Prieto J, Corchado JM. A framework for knowledge discovery from wireless sensor networks in rural environments: A crop irrigation systems case study.

Wireless Communications and Mobile Computing. 2018;**2018**:6089280

[44] Sebastian S, Ray PP. Development of IoT invasive architecture for complying with health of home. In: Proceedings of the International Conference on Computing and Communication Systems (I3CS), Shillong, India. Wireless Communications and Mobile Computing; 9–10 April 2015

[45] Ferrández-Pastor FJ, García-Chamizo JM, Nieto-Hidalgo M, Mora-Pascual J, Mora-Martínez J. Developing ubiquitous sensor network platformusing internet of things: Application in precision agriculture. Sensors. 2016;**16**:16-1141

[46] Merezeanu D, Vasilescu G, Dobrescu R. Context-aware control platform for sensor network integration in IoT and cloud. Studies in Information and Control. 2016;**25**:489-498

[47] Robles T, Alcarria R, Martín D, Morales A, Navarro M, Calero R, et al. An internet of things-based model for smart water management. In: Proceedings of the 28th International Conference on Advanced Information Networking and Applications Workshops, Victoria, BC, Canada. 13–16 May 2014

[48] Ayaz M, Ammad-Uddin M, Sharif Mansour Z, Aggoune E-HM. Internet-of-things (IoT)-based smart agriculture: Toward making the fields talk, special section on new technologies for smart farming 4.0 research challenges and opportunities. IEEE Access. 2019;7: 129551-129583

Chapter 5

Smart Rainwater Harvesting System for Sustainable Agricultural Irrigation and Drainage System

Mohd Hudzari Haji Razali, Abdul Qudus Puteh, Alawi Haji Sulaiman and Mohamad Hakim Mohamad Yatim

Abstract

Nowadays the world population increases, so the demand for clean water is rising. Rain is the faster resource that can recharge compared to ground water. Rainwater harvesting system (RWHS) is one of the traditional and easiest ways of rising fresh water supplies. This system already implements by many countries in the world as a viable decentralized water source. Malaysia can be classified as a country with high annual rainfall and high consumption of domestic water. Malaysia is well and strategically positioned to harvest rainwater for both potable and non-potable uses. This research describes the collaborative and development affordable technology for capturing and retaining runoff starting from rooftop gutter until the tank storage using as a valuable source of water and recharge the percolation well and increase ground water level. The developed system consists of soil sensor, integrated water pump, Arduino controller and water tank harvesting with dynamic mechanical flushing technique which improving filtration method. The results show that the system can be efficiently used for small-scale drip irrigation especially in urbanization farming as nowadays scenario of agriculture demand. Hopefully, this can be helpful as a valuable water source in future.

Keywords: smart irrigation and drainage, smart water tank, rain harvesting system, Arduino microcontroller

1. Introduction

1.1 Background of study

Water is nature's gift for the good purpose of living on the earth. The righteous book of Al-Quran as in Sura Al Anbiya verse 30, mentioned that water is the source for living. To survive, living things, humans and plants need water as an important source. Nowadays, water shortage becomes a major issue. The increasing population causes a high demand for water. To overcome the water issue, water can be collected from rainwater as an alternative where it can be used for daily routine such as bathing, watering plants, vertical farming, and washing the car. Rainwater harvesting system is the best solution and it's appropriate to do it in Malaysia because our country receives 990 billion m3 of rainfall per year. (Department of Irrigation and Drainage Malaysia).

The rainwater harvesting system (RWHS) is an alternative way of agriculture activity. The agriculture sector is the most water use compared to other industries. It can collect rainwater from the rooftop and open space. The scale of the rainwater catchment area is depended on the area of the roof and demand. Even though the rainwater harvesting system has the potential for agriculture like irrigation, but there are still drawbacks to this system, particularly in large-scale projects that cannot be applied to gain potential output and best result.

The aim of this study is to develop a prototype system for sustainable agriculture. The source of water for this system is derived from the original rainwater source. In order to allow the water to flow out from the tank, a sensor remote is needed.

2. Literature review

2.1 Rainwater harvesting system

In ancient times, rainwater was used to persevere. The use of the rainwater harvesting system has two main aims. The first was to provide water for various household activities, especially drinking, bathing, etc. Secondly, rainwater evaporation from some of the domestic ones impounding systems such as pools and reservoirs made some sort of air-conditioning feeling available which has improved more or less on the microclimate, making it cooler and more comfortable.

Rainwater was the primary source of water supply in old days for drinkable and non-drinkable uses because water supply systems are not yet in development. Simplest method used to make a rainwater harvesting system. No treatment was applied to the collected rainwater before it was used. The systems can be categorized as small, medium and large scale [1]. Normally, the size of rainwater harvesting was based on the size of the catchment area [2].

Rainwater has a lot of potential as the main water resource for the future because of its high quality [3]. Clean rainwater collected not only solved the water crisis problem but also reduce the water treatment cost. For the economic benefit. This system can reduce or cut the utility bill for domestic water. So, RWHS is implemented if the water tariff is higher.

2.2 Drivers for RWHS

Rainwater Harvesting System has been used in the water supply system for many years, and has been around for this reason:

1. Water Scarcity

Water scarcity may be due to a lack of adequate surface area and improper control of groundwater resources, polluted land, and groundwater resources and insufficient support for programs to harness and distribute water.

2. Reduce flooding and erosion

This system can help reduce erosion around downspouts and in the garden. It controls the stormwater run-off.

3. Reduce water bills

The use of rainwater can substantially reduce the cost to supply main water services. When needed, the rainwater acts as a substitute for a backup source.

4. For sustainable agriculture

The collection of rainwater can be used to boost plants and gardens too. The salt accumulation can be flushed from plants and soil using collected water. Harvested rainwater is usually free of different forms of pollutants and man-made toxins. Rain is also chlorine-free.

2.3 Benefits of the rainwater harvesting system

The harvesting of rainwater is the element of sustainable agriculture and tends to bring a variety of value not just for users but also environment and government (**Table 1**).

2.4 Working principle of the rainwater harvesting system

Nowadays, the rainwater harvesting system is well known to the public regarding its function and benefits. However, the working process still has the same basic method and component. There are six basic components for a domestic rainwater harvesting system:

- The catchment surface: Surface allowing water to collect from the rain, such as a rooftop. The size of the catchment area determines how much rainwater is harvested.

User	Government	Environment
Rainwater is safe, free from disease and an adequate source of water.	Replace the old system in supply water for infrastructure.	Improve soil moisture level
This method is inexpensive and was easy to maintain.	It may reduce the cost of public water system access.	Reduce the amount of rainwater flow to the drainage that can cause floods to happen.
The system is easy and very user friendly		suitable for ecosystems and plants grow well on rainwater compare to use other sources that may have an element that contaminates plant and soils.
Save cost by cutting the amount of water purchased from domestic systems.		
Away from overuse of water are well.		

Table 1.
Benefits of the rainwater harvesting system.

- Gutter and downspouts: Channel where the water flows into the storage tank.
- Leaf screens: The screens get rid of the waste, dust, dry/dead leaves from the rainwater captured.
- Water storage tanks: where the catchment water is collected before it is distributed to the ground.
- Delivery system: after water passes the filtration water will distribute through remote sensor piping and pump system.
- Treatment: the filtration system to ensure the water captured is clean and secure.

3. Research methodology

3.1 Location of study

The location of study is at University Technology MARA Campus Jasin Melaka. This location is chosen because accessible, safe and complete facilities are available to perform a rainwater harvesting system field experiment.

3.2 Research design

The study utilized data from the experiment and tested the method of collecting rainwater against soil moisture. The work was carried with a guideline and procedure. The water obtained from the rainfall was stored in a small tank where it was installed on the device. To disperse the water to the soil, the water tank already applies with a remote sensor where it detects the soil moisture.

3.3 Development and methodology

This study used a rainwater harvesting system as a source of water for good agriculture practices. To complete the research and run the test, a rainwater harvesting system with a remote sensor is built to detect moisture. To construct the system, a few factors need to be considered in the design of the system:

a. Size of PVC tubing or pipeline Influence the water flow from the tank to the ground.

b. The water tank should have enough water at all times because the water pump is high pressure. This is to avoid running out of water which can interfere with the smooth running of the system.

c. The high placement of the water tank is very important to give an advantage to the pipeline to disperse water with long distance.

3.3.1 Mechanization of rainwater harvesting system prototype

The basic part and component to build a prototype of a rainwater harvesting system are (**Table 2**):

DOI: http://dx.doi.org/10.5772/intechopen.104442

Item	Size	Quantity	Picture
Soil moisture sensor	—	1	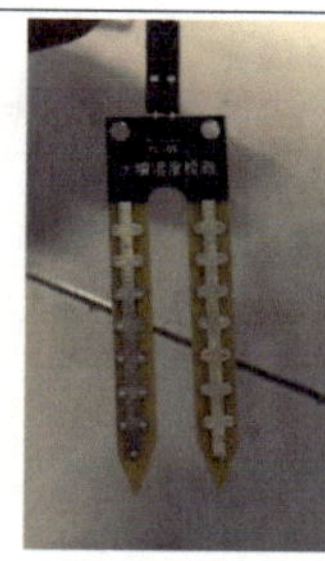
Arduino	—	1	
Relay module	—	1	
Mini Water pump	—	1	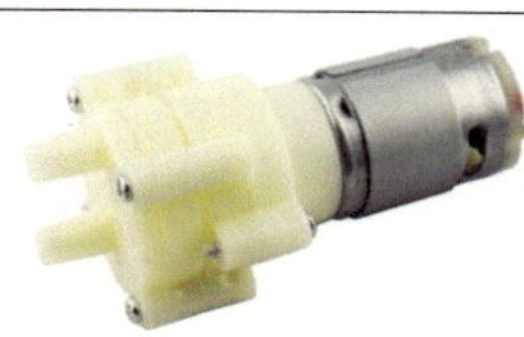
Jump wire	—	10	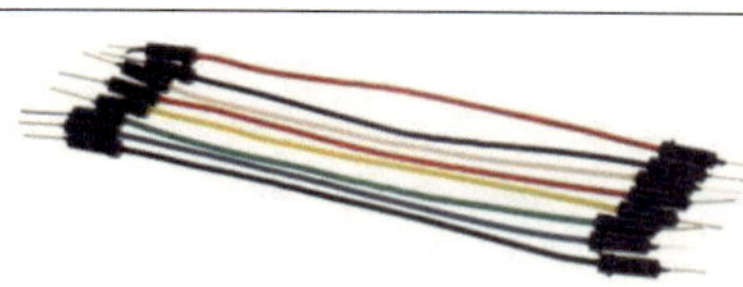
Water container	20 liters	1	
Polyvinyl chloride (PVC) tubing	20 meter	1	

Table 2.
Basic part and component of gravitational drip irrigation system.

3.3.2 Prototype rainwater harvesting installation process

3.3.2.1 Main installation process

The first installation is applying the mini water pump. The function of the water pump is assisted to disperse the water from tank storage to the soil through PVC tubing (**Figures 1–3**).

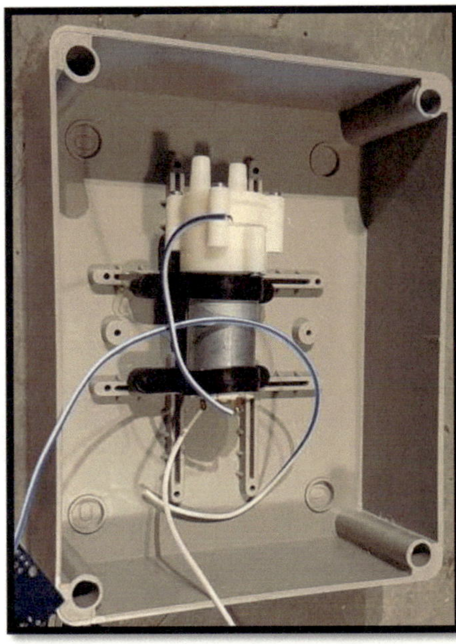

Figure 1.
Installation of mini water pump.

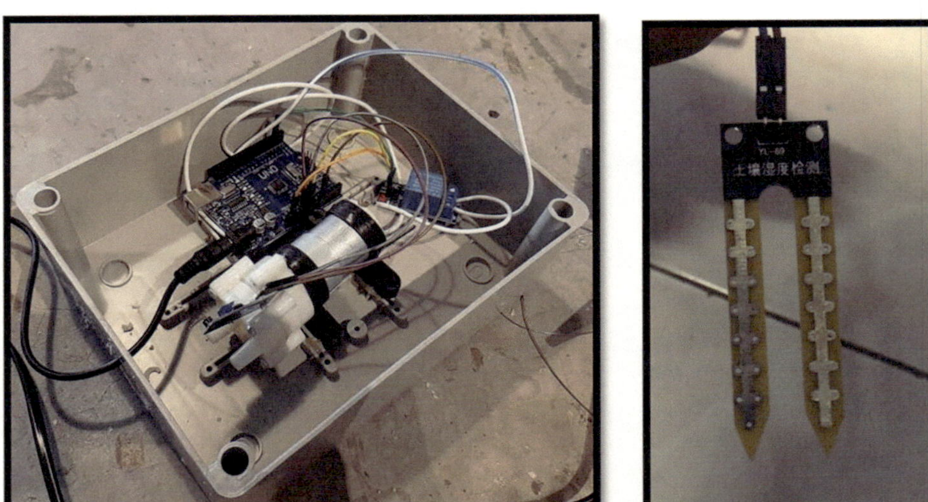

Figure 2.
Arduino UNP R3, relay module and soil moisture sensor.

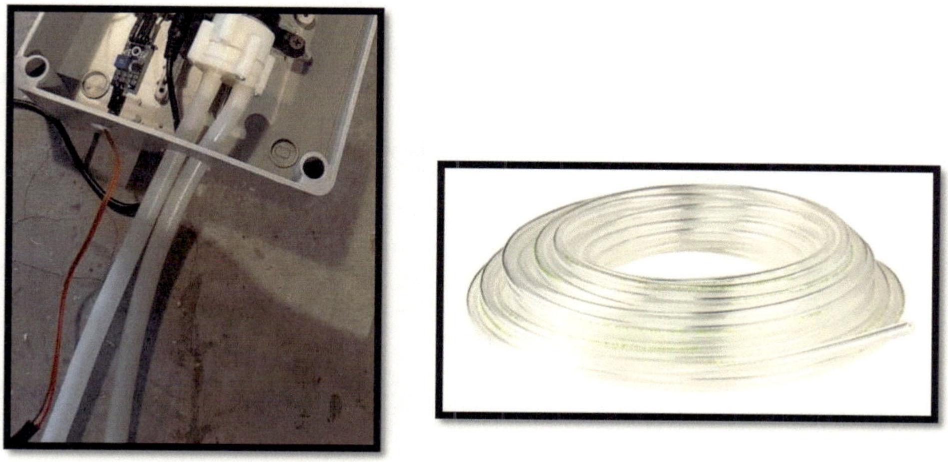

Figure 3.
Installation of PVC tubing.

3.3.2.2 Arduino UNP R3, relay module and soil moisture sensor installation

The second stage is the Arduino UNP R3, relay module and soil moisture sensor installation. At this stage, the Arduino UNP connects to the relay using a jump wire. Soil moisture sensors are installed to detect soil moisture.

3.3.2.3 PVC tubing installation

The installation of PVC tubing is safe transportation of rainwater.

4. Results and discussion

4.1 Introduction

This chapter will outline the project's design and construction process. To complete the project, all equipment structures in this project must be evaluated, and all specifications must be fully implemented.

4.2 General concept

The Arduino microcontroller will be used for the main system in this project. Arduino has been chosen because it has several suitable hardware modules for Arduino boards themselves [4, 5]. There are several Arduino hardware components that can be attached to the Arduino plate, such as micro/mini water pump, soil moisture sensor, and power sources. The equipment and components used in the rainwater harvesting system are:

4.2.1 Soil moisture sensor

The Soil Moisture Sensor is used as the key component to achieve the objective in this analysis. The role of the sensor is to use the capacity to measure soil water

Type	Specification
Model	The model of this sensor is series FC 28
Type of Model	Soil Moisture Sensor
Operating Voltage	The available voltage of this sensor is 3 mA @ 5 V DC
Output Current	0–4.2 V
Operating Temperature	This sensor can withstand temperatures of −40°C to +60°C
Typical Resolution	0.1%
Measuring Range	0 to 45% volumetric soil water content (available at 0 to 100% VWC with alternating calibration)
Accuracy	The level detection accuracy reading is within 4%

Table 3.
Moisture sensor module specification.

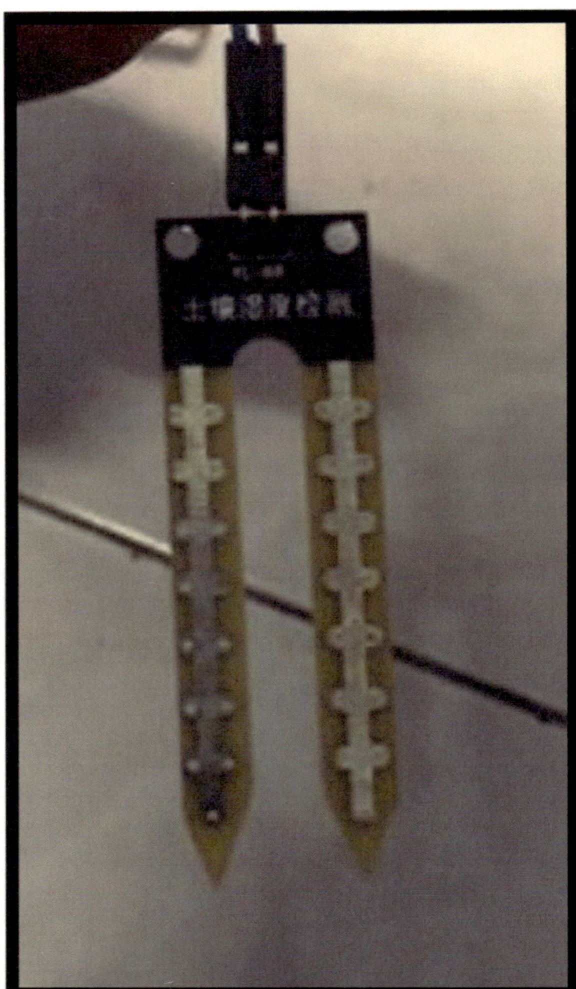

Figure 4.
The moisture sensor.

content and it is very easy to use [6]. Only insert this robust sensor into the soil to be checked, and the volumetric water content of the soil is recorded as a percentage (%). **Table 3** and **Figure 4** show the specification of the Soil Moisture Sensor, as well as the sensor image.

4.2.2 Arduino UNO

Arduino is a microcontroller board which is open-source hardware. The term "UNO" in Italian means "ONE" and was chosen to mark the initial release of Arduino Software. The Uno board is the first in a series of Arduino UNO board based on USBArduino Uno has 14 digital pins, six analog pins, a USB interface, an ICSP header, a power barrel port, and a reset button. This Arduino UNO is used after upload coding to store data, and the project can fully function with a power supply without being connected to a laptop or computer. **Table 4** shows the technical specification of this board. **Figure 5** shows the Arduino UNO board.

Type	Specification
Microcontroller	The microcontroller used is ATmega8U2 model
Operating Voltage	Only 5 volts of operating voltage
Analog Input Pins	This board includes six analog input pins to be used
Flash Memory	32 KB (0.5 KB is used for Bootloader)
Frequency (Clock Speed)	16 MHz
Digital I/O Pins	14 (Out of which 6 provide PWM output)

Table 4.
Arduino UNO technical specification.

Figure 5.
Arduino UNO board.

4.2.3 Jumper wire

Jumper wire is used to connect several components and parts to the Arduino UNO board. For example, Jumper wire act as a bridge to connect the moisture sensor to the Arduino UNO board. **Table 5** shows the jumper wire specification and **Figure 6** shows the picture of jumper wire (**Figure** 7).

4.2.4 Mini water pump

In this study related to the water system, the Mini/Micro Water Pump is used since it integrates easily. The water pump operates with a water suction system that collects the water through its inlet and releases it through the outlet. **Table 6** shows the jumper wire specification and **Figure 8** shows the picture of jumper wire (**Table** 7).

4.3 Maintenance of the system

Every rainwater harvesting system has the same technique and method to do their maintenance. To cut the cost of maintenance, work should be done carefully and properly during the installation stage. The purpose of maintenance is to make sure

Type	Specification
Size	The size of the jumper wire pin header for this project is 0.1 "
Length	The length of every one jumper wire is 180 mm

Table 5.
Jumper wire specification.

Figure 6.
Mini water pump.

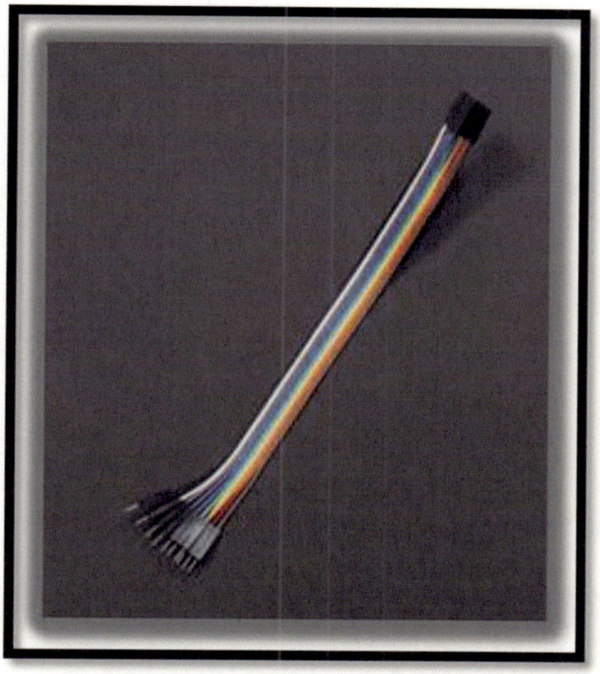

Figure 7.
Jumper wire.

Type	Specification
DC Voltage	3-5 V
Maximum Lift	40-110 cm / 15.75″-43.4″
Flow rate	1.2–1.6 L/H
Working life	500 Hr
Material	engineering plastic
Outside diameter of water outlet	7.5 mm/0.3″
Inside diameter of water outlet	4.7 mm/0.18″
Weight	30 g

Table 6.
The mini water pump specification.

the operating system is running smoothly. Need to ensure the water source free from unwanted things that can cause clogging happen at the water tank or PVC tubing.

Factors need to focus and consider during the maintenance of the system:

a. Water storage tank – Should always check the tank 's cleanliness and ensure that water is always available to the system for smooth operation.

b. Filtration - filters should be maintained periodically to avoid any damage to the system.

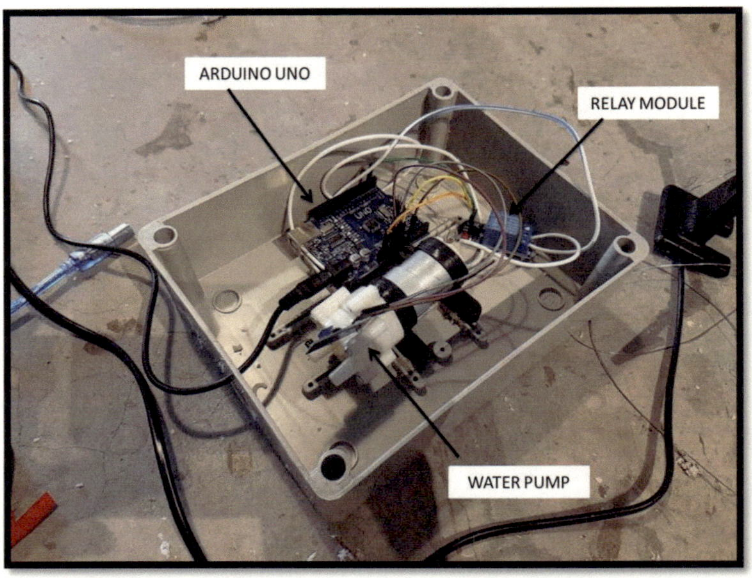

Figure 8.
Project prototype.

c. Transportation and delivery system – All plastic water-supply facilities will last longer. However, periodic inspections are required to ensure all appliances are in good condition.

4.3.1 Prototype project

To complete this a prototype project, a container with soil was used to place a soil moisture sensor in it. Thereafter, the entire project is tested to achieve the project objectives.

Hardware component project	Function (%)
Soil Moisture Sensor	
Sensor detected	100
Analog reading	100
MINI WATER PUMP	
Water flows out of the pump when the sensor detects low humidity	100
Water does not flow out of the pump when the sensor detects low humidity	100

Table 7.
Result of system functionality testing.

4.4 Performance of the developing rainwater harvesting system using soil moisture sensor

The functions of each component within the project are tested and recorded for final results. There are several components that need to be tested such as soil moisture sensors, mini water pumps and PVC tubing to prevent clogging. The ability of these components to work well proves that this prototype has been successfully implemented.

4.4.1 Evaluation function testing project

The prototype will be checked multiple times to ensure the proper functioning of each part within it. To submit information to the water pump acting as output, the soil moisture sensor data will be interpreted. It is intended to prevent and minimize accidents that result in injury during operations [7, 8]. All of the errors may affect the achievement of this study's objectives.

4.4.2 Prototype project unit testing and functional testing

Table 6 shows the complete results of the prototype system for this study. No error occurred at the end of the test after the prototype execution was complete. From this analysis, it can be assumed that it is applied effectively and according to the actual requirements. In this study the most important thing is to ensure that the sensor works properly to detect moisture in the soil. So, it can receive data and transmit information so that other components can effectively perform their tasks. This makes it easier for consumers such as farmers to control the amount of water to crops.

4.5 Discussions

The rainwater harvesting system has been selected as the watering system for agricultural sustainability. The performance of the system has achieved the real objectives of this study. This device is maintained to supply an adequate quantity of water for plant production. So, this system is very relevant to be used by the small-scale farm.

5. Conclusion and recommendations

5.1 Conclusion

The research was conducted to build a rainwater harvesting system prototype. This is also intended to assess the efficiency of the device against soil moisture. The data were obtained at the MARA Campus Jasin University of Technology. Shape the data, which may be linked to the objectives of this stud. The results show the prototype function very well to detect the soil moisture.

Next, proper utilization of this system will help restore the deteriorated aquatic ecosystem and mitigate global climate change. However, research into this knowledge needs to be improved to obtain a greater result, especially for the agricultural sector. A researcher studying irrigation will consider the effective harvesting of rainwater for agricultural irrigation a significant subject.

As a conclusion. Rainwater harvesting has been proved to be efficient in an irrigation system for plant growth. Based on the study, it has been estimated that the RWHS gives a good impact and good vibes to the small-scale farm.

5.2 Recommendation

For the recommendation, to enhance the performance rainwater catchment tanks should be in a high place to facilitate the movement of water. This system required a large size of water storage so the water harvested can hold for a long period and facing no problem during the dry season. It also can improve field performance and capability. For the filtration stage, regular maintenance and check must be done to avoid any clogging.

In many developing countries, this system does not get full attention and is less attractive as installation and maintenance are very costly. The low tariff of domestic water resources also causes the factor less attractive to the system. To implement the rainwater harvesting system's success, the government plays an important role to help give some support, especially during the initial stage. Many kinds of approaches can be done by the government such as doing advertisements through media social, television, and new paper. They also can organize an exhibition regarding this system to expose its function and its benefits.

In Malaysia, to introduce the system need to educate the public through a formal and informal forum such as among students in the school, college and etc. The benefits can be fully highlighted by providing the proper tools where it seems could be effective to encourage the public to save water. The behavior of the public about water use can be changed once the water tariff getting higher. They also find an alternative if they experienced a long drought season and lack water resources.

Acknowledgements

The authors would like to acknowledge the role of the Ministry of Science, Technology and Innovation (MESTECC) and Universiti Teknologi MARA (UiTM) for providing grants funding of 100-IRMI/GOV 16/6/2 (025/2019) and 600-RMC/LESTARI SDG-T 5/3 (077/2019), and all staff in Faculty of Plantation and Agrotechnology, UiTM Melaka, in facilitating the field trials and allow to conduct this research.

Author details

Mohd Hudzari Haji Razali*, Abdul Qudus Puteh, Alawi Haji Sulaiman
and Mohamad Hakim Mohamad Yatim
Faculty of Plantation and Agrotechnology, Universiti Teknologi MARA, Jasin
Campus, Melaka, Malaysia

*Address all correspondence to: hudzari@uitm.edu.my

References

[1] Gould J. Contributions relating to rainwater harvesting. In: Paper Prepared for the World Commission on Dams Secretariat (WCD) Thematic Review IV.3. Cape Town, South Africa: Vlaeberg; 1999

[2] Thamer AM, Megat-Johari MM, Noor AHG. Study on potential uses of rainwater havesting in urban areas. In: Paper Presented at Rainwater Utilization Colloquium on 19 & 20 April 2007 at NAHRIM Mini Auditorium. Selangor, Malaysia: University Putra Malaysia; 2007

[3] Texas Water Development Board, editor. The Texas Manual on Rainwater Harvesting. 3rd ed. Austin, Texas: Texas Water Development Board; 2005

[4] Hudzari RM, Noorman MM, Asimi MNN, Atar MAM, Nashriyah M. Engineering technological in agriculture research and education. Advanced Materials Research. 2013;**705**:493-498

[5] Hudzari RM, Syazili R, Ssomad MAHA, Farizal MSA. Design and development of innovative Highland water filtration system, special issues: Mechatronic system. World Journal of Engineering and Technology. 2016;**04**(03)

[6] Sahit K, Razali MH, Mustapha S. Effectiveness smart sensor devices for sustainable irrigation in agriculture. In: 10th IEEE International Conference on Control System, Computing and Engineering (ICCSCE2020); 21-22 August 2020; Penang, Malaysia. 2020

[7] Mat N, Mohd R, Hudzari, Khandaker, Moneruzzaman M. Effects of irrigation frequencies on aerial agro-morphological parameters of Dioscorea hispida Dennst. (Dioscoreaceae). Journal of Applied Sciences Research. 2014;**8**(9):27-37

[8] Ismail WIW, Razali MH, Tan MY. Development of automation and control design system for lowland tropical greenhouse (Scopus). Pertanika Journal of Science and Technology. 2013;**21**(1):357-366

Chapter 6

Water Sustainability through Drainage Reuse in Agriculture – A Case for Collaborative Wireless Sensor Networks

Huma Zia

Abstract

With increasing prevalence of wireless sensor networks (WSNs) and internet of things (IoT) in agriculture and hydrology, there exists an opportunity for providing a technologically viable solution for the conservation of already scarce freshwater resources. In this chapter, a novel framework is proposed for enabling a proactive management of agricultural drainage and nutrient losses at farm scale where complex models are replaced by *in situ* sensing, communication, and low complexity predictive models suited to an autonomous operation. This is achieved through the development of the proposed Water Quality Management using Collaborative Monitoring (WQMCM) framework that combines local farm-scale WSNs through an information sharing mechanism. In this chapter, we present the design of a framework for facilitating real-time utilization or disposal of agricultural drainage among farms using collaboration among prevalent farm networks. The basic system architecture comprises modules for environmental learning, prediction of the impact of neighboring events in terms of drainage and nutrients losses, and a local decision support mechanism. The overall functionality of the framework is explored in terms of stages of learning, training, and testing. A network learning model is required to identify flow links of a network with neighboring networks.

Keywords: wireless sensor networks, internet of things, collaborative networks, water quality management

1. Introduction

Owing to the coupled impact of traditional farming practices and inherent inefficiency of nutrient uptake by crops (up to 60%), there is an inevitable release of drainage water (35–60% of surface irrigation water) rich in nutrients [1]. For example, in a watershed (190,000 ha) in Western Australia, facing high nutrient fluxes, phosphorous losses are measured to be 140 tonnes per annum (tpa) that is twice its target. It is expected to rise to 1300 tpa in the next 100 years if current practices continue [2]. In another study on the North China Plain, the recovery of fertilizer N by the crop at the conventional N fertilizer rate (300 $KgNha^{-1}$) was approximately 25%, while 30–50% of the applied N was lost [3].

In order to avoid exacerbating the water crisis and to prevent food shortages, an advantageous strategy is the conservation and reuse of agricultural drainage (and the dissolved nutrients) before it ends up in freshwater system [4]. Reusing drainage water and nutrients emanating from one farm in another farm, before they enter the water system, can have huge environmental as well as economic benefits. In particular, reuse reduces the amount of freshwater extracted from the environment, thus lowering its diversion from sensitive ecosystems. In regions where irrigation water supplies are limited, drainage water can be used to supplement them [5, 6]. In addition, agricultural drainage reuse can benefit the farmers (or any stakeholders) by saving cost on not using fresh irrigation water and fertilizer inputs.

The only concern about drainage water reuse is whether or not the water is safe for reuse, that is, does not contain high concentration of salts and pesticides. Highly saline water cannot be used for salt-sensitive crops. However, it can successfully be used for salt-tolerant crops, trees, fodder, and natural wetland and even for salt-sensitive crops at later growth stages [4, 7]. Conjunctive use of saline water with freshwater increases the suitability of drainage water. With regard to pesticides, in areas where strong environmental safeguards exist for pesticide usage, there is little risk associated with the reuse of surface runoff or tail water drainage [7–9]. Hence, drainage water can safely be used if appropriate considerations are taken into account.

Drainage water and dissolved nutrients have been globally utilized for crops and greenhouses. In some intensive farming areas, farmers have begun to test their groundwater for nitrate concentrations and therefore change their nutrient budgets accordingly [10]. In another case, reapplication of N-rich runoff waters provided more than the annual nutrient requirements for that land [11]. In one study, reuse of saline water for salt-tolerant forages has been investigated under varying salinity-level treatments (between 15 and 25 dS/m). For this experiment, sand tanks were used in a greenhouse. Almost all forages showed promise with regard to biomass production, whereas wheatgrass, Bermuda-grass, and paspalum performed particularly well [12].

Some work in local drainage reuse is reported for hydroponic systems maintained in a greenhouse (in which plants are grown in water instead of soil). In one application, high-quality tomato was grown with drainage reuse [13]. During seedling stage, fresh nutrients were supplied with irrigation of which 20–30% overflowed as drainage. At the final stage of ripening, the preserved drainage was reused with no wastage being drained out from the greenhouse. In a similar work based on greenhouses in Australia, drainage reuse was used for growing cucumber and tomato [14]. The study was aimed at investigating the use of drainage water of the greenhouse to increase water and nutrient use efficiency and reduce the environmental impact. Flow meters were installed to gauge the volumes of water applied to the crops. Water samples were taken five times a day for inflows and outflows, and were analyzed for pH, salinity, and concentrations of nutrients. The results indicated 33% reduction in freshwater usage for irrigation. Furthermore, it was determined that drainage water collected from the greenhouse contained 59% of applied N and 25% of applied P. These studies, though small scale and based on local drainage reuse, are very encouraging.

Existing work though promising is based on spatially and temporally limited manual sampling of soils and waters and on hypothetical guesses as to the processes involved in the N cycling. Furthermore, various resource constraints and farmer's concerns regarding real time availability of information on volumes, timings, and quality of discharges that will be delivered to the farms [15, 16] restricts wide

DOI: http://dx.doi.org/10.5772/intechopen.106486

adoption of this mechanism in agriculture. Despite tremendous promise of the benefits of drainage water reuse and technological advancements, implementation of an intelligent and autonomous management mechanism has not kept pace with the deteriorating water situation. Some of the reasons as outlined in a detailed study [17] are as follows: (i) insufficient awareness of available technology, (ii) unavailability of soil, weather, and crop data, (iii) inappropriate model selection which inadequately capture the system details, and (iv) gap between decision makers and scientists. The integration of useful and relevant scientific information is necessary and critical to enabling informed decision making for drainage reuse or disposal [18]. Recent adoption of WSNs in agriculture and hydrology presents huge promise for improving water management, and the next section discusses the applications of WSNs for water quality monitoring and agriculture and identifies huge opportunities available with real time, dense, and remote data availability.

This chapter presents the architecture of water quality management using collaborative monitoring (WQMCM), which uses existing networked farms and water systems and low-complexity predictive models to enable real-time drainage water management [19–21]. The functional overview of the WQMCM framework with the design of a modified drainage network is discussed.

2. Function overview of WQMCM

WQMCM is an integrated control and management strategy, which requires that individually targeted monitoring units or local networks, representing different stakeholders in a catchment, for example, a farm, should be able to share information with each other about runoff, drainage, or nutrient fluxes. These events may be intense but are short-lived and so information sharing becomes important as they may be very fast, and so may normally be missed with the usual sampling rate. Allowing event information to be transmitted across multiple networks as they are detected will allow prediction of when the repercussions of that event might be seen downstream, allowing other stakeholder networks in the vicinity to adjust their monitoring and management strategy. This will include taking decisions about reusing or disposing the drainages, or increasing their sample rate to catch transient events. As emphasized in the literature, drainage reuse strategy reduces the overall stress of nutrient losses to the water system and provides economic benefit as well by reducing fertilizer usage. The proposed framework enables stakeholders to manage and benefit from this reutilization by sharing information about their availability and presence. Such a de-centralized approach comprising autonomous networks presents a flexible methodology where independent networks, in addition to local monitoring objectives, seek to opportunistically utilize neighboring events.

To demonstrate the mechanism of the proposed framework with respect to agricultural drainage reuse, a modified drainage network is designed. **Figure 1** illustrates an example irrigation and drainage system in which various farms and drainage regions are linked with each other through water flow paths. The figure shows an additional bay, drainage reuse bay, linked with individual farm's irrigation and drainage bay to implement the drainage reuse mechanism. Each farm would have the option to either use drainage from another farm or fresh irrigation water for irrigation. As mentioned earlier, the WQMCM framework aims to combine local individual networks into an integrated mechanism; therefore, it is assumed that these farms are monitored by individual networks with local application objectives. These objectives are to facilitate farming decisions with regard to, for example, irrigation or pesticides scheduling by monitoring microclimate (soil moisture, crop cover, and soil temperature) of the field. For implementing the framework, an additional network on the water system, the drainage reuse bay in this case, is required to monitor drainage and nutrient contributions by each farm. As shown in **Figure 1**, individual sensors in the drainage network are deployed at the outlet of each farm to monitor its drainage outflow. Other nodes monitor base flow in the drainage bay. This network will be either deployed by an official governing body working toward maintaining water quality or by local farmers for a collaborative cause.

These networks, under the proposed WQMCM framework, share information about the start of a daily event with each other, for example, an irrigation event in a farm or high pollutant drainage discharge from drainage bay. When event information is received from a farm network (e.g., farm A), the drainage network node associated with that farm uses on-node predictive models to forecast the values for expected drainage and nutrient dynamics as a result of that event. The forecasting of

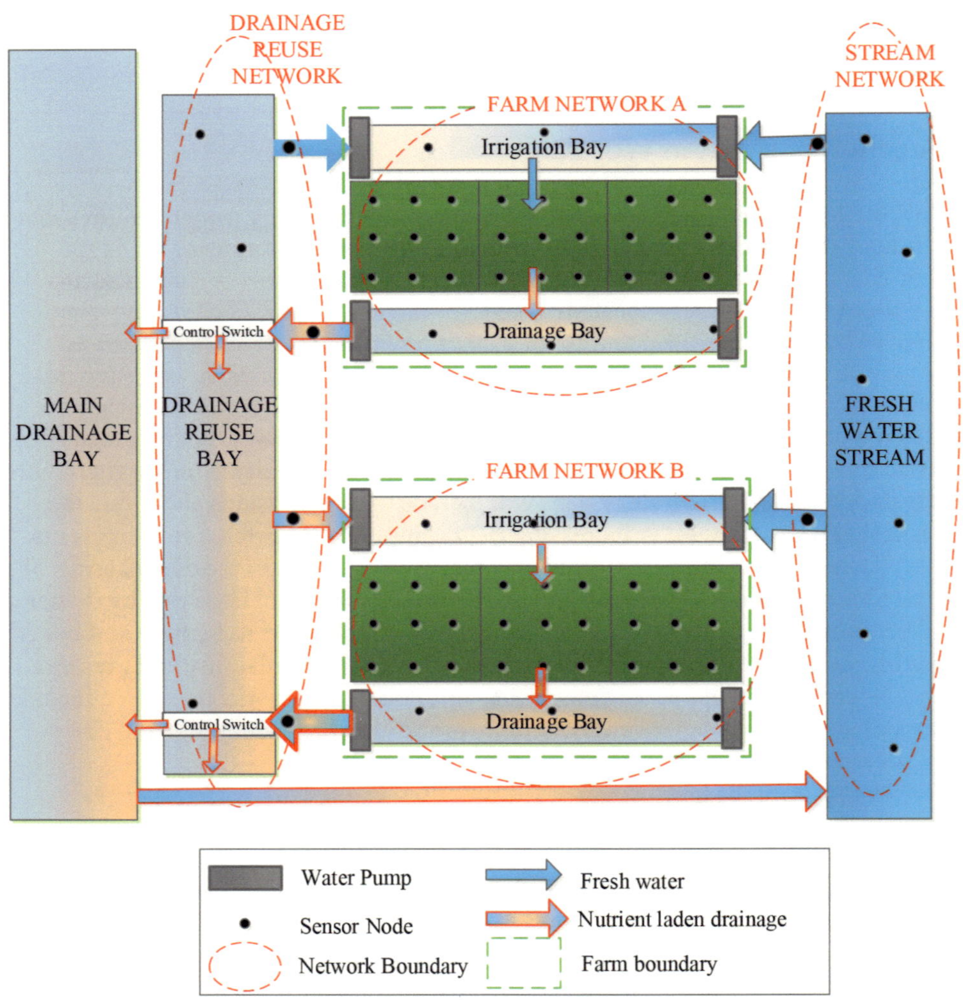

Figure 1.
A modified drainage network design to implement drainage reuse for the WQMCM framework.

drainage dynamics is undertaken by the drainage network for the following reasons. Firstly, because drainage network links all the farms and the stream networks, it is ideal to have the drainage network disseminate the predicted information about drainage to all the other farms and stream network for reuse, treatment, or disposal. Secondly, the main drainage bay could be distantly located from the drainage ditch of a farm; hence, volumes of actual drainage outflows received by a drainage bay from a farm may change owing to evapotranspiration and absorption during its transport. Additionally, running predictive models is a computational overhead, which should naturally be taken up by the network responsible for decision making.

Figure 2 illustrates the format of information shared by a farm and the parameters predicted by a drainage network. The shared event information packet from a farm includes network and event details. To identify a network, information such as network id, type, and location is included. Network type is related to whether it is a farm, drainage, or a stream network, which helps filter out received messages. For instance, a farm network may only want to receive information from drainage or stream network, or a drainage network may only be interested in information coming from farms for obvious reasons. Network location filters out geographically dislocated networks or the ones located downstream, which are unlikely to impact upstream networks. Further to that, event detail in the information packet includes event depth/volume, event duration, fertilizer quantity applied. Any additional event information will be governed by the requirements of a predictive model, which is discussed in the next chapter. As far as the predicted parameters for expected drainage are concerned, as discussed in chapter 1, the relevant information necessary to implement a proactive monitoring and management system is drainage depth/volume (Q), fertilizer loads in the drainage (TON), start time (t_1), and duration (t_d) of the drainage.

Predicted values of drainage and nutrient dynamics by the sensor node are transmitted to the gateway of drainage network, from where it is relayed to the neighboring farm and stream networks. The farm networks (e.g., farm B) uses the predicted information and local decision support model to decide whether to reuse the drainage or not, and transmit a reuse acknowledgement to the drainage network. In the former case in which network B intends to reuse the drainage, the drainage water, once available from farm A, is allowed to drain into the drainage reuse bay (through a control pump) instead of the main drainage bay. From the reuse bay, the drainage is then pumped into the irrigation bay of farm B. In case none of the networks send reuse

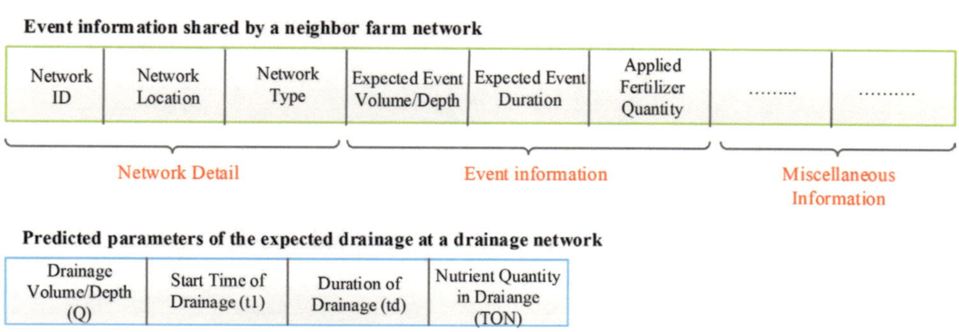

Figure 2.
Parameters related to upcoming event shared by neighbor networks and the predicted variables for the resulting drainage event.

acknowledgements, the drainage would be drained into the main drainage bay. The stream network can then decide, based on the predicted values for nutrients, whether to divert the flow in case of high-nutrient outflows or to otherwise allow it to enter the stream.

3. Architectural detail

The fundamental part of the WQMCM framework is that individual networks learn their environment to predict the impact of events elsewhere in the catchment on their own zone of influence. The predicted drainage information can be beneficial for adjusting management strategy in a farm or in a stream network accordingly, by adopting drainage reuse, disposal, or treatment. For managing agricultural reuse, the overall architectural detail comprises of various modules encompassing drainage, farm, and stream networks, as illustrated in **Figure 3**. For enabling forecasting of drainage dynamics expected as a result of an event in a neighboring farm, two key modules are developed in the drainage network: neighbor linking model, and drainage and pollutant dynamics module. The neighbor linking model uses neighbor event information and sensed drainage data to link the impactful neighbors. The predictive module further comprises individual models to predict Q, t_1, t_d, and TON. The predicted drainage information is used by decision support models in farm and stream network to enable decision making about its reuse or disposal. Furthermore, this information is further used to adjust sampling rate of the sensors, to capture the approaching drainage flow, at the predicted response time.

It has been emphasized in this chapter that due to inevitable drainage and nutrient losses despite adopting BMPs, it is important to enable mechanism for their reutilization. Therefore, a simplified decision support model is developed just as an example to illustrate

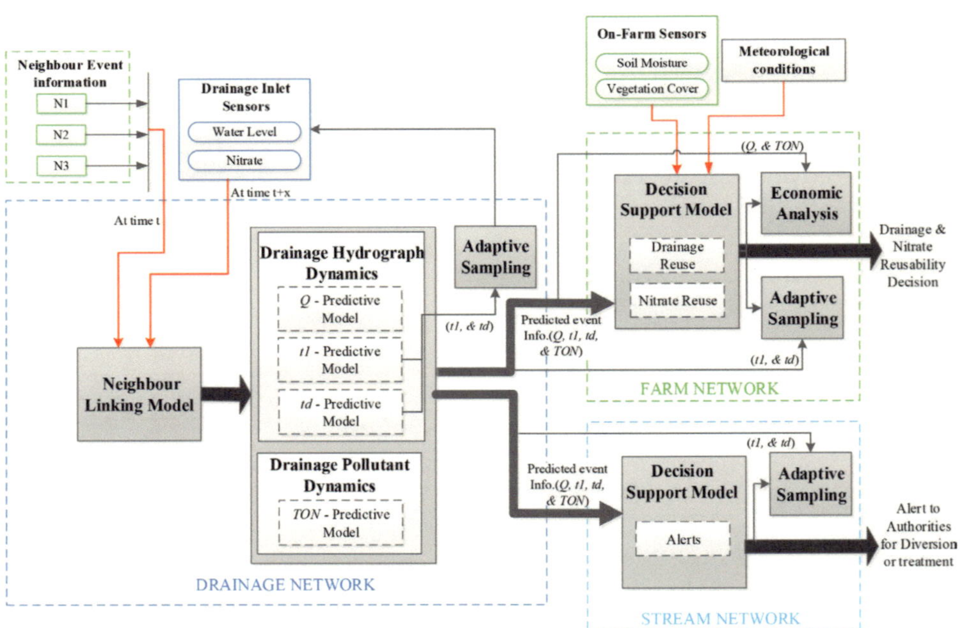

Figure 3.
Block diagram of the WQMCM framework architecture.

the utilization of predicted information for enabling reuse mechanism. The modules of drainage network and farm network blocks are briefly introduced. **Figures 4** and **5** found illustrates the functional flow of these modules for both the blocks.

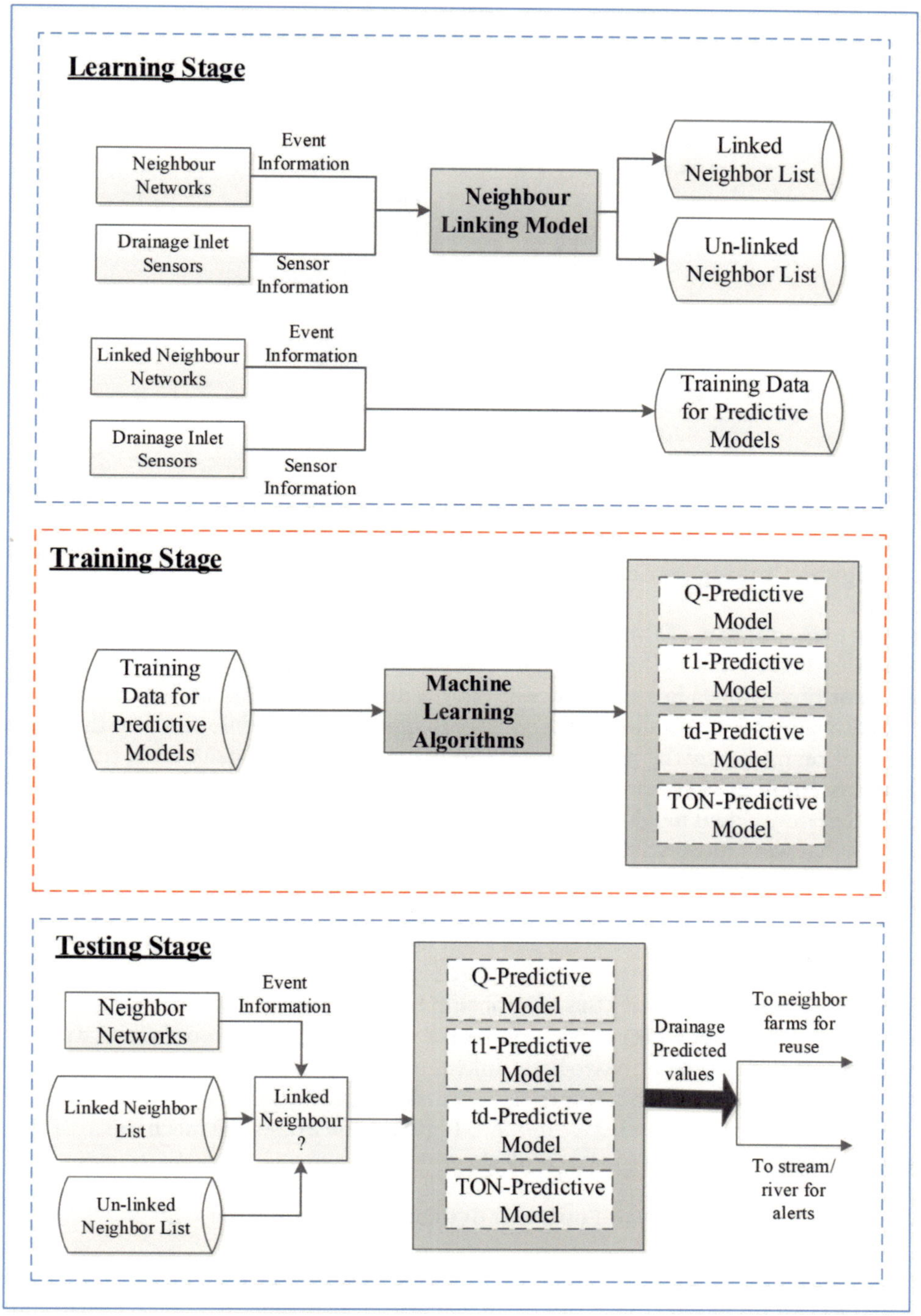

Figure 4.
Functional stages of the drainage network modules, under the WQMCM framework, such as "Learning," "Training," and "Testing."

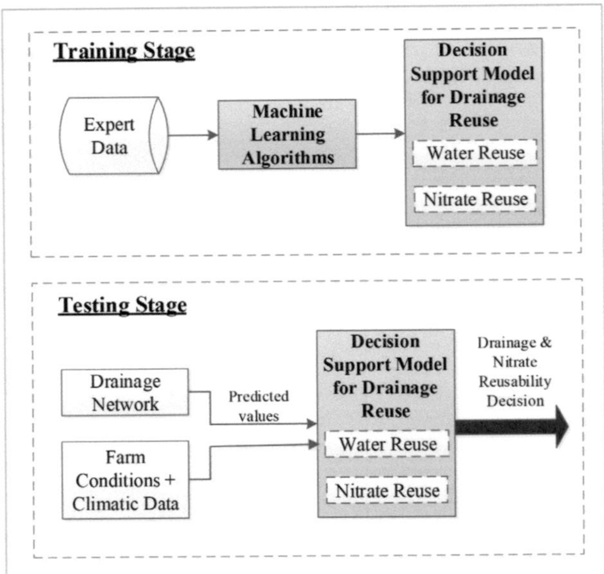

Figure 5.
Functional stages of a decision support model at a farm network, which intends to reuse drainage water, under the WQMCM framework.

3.1 Neighbor linking model

The main purpose of this module in the drainage network is to identify the farm networks that drain into this drainage network. These links are identified over a period of time using a learning process. Firstly, dislocated networks (e.g., located at lower altitudes of the catchment) are filtered out using network location in the shared information packet by the neighbors. Secondly, for the filtered neighbors, training dataset is acquired over a period of time, as respective event information is received from these individual neighbors. Training data consists of (i) the event information packets that were received that particular day, for example, at time t, and (ii) the sensed values of the received drainage and nutrients at the drainage bay at time $t + x$. Here, x refers to the time it takes for drainage to appear in the drainage bay from the time event information is received, and t is ranged between 00 hrs and (x-1) hrs.

For each set of the acquired training data for a particular neighbor, a linear regression model is used to identify the relationship between the sensed drainage and the received event details. This process divides the neighbors, which sent information over the learning phase, into two lists: linked neighbors and un-linked neighbors. Later, for the linked neighbors, sufficient training data are acquired to provide for the development of the predictive models. **Figure 4** illustrates the mechanism of the neighbor linking model in the learning stage section.

3.2 Drainage hydrograph and nutrient dynamics predictive models

Once training dataset is acquired for the linked neighbors, the next step is to develop the models for predicting the drainage hydrograph and nutrient dynamics. Constraints on network nodes (battery life, computing power, availability of sensors, etc.) require a simplified underlying physical model, and a simple machine learning model based on fewer and, ideally, real-time field parameters acquired autonomously and shareable across neighboring farms. Ideally, the model should be based on minimal training samples so that the model can be implementable soon after the deployment of the network. Such models are local in the sense of being valid for a given site (farm in this case). Once developed at the gateway, these models are deployed on the relevant node associated with the particular farm. This facilitates distributed computing where individual nodes of the drainage network, deployed at the outlets of farms, run the learned predictive models for forecasting drainage from those farms. These models can then generate expected drainage hydrograph and nutrients dynamics, which are transmitted to the gateway for further action regarding transmission to neighboring networks.

These models intrinsically self-calibrate because the evolving record of the observations allows them to adapt to the latest condition. This creates portability from one season to the next and from one climate regime to the next. With new data regarding a farm, the models are calibrated at the gateway and re-deployed at the relevant node. However, it is important that a model must maintain a balance between the complexity of the model and the predictive accuracy of the model.

Existing state-of-the-art predictive models are used as a basis to derive low-complexity models for Q, t_1, t_d, and *TON*. A machine learning algorithm, M5 tree, is then used to train the individual models as shown in the training stage of **Figure 4**. Once the models are trained with acceptable prediction performance, the drainage network progresses to the testing stage. In the testing stage, neighbor event information is firstly interpreted using developed neighbor linking lists and then used to predict drainage dynamics using the predictive models, in case of a linked neighbor as illustrated in the testing stage of **Figure 4**. As mentioned earlier, the model accuracy can be continuously improved by learning the evolving instances in the testing stage.

The algorithmic flow of these stages for a drainage network is illustrated using a flow diagram in **Figure 6**. When information is received at the data sink of the drainage network by either a drainage network sensor or a neighbor farm, firstly it is checked whether the network is in the learning stage or not. In such a case, the information is passed on to the neighbor linking model. If the model is in testing stage, then in case the received data packet is from a neighbor, it is checked if the neighbor ID is in the linked neighbor list. If it is an already linked neighbor, then the relevant trained predictive models for that particular neighbor are used to predict the event values. Otherwise, it is determined if the neighbor is an un-linked neighbor, in which case the event packet information is disregarded. In case the received data packet is from the drainage sensor, then the data within the packet are linked with the relevant neighbor information and saved for improving the models later.

3.3 Decision support model (for drainage reuse in a farm)

For the decision support model, the challenge lies in designing a model which takes into account local field conditions, predicted dynamics, and expert knowledge. Unlike the predictive models in the drainage network, this model essentially runs on the gateway of the farm network. The model complexity can substantially vary depending upon the requirements set by the farmer. For example, the farmer may want the model to advice on the possible repercussions of drainage reuse on crop. Furthermore, in case the available drainage is not enough or high in N, the model may also advise on mixing drainage and freshwater for irrigation to fulfill its requirements or to disregard the excess nutrients in the drainage which the farm may not want to reuse. These complexities are highly scenario dependent and require sufficient expert knowledge and data to address. In this chapter, a simplified decision support model is

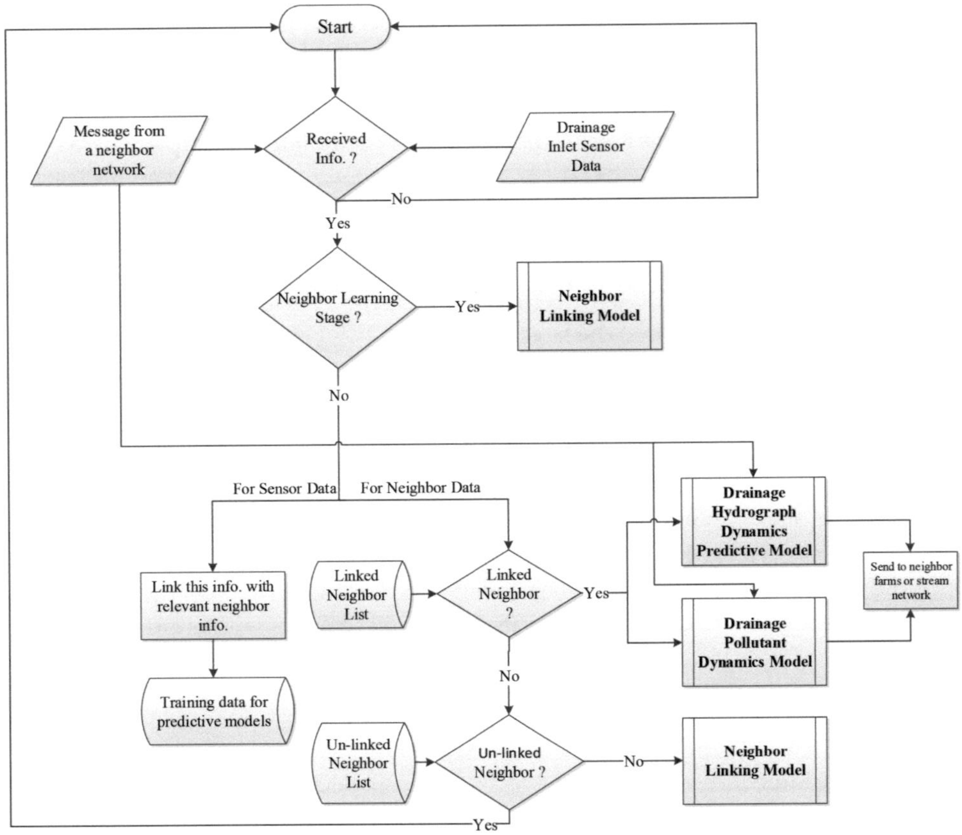

Figure 6.
Basic algorithm running at a sensor node of a drainage network under WQMCM framework.

developed as an example to demonstrate the utilization of predicted information for enabling the reuse mechanism.

Figure 6 illustrates the functional stages of the development and use of the decision model for drainage reuse in a farm. In the training stage, expert knowledge and machine learning algorithms are used to implement simplified models for drainage and nutrient reuse. Once the model is trained, predicted drainage information received from a drainage network, and local field conditions and climatic data are used to classify the usability of the drainage water and nutrients (as shown in the testing stage of **Figure 6**).

4. Conclusions

This chapter presented the water quality monitoring, control, and management framework for a collaborative control and management of agricultural drainage water for addressing the issue of prevalent water crisis. The framework leverages individual networked farms and streams into an integrated water management mechanism. Such a monitoring system should enable each farm to share information about its drainage flow with neighbor networks, for example, with a drainage bay network, which can then process the information for timely treatment, disposal, or reuse of the drainage.

To implement the drainage management, the architecture of the WQMCM framework comprises various modules. Modules for a drainage bay network include neighbor linking model, and predictive models for drainage and pollutant dynamics, whereas, for a farm network, a decision support model is used to ascertain the reusability of the predicted drainage event. The overall functionality of the framework is explored in terms of stages of learning, training, and testing. In the learning stage, neighbor linking model is used to determine the correlation of events in various farm networks with the events received in the drainage bay by the drainage network. The model results in identifying linked and un-linked farm networks by using a combination of geographical filtering and linear regression methods. For the linked networks, training dataset is acquired to provide for the development of the predictive models for drainage dynamics and nitrate losses. When the drainage network has learned the environment and the predictive models for individual farms, it is brought into a testing stage. In this stage, neighbor event information is firstly interpreted using developed neighbor linking lists and then, in case of a linked neighbor, used to predict drainage dynamics. These predicted values are transmitted by a drainage network to other farms and stream networks so that they can take a decision for the reuse, disposal, or conservation of the expected drainage.

Acknowledgements

I would like to express my gratitude to Dr. Nick Harris and Dr. Geoff Merrett (University of Southampton, United Kingdom), and also acknowledge Dr. Mark Rivers and Dr. Keith Smetten (University of Western Australia, Australia) for their support and valuable advice on this research work.

Author details

Huma Zia
Computer Engineering, College of Engineering, Abu Dhabi University, Abu Dhabi, United Arab Emirates

*Address all correspondence to: huma.zia@adu.ac.ae

References

[1] Eisenhauer DE. Irrigation Efficiency and Uniformity, and Crop Water Use Efficiency. The Board of Regents of the University of Nebraska - Lincoln Extension. 2011. Contract No.: EC 732

[2] Rivers M, Weaver D, Smettem K, Davies P. Estimating future scenarios for farm–watershed nutrient fluxes using dynamic simulation modelling. Physics and Chemistry of the Earth, Parts A/B/C. 2011;**36**(9):420-423

[3] Liu X, Ju X, Zhang F, Pan J, Christie P. Nitrogen dynamics and budgets in a winter wheat-maize cropping system in the North China Plain. Field Crops Research. 2003;**83**(2):111-124

[4] Tanji KK, Kielen NC. Agricultural drainage water management in arid and semi-arid areas: Food and Agriculture Organization of the United Nations; 2002

[5] Rhoades J. Intercepting, isolating and reusing drainage waters for irrigation to conserve water and protect water quality. Agricultural Water Management. 1989;**16**(1):37-52

[6] Assar W, Ibrahim MG, Mahmod W, Allam A, Tawfik A, Yoshimura C. Effect of water shortage and pollution of irrigation water on water reuse for irrigation in the Nile Delta. Journal of Irrigation and Drainage Engineering. 2020;**146**(2):05019013

[7] Madramootoo CA, Johnston WR, Willardson LS. Management of agricultural drainage water quality: Food & Agriculture Organisation; 1997

[8] El-Sayed A, Shaban M. Developing Egyptian water quality index for drainage water reuse in agriculture. Water Environment Research. 2019;**91**(5): 428-440

[9] Sharifipour M, Liaghat A, Naseri A, Nozari H, Hajishah M, Zarshenas M, et al. Drainage water management of irrigation and drainage networks of South West Khuzestan. Iranian Journal of Soil and Water Research. 2020;**51**(2):525-539

[10] Adelman DD. Simulation of irrigation reuse system nitrate losses and potential corn yield reductions. Environmental Science & Policy. 2000;**3**(4):213-217

[11] Harper HH. Impacts of Reuse Irrigation on Nutrient Loadings and Transport in Urbanized Drainage Basins. Environmental Research & Design, Inc.; 2012

[12] Grattan SR, Grieve CM, Poss JA, Robinson PH, Suarez DL, Benes SE. Evaluation of salt-tolerant forages for sequential water reuse systems: I. Biomass production. Agricultural Water Management. 2004;**70**(2):109-120

[13] Okano K, Sakamoto Y, Watanabe S, editors. Reuse of drainage water for the production of high quality fruits in single-truss tomato grown in a closed hydroponic system. In: XXV International Horticultural Congress, Part 1: Culture Techniques with Special Emphasis on Environmental Implications. 1998

[14] Grewal HS, Maheshwari B, Parks SE. Water and nutrient use efficiency of a low-cost hydroponic greenhouse for a cucumber crop: An Australian case study. Agricultural Water Management. 2011;**98**(5):841-846

[15] Carr G, Potter RB, Nortcliff S. Water reuse for irrigation in Jordan: Perceptions of water quality among farmers. Agricultural Water Management. 2011;**98**(5):847-854

[16] Oster J, Grattan S. Drainage water reuse. Irrigation and Drainage Systems;**16**(4):2002, 297-2310

[17] Bastiaanssen WGM, Allen RG, Droogers P, D'Urso G, Steduto P. Twenty-five years modeling irrigated and drained soils: State of the art. Agricultural Water Management. 2007;**92**(3):111-125

[18] Liu Y, Gupta H, Springer E, Wagener T. Linking science with environmental decision making: Experiences from an integrated modeling approach to supporting sustainable water resources management. Environmental Modelling & Software. 2008;**23**(7):846-858

[19] Zia H, Harris N, Merrett G, editors. Collaborative Catchment-Scale Water Quality Management using Integrated Wireless Sensor Networks. EGU General Assembly Conference Abstracts; 2013.

[20] Zia H, Harris NR, Merrett GV, editors. Water Quality Monitoring, Control and Management (WQMCM) Framework using Collaborative Wireless Sensor Networks. 11th International Conference on Hydroinformatics 2014; New York City, USA

[21] Zia H, Harris NR, Merrett GV, Rivers M, Coles N. The impact of agricultural activities on water quality: A case for collaborative catchment-scale management using integrated wireless sensor networks. Computers and Electronics in Agriculture. 2013;**96**(0):126-138

Chapter 7

Improvement of Tertiary Irrigation Networks, Changes in Cropping Patterns, and Increasing Cropping Index at Kendal Indonesia

Meinarti Norma Setiapermas, Anggi Sahru Romdon and Yulis Hindarwati

Abstract

Increasing food crop production remains a top priority for the Indonesian government, as demand increases as the population grows. One of the obstacles faced in increasing production is climate change. One of the adaptations to climate change in agriculture is to establish policies for the development and modification of infrastructure that can save water resources management and establish institutions involved in the planning and implementation of water resources management. Kendal Regency, Central Java Province, is one of the regions that contributes to food availability in Indonesia. These conditions support the development of food crops, especially rice, corn, and soybeans. Since 2015, the government and farmers have made improvements to the tertiary irrigation network in paddy fields. This activity aims to increase the cropping index and achieve cropping patterns for one year. In the implementation of irrigation network improvement activities, the community of farmers using water usually works together in determining the location for repairs and making suggestions for improvements to the government. Improvements to irrigation networks were able to increase the cropping index by 0.37 from 1.85 in 2015 to 2.22 in 2016 or equivalent to an area of 8,880 ha (standard area ±24,000 ha).

Keywords: Kendal Regency Central Java Indonesia, irrigation network, increasing food crop production

1. Introduction

Raising food crop production remains the top priority for the government. This is done to meet the increasing food needs and the expanding number of people in Indonesia. One of the hindrances faced in increasing production is climate change. The world of agriculture is faced with more significant challenges, mainly due to global climate change, which impacts the area of sub-optimal land with more severe stress levels [1].

Climate change will have negative impacts on water resources, agriculture, forestry, health, and the vulnerability of public infrastructure, as well as the extinction of various species [2]. The area under rice cultivation that is nationally threatened with drought in the next one or two decades will increase from 0.3–1.4% to 3.1–7.8%, while the area that is unharvested due to drought will also increase from 0.04–0.41% to 0.04–1.87% [3]. On the other hand, the La Nina impact caused an increase in flood-prone cropping areas from 0.75–2.68% to 0.97–2.99%, and the area that is unharvested from 0.24–0.73% to 8.7–13.38%. As a result, the risk of decreasing food production, due to floods and droughts, will rise from 2.4–5.0% to more than 10% [4].

Survey activities for the improvement of tertiary irrigation networks intensively in irrigated rice fields in the Kendal Regency were carried out in 30 (thirty) villages spread over 11 (eleven) subdistricts. The survey location is in the form of agroecosystems of rice fields irrigated with technical irrigation networks and semi-technical irrigation networks.

2. River flow areas basic irrigation network improvement

2.1 Irrigation network to increase food production

Replace in anticipating the impacts of climate change, the government has set policies, such as developing and modifying infrastructure, that can save the management of water resources and biodiversity, taking into account the emissions trading, reducing greenhouse gases, and establishing institutions involved in planning and implementing water resources management, as well as innovations on the application of technology that has a good impact on water resources and other natural resources [5].

The irrigation system includes irrigation infrastructure, irrigation water, irrigation management, irrigation management institutions, and human resources [6]. The Important infrastructure to support the availability of water is the irrigation network. The availability of water will determine the success of increasing production. The water sufficiency for plants will provide optimal results for both growth and production [7]. Regarding food crops, the availability of water from technical irrigation networks (weirs, culverts, and reservoirs), as well as other water sources (dams/spring ponds, long storage, and others) is the basis for accelerating the increase in food and agricultural production [8].

Storage and utilization of abundant water is an applied operational strategy to increase food production in rain-fed rice fields [9]. In addition, through water management (utilizing rainwater and other water resources as much as possible), increasing food production in 4 million ha of distributed rain-fed rice fields in various islands can be done, including by increasing the cropping index from one to two or three times per production in a year [10, 11]. Moreover, the tertiary irrigation network restoration of rice fields can also increase the cropping index and production [12].

The rehabilitation of irrigation networks to optimize the water resources is mainly to support the establishment of food security, increase added value, and at the same time increase farmers' welfare [13]. The Indonesian government has rehabilitated the irrigation network with the realization of around 3.05 million hectares of rice fields that are spread across all agricultural areas in Indonesia [6].

Generally, the type of channel in the irrigation network is divided into primary, secondary, tertiary, quarter, and drainage channels [14]. Primary and secondary irrigation networks become the responsibility of the central, provincial, and district governments. Meanwhile, the maintenance of tertiary irrigation networks is accomplished by farmers or water management groups. The provision of optimal tertiary irrigation networks in irrigated rice fields is a direct and adaptive operational action that can be taken by farmers in dealing with drought or water needs. One of the main culprits of the food and agricultural production opportunities loss is the poor condition of supporting infrastructure.

The deterioration of the irrigation system network will threaten the increase in food production [6]. Restoration of damaged irrigation networks covering an area of 3 million ha or 52% of the total irrigation area, can expand irrigation services to increase the cropping index by 0.3 or more [8]. The network restoration is gradually carried out by the government and farmer groups. Irrigation management requires institutions, namely, management, members, and various accompanying regulations to be efficient in their use and remain sustainable [6]. The results of research in Central Sulawesi Indonesia recommend a more efficient and useful technology in dealing with climate change in food crop production to increase farmers' knowledge, provide user-oriented features, and various institutions to provide collaborative programs in facing the challenges of climate change [15].

2.2 Distribution of watersheds in Kendal Regency

Kendal Regency, Central Java Province, is one of the regions that contributes to food availability in Indonesia. Generally, Kendal Regency has an area of 23,270.07 ha of irrigated rice fields and 809.6 ha of rain-fed rice fields [16]. These conditions support the development of food crops, especially rice, corn, and soybeans. Efforts to increase food crop production can run optimally; therefore, since 2015, restorations have been made to tertiary irrigation networks in rice fields. Through the restoration of the irrigation network, it expects an integrated increase in IP and resource management with a specific location, as well as the application of technological innovations according to technical standards can also be carried out. Finally, the target of increasing production, productivity, and rice crop index in a year can be realized. One of the efforts made by the government to fulfill food needs is by increasing food production in agriculture. The most productive agricultural effort is the use of water for irrigation. Based on these conditions, proper, regular, and sufficient water supply and management is a must [13].

Irrigation can be interpreted as an effort to bring in water by making buildings and channels to drain water for agricultural purposes, distributing water to rice fields or farms regularly and in sufficient quantities, and then disposing of the unnecessary water is required [17]. Generally, irrigation areas are closely related to watersheds as sources of irrigation in the form of large rivers, springs, and others [18].

This watershed is usually associated with the local climate of a place, which is a rainfall area. In Kendal Regency, there are four groups of rainfall areas (**Figure 1**), namely, 500 mm/year rainfall, 1000 mm/year rainfall, 1500 mm/year rainfall, and 2000 mm/year rainfall. The highest rainfall (2000 mm/year) is in two subdistricts, namely, Plantungan and a small part of Sukorejo, 1500 mm/year is in Limbangan District, 1000 mm/year is in Boja, Singorojo, Patean, and parts of Sukorejo. The remaining 500 mm/year of rainfall occurs in most subdistricts in the Kendal Regency. These conditions determine the watershed and the construction of irrigation networks. For example, the Cepiring subdistrict is an irrigation area originating from the Bodri watershed or Sidomukti irrigation area (**Figure 2**).

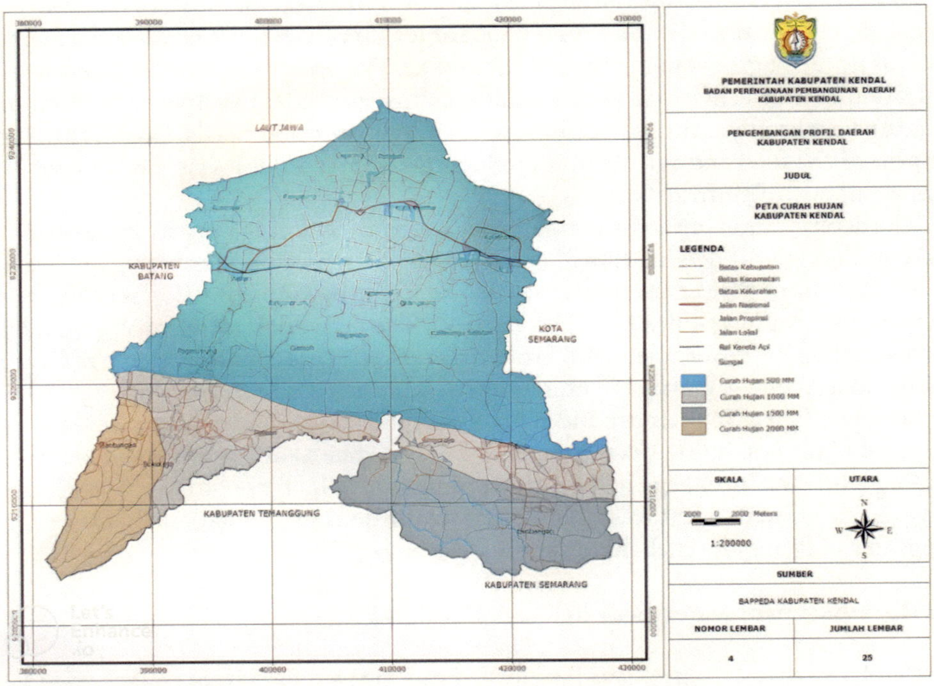

Figure 1.
Map of rainfall in Kendal Regency [19].

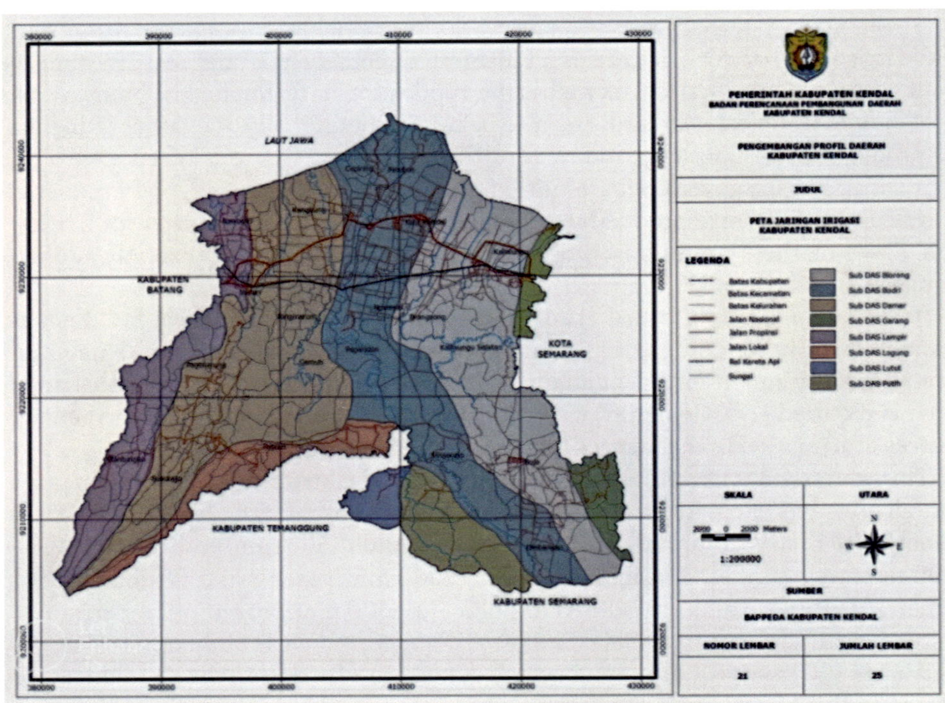

Figure 2.
Irrigation network in Kendal Regency [19].

Based on the type of rainfall and watershed in the Kendal Regency, the construction of a tertiary irrigation network is displayed in **Table 1**.

Table 1 shows that there are 30 locations for the construction of tertiary irrigation channels in the Kendal Regency. These locations are spread over 11 subdistricts with different sources of irrigation water. Based on the table, it is known that seven sources of irrigation water or watersheds become the source of irrigation in the Kendal Regency.

No	Subdistrict	Village	Benefit recipients	Name of irrigation area
1	Boja	Bebengan	Rukun Tani	Sojomerto
2		Ngabean	Tani Mulyo II	Sojomerto
3		Trisobo	Tuk Mandiri I	Sojomerto
4	Brangsong	Brangsong	Gempolsari	Sojomerto
5		Kertomulyo	Sidomulyo	Sojomerto
6		Sidorejo	Sido Makmur	Sojomerto
7		Tosari	Tirto Arum	Sojomerto
8	Cepiring	Juwiring	Sido Maju	Sidomukti
9		Sidomulyo	Sido Maju	Sidomukti
10	Gemuh	Galih	Tirto Rahayu	Kd. Pengilon
11		Poncorejo	Lumintu	Kd. Pengilon
12		Sedayu	Tirto Sido Ayu	Kd. Pengilon
13		Triharjo	Sumber Rejeki	Kd. Pengilon
14	Kaliwungu Selatan	Plantaran	Ngudi Makmur III	Bodri Trompo
15		Sidomakmur	Tani Barokah	Bodri Trompo
16		Sukomulyo	Tirto Lestari	Bodri Trompo
17	Limbangan	Gonoharjo	Darmatirta	Sidomukti
18		Tamanrejo	Sumberrejo	Sidomukti
19	Ngampel	Dempelrejo	Sudi Makmur	Blimbing
20		Ngampel Kulon	Mugi Rahayu	Blimbing
21		Ngampel Wetan	Mugi Rahayu	Blimbing
22	Patean	Mlatiharjo	Mlati Tirto Mulyo	Mudal
23		Plososari	Tani Luhur	Mudal
24	Patebon	Donosari	Maju Makmur	Jlegong
25		Pidodo Kulon	Ngudi Makmur	Jlegong
26	Pegandon	Gubugsari	Gemah Ripah	Bodri Trompo
27		Karangmulyo	Tirto Mulyo	Bodri Trompo
28		Pesawahan	Sido Rukun	Bodri Trompo
29	Singorojo	Kaliputih	Ngudi Makmur	Bodri Trompo
30		Singorojo	Tirto Tlogosari	Bodri Trompo

Table 1.
Tertiary irrigation channels in Kendal Regency, Central Java Province.

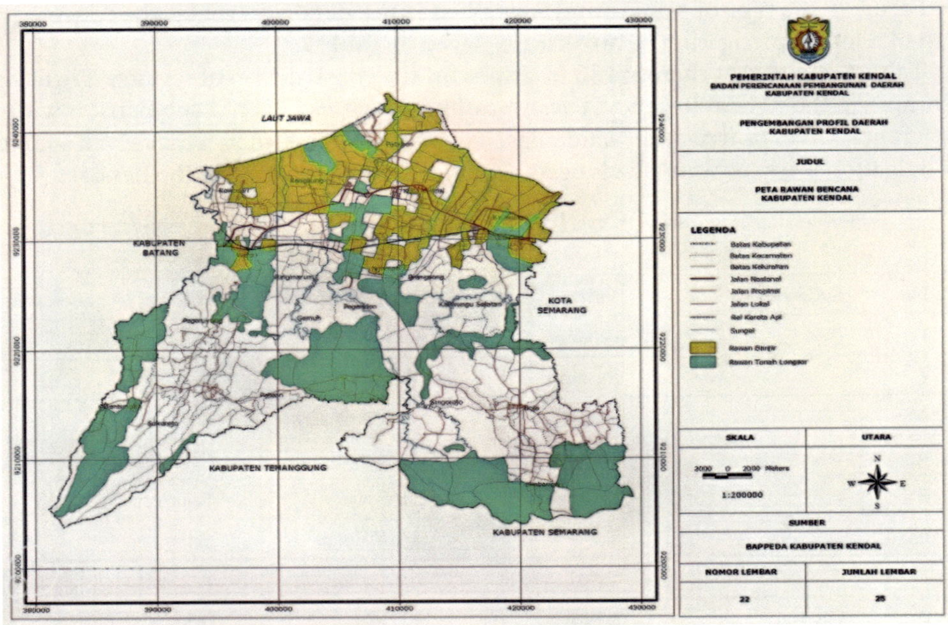

Figure 3.
Disaster-prone area in Kendal Regency [19].

Regarding the restoration of irrigation networks, the determination of the point of the tertiary irrigation network restoration is based on the watershed and the irrigation area that has the potential to increase the cropping index. The restoration of the tertiary irrigation network is not carried out in locations that are prone to flooding, whether it is rainwater puddles or seawater intrusion.

Figure 3 depicts that the northern coast (yellow) is a flood-prone area. Therefore, in the first planting season, the related agency suggested farmers plant rice varieties tolerant of flooding. If **Figure 3** is combined with the position of the irrigation network in **Figure 2**, the Kendal Regency area is an area where the condition of the availability of irrigation water is very sufficient.

The water requirement in rice cultivation is divided into several stages. Tillage is the stage that requires the most water. Optimal tillage is the beginning of preparation for plant life that affects the growth and production of rice plants. Therefore, water is absolutely necessary. Water requirements in each growing season are different. The water requirement for early maturing rice plants is highest in the second planting season because rainwater begins to decrease. The largest available water discharge occurs at the end of February and the potential for water availability is relatively small [20]. Therefore, in the second and third planting seasons, the expansion of water discharge through the tertiary irrigation network significantly affects the increase in the cropping index and the enlargement of rice fields, particularly in technically irrigated rice fields [13].

Water sufficiency for crops in a rotation pattern is the foremost step in considering whether a crop rotation is possible to be applied to an area. The irrigation planning criteria offer an effective rain calculation based on rainfall measurement data at the nearest station. In areas that have irrigation network facilities or irrigated rice fields, the water source is better used as chief support to irrigate crops than rain. A rational rotation pattern is selected based on the following criteria: (1) the need and sufficiency of water; (2) the highest economic profit each season per year; and (3) other considerations, such as market demand and government policies [21].

2.3 The restoration of the tertiary irrigation network in Kendal District

Irrigation network facilities and infrastructure restoration is divided into two activities, namely, development and maintenance. Related to these activities, there are many locations in irrigation areas that require first handling. The results of the research in Yogyakarta showed that the tendency of tertiary irrigation networks in 10 irrigation areas was that the higher the command area, the higher the priority ranking for the development or management of a location. If there are several proposed locations with similar conditions, prioritization can be determined based on the command area [22].

In Kendal Regency, the restoration of the tertiary irrigation network was conducted to support the enlargement of the planted area for food crops. The irrigation network restoration was conducted in 57 locations (**Table 2**), including in the Districts of Boja, Kaliwungu Selatan, Pegandon, Singorojo, Limbangan, Brangsong, Ngampel, Cepiring, Patebon, Patean, and Gemuh. The estimated area of irrigated rice fields is about 3500 ha. The restoration of the tertiary irrigation network is a shared commitment between the government and the water management groups. However, in practice, the *officers of the union of water-user farmers* usually work together in determining the location for repairs and making suggestions for restorations to the government.

The union of water-user farmers in the Bantimurung Irrigation Area, Maros Regency, South Sulawesi, Indonesia, has moderate authority in the utilization, development, and management of irrigation water [23]. Whereas in Morocco, farmer organizations formed to intervene in water management and sugar production appear to be inactive or have weak relationships with their constituents. Therefore, irrigation managers and the sugar industry continue to interact directly with farmers in a centralized manner [24].

The restoration of the tertiary irrigation network in the Kendal Regency was conducted in the areas of lowland, medium, and highland irrigated rice fields. In **Figures 4** and **5**, some examples of tertiary irrigation network restoration locations in the lowlands include Patebon Subdistricts (Maju Makmur water user farmer association covering an area of 60 ha), Pegandon (Gemah Ripah water user farmer association covering an area of 80 ha) and Gemuh (Tirto Rahayu water user farmer association covering an area of 80 ha and Sumber Rejeki water user farmer association covering an area of 80 ha).

The restoration of the lowland tertiary irrigation network was also conducted in Weleri Subdistrict (Sumber Agung Village and Karanganom Village) and Kangkung Subdistrict, Gebanganom Village. The restoration of the tertiary irrigation network in Sumber Agung Village is an irrigation network from the Timbang Weir (Weleri District), which is a transfer from the Damar river. The cropping pattern that is usually done by farmers around the restoration of irrigation networks, in general, is paddy-paddy-paddy. But some farmers plant with paddy-paddy-tobacco or paddy-paddy-horticultural (vegetable) cropping patterns. Meanwhile, irrigation in Karanganom Village, Weleri District, comes from the Sasem Weir (from Grinsing District, Kendal Regency). The usual cropping pattern is paddy-paddy-paddy (**Figure 6**).

For the area of Tani Makmur water-user farmer association Gebanganom Village, Kangkung District (**Figure 7**), the restoration of the tertiary irrigation network

2015 State Budget					2015 Revised State Budget			
Sl. No.	Subdistrict	Village	Name of water-user farmer association	Area (Ha)	Subdistrict	Village	Name of water-user farmer association	Area (Ha)
1	Boja	Bebengan	Rukun Tani	35	Pageruyung	Bangunsari	Taru Martani	50
2		Ngabean	Tani Mulyo	30	Plantungan	Bendosari	Ngudi Sejahtera	80
3		Trisobo	Tuk Mandiri I	25	Ringinarum	Pagerdawung	Tirto Sebrumbun	70
4	Kaliwungu selatan	Plantaran	Ngudi Makmur III	80		Purworejo	Ringin Wangun II	70
5		Sidomakmur	Tani Barokah	70		Mojo	Tirto Agung	70
6		Sukomulyo	Tirto Lestari	80		Kedungsari	Sumber	70
7	Pegandon	Pesawahan	Sido Rukun	70	Cepiring	Podosari	Tirtosari	40
8		Karangmulyo	Tirto Mulyo	80		Korowelangkulon	Ngudi Rejeki	60
9		Gubugsari	Gemah Ripah	80	Gemuh	Sojomerto	Joyo Klantung	70
10	Singorojo	Singorojo	Tirto Tlogosari	70	Singorojo	Getas	Usaha Maju	60
11		Kali Putih	Ngudi Makmur II	80	Boja	Karangmanggis	Ngudi Rahayu	25
12	Limbangan	Gonoharjo	Sido Makmur	40		Blimbing	Dewi Sri Makmur	25
13		Tamanrejo	Sumberrejo	50		Bebengan	Tanjungsari	30
14	Brangsong	Kertomulyo	Sido Mulyo	70	Ngampel	Winong	Pengilon	70
15		Sidorejo	Sido Makmur	70		Sudi Payung	Sudi Makmur	70
16		Brangsong	Gempolsari	70	Patebon	Tambakrejo	Tirto Sari	40
17		Tosari	Tirto Arum	70		Pidodowetan	Mugi Langgeng	50
18	Ngampel	Dempelrejo	Sudi Makmur	80	Pegandon	Puguh	Tirto Panguripan	60
19		Ngampel Kulon	Mugi Rahayu I	80		Tegorejo	Dadi rejo	80

2015 State Budget					2015 Revised State Budget			
Sl. No.	Subdistrict	Village	Name of water-user farmer association	Area (Ha)	Subdistrict	Village	Name of water-user farmer association	Area (Ha)
20		Ngampel Wetan	Mugi Rahayu II	70	Weleri	Ngasinan	Tirto Arum	70
21	Cepiring	Sidomulyo	Sido Maju	70		Karanganom	Loh Jinawi	70
22		Cepiring	Sido Maju	70		Sumberagung	Ngudi Luhur Sejati	70
23	Patebon	Donosari	Maju Makmur	60		Karangdowo	Tirto Asri	70
24		Pidodo Kulon	Ngudi Makmur	80	Kangkung	Gebanganom	Tani Makmur	60
25	Patean	Ploso Sari	Tani Luhur	50		Jungsemi	Sido Kabul	70
26		Mlatiharjo	Mlati Tirto Mulyo	50				
27	Gemuh	Triharjo	Sumber Rejeki	80				
28		Sedayu	Tirto Sido Ayu	80				
29		Poncorejo	Lumintu	80				
30		Galih	Tirto Rahayu	80				
Area of irrigated rice field				2000				1500

Table 2.
Tertiary irrigation network repaired in the Kendal Regency in 2015.

Figure 4.
Secondary irrigation channels in Gubugsari Village, Pegandon District (not yet repaired), and tertiary irrigation networks in Gubugsari Village, Pegandon District.

Figure 5.
Secondary (no restoration) and tertiary channels in Galih Village, Triharjo Village, Gemuh District.

Figure 6.
Restoration of the Tertiary Irrigation Network at Ngudi Luhur water user farmer association Sejati, Sumberagung Village, Weleri District, and Timbang Weir, Weleri District.

Figure 7.
Tertiary irrigation network rehabilitation at P3A Tani Makmur, Gebanganom Village, Kangkung District.

is adjacent to the secondary channel from the Juwero Weir (in Gemuh District). The initial cropping pattern was paddy-pady-corn. With the restoration of the tertiary irrigation network, it is expected that the cropping pattern will become paddy-paddy-paddy.

Restorations of the tertiary irrigation network in the highlands were conducted in the district of Boja (in **Figure 8**), water-user farmer association of Bebengan Village for an area of 35 ha, Tani Mulyo water-user farmer association (Ngabean Village) for an area of 30 ha and in Trisobo Village Tuk Mandiri I water-user farmer association for an area of 25 ha. In addition, restorations of the irrigation network were carried out at the location of Taru Martani water-user farmer association (in **Figures 9** and **10**, Bangunsari Village, Pageruyung District), as well as repairing the Kreon weir from the Kuto River, which was an aspiration fund. The cropping index around Kreon Weir is not rice, but annuals. It can be said that the restoration of the irrigation network around the Kreon Weir is to improve the irrigation of the rice fields under the Kreon Weir. Restorations to the irrigation network were executed in two places, namely, around the Kreon Weir and near the rice fields.

Besides Pageruyung (**Figure 11**), other highland rice fields are located in Plantungan District. The restoration was carried out at Ngudi Sejahtera water-user farmer association, Bendosari Village, Plantungan District. This location is quite far and close to Batang Regency. The flow of water comes from the Bulus river, which is runoff from the Kuto River. The cropping pattern of irrigated rice fields is paddy-paddy-paddy. It can be said that the restoration of the irrigation network is to expand the area of rice fields that can be irrigated.

The restoration location is quite far from the rice fields. However, the land around the irrigation network, which was previously planted with annual crops, is expected to be planted with rice in the third planting season of the following year.

Figure 8.
Semi-technical tertiary irrigation channels in Bebengan Village and Ngabean Village, Boja District.

Figure 9.
The secondary irrigation channel has a disconnected position with a tertiary channel in Trisobo Village, Boja District.

Figure 10.
Kreon weir, irrigation network, and irrigated rice fields in Bangunsari Village, Pageruyung District.

Figure 11.
The location of the irrigation network is in Bendosari Village, Plantungan District.

Based on the results of field observations and focused discussions with the agriculture, plantation, livestock, and forestry service of Kendal Regency (coordinator of instructors, agricultural assistants, instructors, and agricultural infrastructure staff) and farmers, information was obtained that the restoration of the tertiary irrigation network can reduce water loss due to leakage during irrigation. Therefore, the area of irrigated rice fields increases, especially in the second season of rice planting. In addition, farmers hope that the rice fields can be planted with food crops after the restorations of the tertiary irrigation network in the third planting season.

The implementation of irrigation network infrastructure development in Bengkulu Tengah Regency, Bengkulu Province, Indonesia, is a priority in supporting the provision of water resources, functioning to support dynamic and interactive rural development, as well as irrigated agriculture in the future. The utilization of water resources is fully aimed at improving the economy and welfare of farmers. By using technology it supports infrastructure development, as well as water management to produce a more effective irrigation network [25].

Small irrigation systems with a command area of less than 500 ha are the backbone of family food security, which, in turn, leads to food security at the national level. The deterioration of the irrigation system network will threaten the increase in food production. In the future, irrigation infrastructure must be supervised better. Therefore, the agricultural sector can realize agricultural diversification. The wider the conservation, local wisdom, and social capital in irrigation management can be maintained. Irrigation expansion and development should include the participation of farmers and water-user farmer association through self-management, not a tender system (auction), and target-oriented. In addition to better performance, the self-management system fosters a high sense of ownership and responsibility by the water-user farmer association [6].

Self-management will be more efficient if it is combined with a mutual cooperation system, therefore, restoration targets can be achieved and even exceeded. In addition, supervision from farmers and even cross-control between group members will automatically occur. In the future, the role of the water-user farmer association will be improved to increase the function of the development and management of irrigation, especially in small irrigation. Good small irrigation management involving the role of government and stakeholders is expected to improve farmers' income levels, expand job opportunities in farming and outside farming, food resources, soil and environmental damage prevention, and ownership of productive assets [6].

Participatory irrigation, where farmers are given greater control and management responsibility, has been a topic of controversy for many years. Initially seen as a panacea for dealing with weaknesses in state-run irrigation, participatory irrigation has generated mixed results, especially in south Asia. Part of the challenge in understanding the conditions that elevate and undermine participatory irrigation is that it is rarely deployed in the same way. For example, irrigation fees collected by farmers are not handled collectively, even within a single country. In some instances, a large amount of collected fees is retained locally. Only a small amount is kept for local use. In this paper, we use game theory to consider how the portion of irrigation fees retained locally might impact the effectiveness of participatory irrigation. We show that there are multiple plausible equilibria and that allowing farmers to retain more funds locally might shift behaviors from an uncooperative equilibrium to a cooperative outcome. However, we also find that it is unlikely to be a singular fix. We use empirical evidence to demonstrate the conundrums of making participatory irrigation sustainable [26]. Irrigation water can increase rice production, productivity, and farmers' revenue through the restoration of irrigation infrastructure, utilization of biological fertilizers, and proper management of irrigation systems [27].

3. Food production and tertiary irrigation improvement

The impact of constructing a tertiary irrigation network in the Kendal Regency can be seen in two things, the cropping index and productivity. The construction of irrigation networks can affect productivity and production [28]. Tertiary irrigation construction not only increases production by 22.19% but also increases the cropping index [29]. Generally, the impact of tertiary irrigation construction on cropping index and production in the Kendal Regency is illustrated in **Figures 12** and **13**.

The picture above depicts that there was an increase in the cropping index of rice plants in 2016. This increase occurred after the restoration of the irrigation network in 2015. The increase in the cropping index occurred by 0.37 from 1.85 in 2015 to 2.22 in 2016 or equivalent to an area of 8.880 ha (standard area ±24,000 ha). However, in the following year until 2019, the rice planting index decreased to 0.73. Exploration of further information related to the decline in the index of rice cultivation was caused by the shift of commodities, which are usually paddy to corn, in several subdistricts, such as Brangsong, Patebon, Gemuh, and Ringinarum. Farmers said that the corn

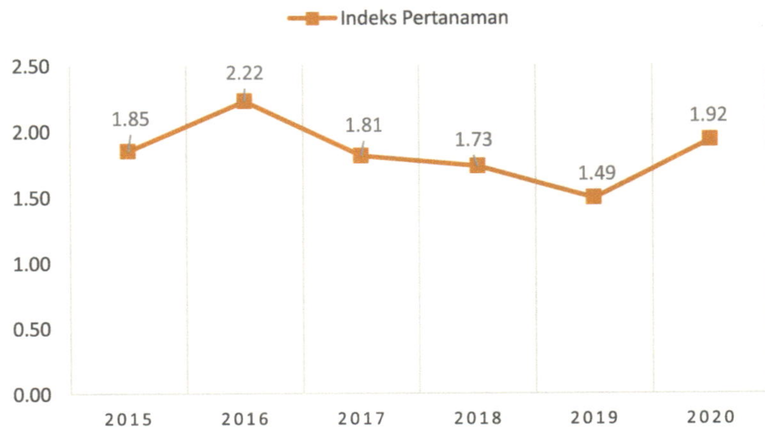

Figure 12.
The effect of tertiary irrigation network construction on increasing IP in Kendal Regency in 2015–2020.

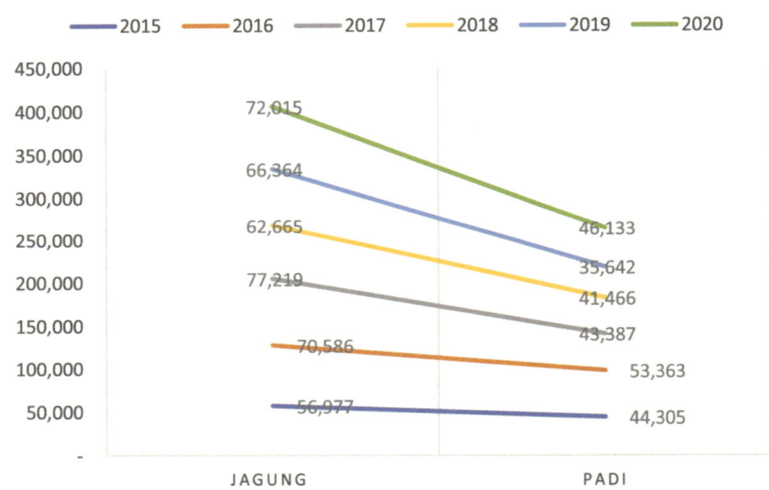

Figure 13.
The effect of tertiary irrigation network construction on rice and maize production in 2015–2020.

commodity in recent years was more promising both in terms of production and price. Therefore, corn was more profitable. The transfer of commodities conducted by farmers is in line with corn production in Kendal Regency as shown in **Figure 13**.

In **Figure 13**, rice production increased from 2015 was 44,305 tons to 53,363 tons (9.058 tons). However, from 2017 to 2019, there was a decrease to 35,642 tons. On the other hand, corn in 2016 increased in production by 13,609 tons, from 56,977 tons to 70.586 tons. Furthermore, corn production increased to 77,219 tons in 2017, in 2018 by 62,665 tons, in 2019 by 66,364 tons, and in 2020 by 72,015 tons. This condition is contrary to rice production which continued to decline until 2019 and rose again in 2020.

4. Conclusions

Improvement of the tertiary irrigation network is one of the adaptations to climate change for the farmer level. The improvement of the tertiary irrigation network is very much needed by farmers with further management carried out by the water-user farmer association. Improvements to the irrigation network have been carried out at 57 points and spread across all subdistricts from the lowlands, medium to highlands. The improvement of the tertiary irrigation network was able to increase the cropping index by 0.37, from 1.85 in 2015 to 2.22 in 2016, or equivalent to an area of 8,880 ha (standard area ±24,000 ha). The construction of irrigation networks was also able to increase production by 9,058 tons, from 44,305 tons in 2015 to 53,363 tons in 2016. In the following year (2017–2019), rice production decreased because farmers in several subdistricts switched commodities to corn. The role of water-user farmer associations in the improvement and optimal management of tertiary irrigation networks is very important. Thus, the management of water-user farmer associations must be regulated and functioned professionally in sustainable agricultural development in Indonesia.

Acknowledgements

Thank you to the staff of the Department of Agriculture, Plantation and Forestry, Kendal Regency, Central Java Province and farmers who have provided information about the construction and improvement of tertiary irrigation networks. Likewise, we would like to thank all the staff of the Central Java of Assessment Institute for Agricultural Technology who have assisted in this survey, especially Mrs. Endah Winarni.

The survey activity for the improvement of the tertiary irrigation network on irrigated rice fields in Kendal Regency, Central Java was financed by the Indonesian Agency for Agricultural Research and Development, Ministry of Agriculture.

Conflict of interest

The authors declare no conflict of interest.

Appendices and nomenclature

Indonesian farmers really need adaptation technology to climate change, especially farmers in paddy fields. One of the technologies for adapting to climate change is the improvement of tertiary irrigation networks in irrigated rice fields. The Indonesian government, especially the agriculture office, has data on irrigation networks (primary, secondary, and tertiary) and points for repairing tertiary irrigation networks. Determination of the point of improvement of the tertiary irrigation network is based on the possibility of primary and secondary irrigation networks, available water sources, and collaboration of water-user farmer associations and the government. Surveys of farmers using water after the improvement of the tertiary irrigation network show that the water source to meet the irrigation network greatly determines the commodities to be planted, the area of irrigation coverage, and the cropping index.

Author details

Meinarti Norma Setiapermas[1*], Anggi Sahru Romdon[2] and Yulis Hindarwati[1]

1 Research Center for Food Crops, Research Organization for Agricultural and Food, National Research and Innovation Agency, Semarang, Indonesia

2 Researh Center for Behavioral and Circular Economics, Research Organization for Governance Economy and Community Welfare, National Research and Innovation Agency, Semarang, Indonesia

*Address all correspondence to: meinarti.ns@gmail

References

[1] Soepandi D. Fisiologi Adaptasi Tanaman Terhadap Cekaman Abiotik Pada Agroekosistem Tropika. Bogor: Institut Pertanian Bogor Press; 2013. p. 228

[2] Runtunuwu E, Amien I. Dampak dan Antisipasi Perubahan Iklim Sektor Pertanian: Buletin Hasil Penelitian Agroklimat dan Hidrologi. Bogor: Balai Penelitian Aroklimat dan Hidrologi; 2010

[3] Anonim. Laporan Akhir Kegiatan 2008-2009. Jakarta: Badan Penelitian dan Pengembangan Pertanian; 2009

[4] Las I. Teknologi Pertanian Siasati Perubahan Iklim. Balai Besar Penelitian dan Pengembangan Sumberdaya Lahan Pertanian. Badan Penelitian dan Pengembangan Pertanian. Kementerian Pertanian. (Unpublished). 2010

[5] Dwiratna SNP, Suryadi E, Kamaratih KD. Optimasi pola tanam pada lahan sawah tadah hujan di kecamatan cimanggung kabupaten sumedang. Teknotan, Jurnal Industri Teknologi Pertanian, Universitas Padjadjaran Bandung dan PERTETA Indonesia. 2016;**10**(1):37-45. P-ISSN: 1978-1067; E - ISSN: 2528-6285

[6] Purwantini TB, Suhaeti RN. Irigasi Kecil : Kinerja, Masalah dan Solusinya. Forum Penelitian Agro Ekonomi. 2017;**35**(2):91-105

[7] Setiapermas MN, Romdon AS, Winarni E. Dukungan Sarana Prasarana Pertanian pada Progam Upsus di Kabupaten Kendal. In: Dalam: Hermawan A dan Komalawati, editor. Upaya Khusus Peningkatan Produksi Pangan Potensi. Banten (ID) Loka Aksara: Implementasi dan Dampak; 2021

[8] Sulaiman AA, Simatupang P, Las I, Jamal E, Hermanto K, Syahyuti IK, et al. Kebijakan dan Strategi Kedaulatan Pangan dan Kesejahteraan Petani. In: Sukses Swasembada Indonesia Menjadi Lumbung Pangan Dunia 2045. 2017. pp. 81-106

[9] Haryoto I, Raharjo N. Kajian Pendahuluan Sistem Pemanfaatan Air Hujan. JAI. 2015;**8**(1):105-114

[10] Pirngadi K, Makarim AK. Peningkatan Produktivitas Padi pada Lahan Sawah Tadah Hujan melalui Pengelolaan Tanaman Terpadu. Jurnal Penelitian Pertanian Tanaman Pangan. 2006;**25**(2):116-123

[11] Dwiratna SNP, Suryadi E, Kamaratih KD. Optimasi Pola Tanam pada Lahan Sawah Tadah Hujan di Kecamatan Cimanggung Kabupaten Sumedang. Jurnal Teknotan. 2016;**10**(1):37-45

[12] Damayanti L. Pengaruh Irigasi Terhadap Kesempatan Kerja, Kemiskinan dan Ketahanan Pangan Rumah Tangga Tani di Daerah Irigasi Parigi Moutong. Desertasi. Yogyakata (ID): Universitas Gajah Mada; 2012

[13] Zakaria AK. Dampak Rehabilitasi Jaringan Irigasi Perdesaan terhadap Adopsi Teknologi Budi Daya Padi. Penelitian Pertanian Tanaman Pangan. 2014;**33**(2):102-108

[14] Noerhayati E, Suprapto B. Perencanaan Jaringan Irigasi Saluran Terbuka. Cita Intrans: Selaras; 2018

[15] Purwanto MYJ, Anika EN. Peningkatan efisiensi dan produksi pangan dengan pembangunan sistem irigasi pipa di ingkat tersier. Kementerian Pekerjaan Umum Indonesia: Jurnal Irigasi. 2012;**7**(2):99-109.

[16] Badan Pusat Statistik Kabupaten Kendal. Kabupaten Kendal dalam Angka. BPS Kabupaten Kendal. 2022

[17] Available from: https://distan.bulelengkab.go.id/informasi/detail/artikel/pengairan-53 [Diakses: April 30, 2022]

[18] DPUPKP Kulonprogo. Pengetahuan Umum Teknik Irigasi. 2020. Available from: https://dpu.kulonprogokab.go.id/detil/247/pengetahuan-umum-teknik-irigasi [Diakses: April 29, 2022]

[19] Available from: https://www.kendalkab.go.id

[20] Suripto. Kebutuhan Air Irigasi Untuk Tanaman Padi Genjah. Politeknologi;**17**(1):81-87

[21] Makarim AK, Ikhwani MJMMJ. Rasionalisasi Pola Rotasi Tanaman Pangan Berbasis Ketersediaan Air. Iptek Tanaman Pangan. 2017;**12**(2):83-90

[22] Attamimi FF, Buyang CG, Kalalimbong A. Perencanaan Saluran Irigasi Samal Kiri di Kabupaten Maluku Tengah. Jurnal Simetrik. 2021;**11**(2):462-468

[23] Las I. Teknologi Pertanian Siasati Perubahan Iklim. Balai Besar Penelitian dan Pengembangan Sumberdaya Lahan Pertanian. Badan Penelitian dan Pengembangan Pertanian. Kementerian Pertanian Indonesia . 2010. (Unpublished)

[24] Faysse N, Errahj M, Kuper M, Mahdi M. Learning to voice? The evolving roles of family farmers in the coordination of large-scale irrigation schemes in Morocco. Water Alternatives. 2010;**3**(1):48-67

[25] Hidayat R. Evaluasi Pembangunan Infrastruktur Jaringan Irigasi di Kabupaten Bengkulu Tengah. Ekombis Review: Jurnal Ilmiah Ekonomi Dan Bisnis. 2022;**10**(1):334-347

[26] Hone S, Crase L, Burton M, Bethany B, Gandhi VP, Ashfaq M, et al. Farmer cooperation in participatory irrigation in South Asia: Insights from game theory. Water. 2020;**12**:1329

[27] Nurkhaida R, Hamdani A, Suriadi A, Heryani N. Increasing rice productivity and profitability through irrigation water management and bio-fertilizer in West Nusa Tenggara. Web of Conferences. 2021;**306**:04012

[28] Triasni AAR. Analisis Dampak Pembangunan Jaringan Irigasi Tersier Tingkat Usaha Tani Terhadap Peningkatan Produksi dan Produktivitas Padi Sawah di Kelurahan Lapajang Kecamatan Labata Kabupaten Soppeng. Jurnal Ilmiah Agrotani. 2019;**1**(1):16-22

[29] Purwanto MYJ, Erizal N. Peningkatan Efisiensi dan Produksi Pangan dengan Pembangunan Sistem Irigasi Pipa di ingkat Tersier. Jurnal Irigasi. 2012;7(2):99-109

Chapter 8

Irrigation Scheduling Methods: Overview and Recent Advances

Younsuk Dong

Abstract

Applying irrigation at the right time and the correct amount is a challenge. Irrigation scheduling is a method of determining the appropriate amount of water to be applied to a crop at the correct time to achieve full crop production potential. Scheduling irrigation based on the weather, soil moisture, and plant data are reviewed. The advantages and challenges of each irrigation scheduling method are also discussed. In addition, innovative irrigation scheduling methods such as internet of things (IoT)-based on using wireless communication and smartphone app-based are described. In conclusion, the irrigation scheduling method has been continuously developed to be more accurate and precise. Integration of innovative technologies and techniques, such as IoT and machine learning, could be used to take the scheduling method to the next level.

Keywords: irrigation scheduling, agricultural water management, IoT, soil moisture sensors, water efficiency

1. Introduction

Water is an essential component of growing crops. Even in humid climates, precipitation is not enough to meet plant water requirements. Thus, additional water is applied through irrigation. Irrigation management can be complicated with unpredictable precipitation patterns. Not watering at the right time and correct amount can result in plant water stress and reduce the quality and yields of crops. On the other hand, over-watering can increase the risk of nutrients leaching below the root zone, waste resources (water, energy, and nutrients), and environmental impacts. Therefore, it is important to apply irrigation at the right timing and correct amount. Determining the appropriate amount of irrigation and the optimal timing of irrigation are challenging due to unpredictable weather conditions and climate changes.

Irrigation scheduling is a method of determining the appropriate amount of water to be applied to a crop at the correct time to achieve full crop production potential. Scheduling irrigation water has been based on the soil moisture measurement and/or weather data that are estimates of evapotranspiration.

This chapter reviews the various existing and recent advances in irrigation scheduling methods. The irrigation scheduling methods are:

- Feel and appearance.

- Gravimetric Method.
- Weather-based irrigation scheduling.
- Sensor-based irrigation scheduling.
- Plant-based irrigation scheduling.
- IoT sensor technology.
- Smartphone APP.

2. Irrigation scheduling method

2.1 Feel and appearance

The most popular and quickest method is based on the feel and appearance of the soil. A soil probe is typically used to take soil samples. **Table 1** shows the soil moisture and appearance relationship. **Table 1** shows an approximate relationship between field capacity and wilting point. The top of each soil type corresponds to the condition of zero soil moisture deficiency, also known as field capacity. The bottom of each soil type corresponds to the condition of maximum soil moisture deficiency, also known as wilting point. The soil moisture deficiency also presents the available moisture range of the soil. The table provides general numbers for a specific group of soils and may not apply to all soil groups. This method is not quantitative and is judged by the individual, which lacks precision.

2.2 Gravimetric method

Soil moisture content is an important parameter for understanding the water movement in the soil. Taking soil samples is the direct method to measure the actual soil moisture level. This method requires weighing a sample of a known volume of soil and then reweighing it after drying in an oven at 105°C to calculate the mass of water lost by drying [2]. This method allows for calculating gravimetric water content (g/g) and soil bulk density (g/cm^3). Multiplying the gravimetric water content by the soil bulk density allows for calculating the volumetric water content (cm^3/cm^3) [3]. The equation of volumetric water content is described in (Eq. (1)). Soil sample collection method is accurate, but it requires intense labor, time, and soil disturbance. Therefore, continuous soil moisture monitoring through soil sample collection on farmland can be difficult and limited.

$$\theta_v = \frac{\theta_g * \rho_{soil}}{\rho_{water}} = \frac{\left(\left(\frac{(M_{water} - M_{dry})}{M_{dry}}\right) * \left(\frac{M_{dry}}{V_{soil}}\right)\right)}{\rho_{water}} \quad (1)$$

Where,
θ_v = Volumetric water content (cm^3/cm^3).
θ_g = Gravimetric water content (g/g).

Moisture deficiency in/ft	Loamy sand	Sandy loam	Loam	Clay loam
0	Leaves wet outline on hand when squeezed (field capacity).	Appears very dark, leaves wet outline on hand; makes a short ribbon (field capacity).	Appears very dark; leaves a wet outline on hand; will ribbon out about one inch (field capacity).	Appears very dark; leaves slight moisture on hand when squeezed; will ribbon out about two inches (field capacity).
0.2	Appears moist; makes a weak ball.	Quite dark color; makes a hard ball.	Dark color; forms a plastic ball; slicks when rubbed.	Dark color; will slick and ribbon easily.
0.4	Appears slightly moist sticks together slightly.	Fairly dark color, makes a good ball.	Quite dark, forms a hard ball.	Quite dark, will make a thick ribbon; may slick when rubbed.
0.6	Very dry, loose; flows through fingers. (Wilting point)	Slightly dark color, makes a weak ball.	Fairly dark, forms a good ball.	Fairly dark, makes a good ball.
0.8		Lightly colored by moisture, will not ball.	Slightly dark, forms a weak ball.	Will ball, small clods will flatten out rather than crumble.
1		Very slight color due to moisture. (Wilting point)	Lightly colored; small clods crumble easily.	Slightly dark, clods crumble.
1.2			Slight color due to moisture, small clods are hard (Wilting point)	Some darkness due to unavailable moisture, clods are hard, cracked. (Wilting point)

Table 1.
Relationship between soil moisture and appearance [1].

ρ_{soil} = Soil bulk density (g/cm^3).
ρ_{water} = Density of water = 1 (g/cm^3).
M_{water} = Weight of wet soil (g).
M_{dry} = Weight of dry soil (g).
V_{soil} = Volume of soil sample (cm^3).

2.3 Weather-based irrigation scheduling method

Weather-based irrigation scheduling method is based on the weather condition. Four major weather parameters determine evapotranspiration (ET), which drives the weather-based irrigation scheduling method. The weather parameters are solar radiation, air temperature, relative humidity, and wind speed. Higher the solar radiation, the greater ET. This is because sunlight is the main energy source for evaporating water. The warmer the air, the greater ET, because it can hold more water vapor. The drier the air, the greater the ET, because there is less water vapor it already holds.

The greater wind, the greater the ET. In humid climate regions, solar radiation and air temperatures play a significant role in determining daily ET.

ET can be estimated in several ways. One method that is accepted as an international standard is the Penman–Monteith Equation, which is used to calculate the reference potential ET (rPET) using comprehensive weather data. The weather data includes net radiation, soil heat flux, average air temperature, wind speed, saturation vapor pressure, actual vapor pressure, the slope of vapor pressure curve, and psychrometric constant. rPET assumes a four grass-covered surface that is well-watered and unshaded. The actual ET of a crop at any given time depends on the amount of leaf area and the developmental stage, so to calculate the ET for a specific crop type, at a specific developmental stage, the rPET values must first be multiplied by a crop. Kc changes with crops as they grow, for example, the Kc of fruit trees increases rapidly in the spring as the trees leaf out to full canopy. **Figure 1** shows the change of Kc as a soybean grows. To estimate actual crop ET, the rPET is multiplied by the crop coefficient Kc to determine the actual water lost from the crop via ET (see (Eq. (2)).

$$ET_C = K_C * rPET \tag{2}$$

Where,

ET_C =Actual Crop Evapotranspiration (in/day).

K_C =Crop Coefficient (unitless multiplier).

$rPET$ =Reference Potential Evapotranspiration (in/day).

Based on each day of the actual crop evapotranspiration, the suggested irrigation amount can be calculated to ensure that the soil has adequate moisture for plant growth and improve irrigation water use efficiency. For example, if last week's cumulative actual crop evapotranspiration was 2.5 cm, the farmer should apply 2.5 cm of irrigation to maximize crop production and minimize environmental impacts.

2.4 Soil moisture sensor-based irrigation scheduling method

An alternative way to measure soil moisture content is using a soil moisture sensor. A typical soil moisture sensor estimates the volumetric water content (cm3/cm3) in soils. Soil moisture sensors allow monitoring the changes of soil moisture level over time without disturbing the soil. The sensors can be installed at multiple depths of soil to monitor the water flow in soil. In general, there are two types of soil moisture sensors. Soil tension sensors measure the required force for roots to pull water out of the soil. Volumetric water content sensor based on the electrical properties of the soil

Figure 1.
Crop coefficient (kc) changes as the soybean grows.

to estimate the soil moisture level. Knowing some common terminologies used in soil moisture sensor-based irrigation scheduling would be helpful in interpreting these sensor data.

The descriptions of some useful terminologies follow:

Saturation: All soil pore spaces are filled with water.

Field Capacity (FC): Maximum amount of water that soil can hold after drainage.

Wilting point (WP): Soil moisture level where there is no available water for the crop.

Available Water Capacity (AWC): The difference between FC and WP, is often expressed in inches of water per foot of soil. AWC of a soil is primarily related to the soil texture, organic matter content, and bulk density. The equation of AWC is shown in (Eq. (3)).

$$AWC = \frac{\rho_{soil} * T * P_W}{\rho_{water} * 100} \tag{3}$$

Where,

AWC = Available water capacity in inches.

ρ_{soil} = Soil bulk density.

T = thickness of soil horizon under consideration in inches.

P_W = Moisture content between field capacity and wilting point in percentage by weight.

ρ_{water} = Density of water = 1 (g/cm^3).

Maximum Allowable Depletion (MAD): The amount of available water that can be safely depleted without causing drought stress, which depends on the crop and the growth stage.

Soil Matric Potential (SMP): Physical force required for the plant to move water into its root system.

2.4.1 Type of soil moisture sensors: Soil tension sensors

Soil moisture can be estimated by electrical resistance instruments. As the soil moisture content changes, the electrical properties of the soil also change. Electrical resistance sensing devices are sensitive to salts and fertilizer in the soil. A typical soil tension sensor is a solidstate electrical resistance sensing device that also measures soil matric potential (SMP). This type of sensor reads the resistance changes as the soil tension changes, which depends on the soil moisture content. A common soil tension sensor is WATERMARK, manufactured by the IRROMETER company (Riverside, CA, USA). The sensor measures from 0 to 239 kPa. The value of 0 kPa indicates that the soil reached saturation. The measurement of 239 kPa indicates that the soil is dry. Based on the tension (kPa) values, depletion in water holding capacity can be estimated. An example of depletion in water holding capacity for different soil types is described in **Table 2**. From the depletion values, irrigation recommendations and timing can be estimated to ensure the soil moisture level is at optimal condition for plant growth.

2.4.2 Type of soil moisture sensors: Frequency domain reflectometry

Frequency domain reflectometry (FDR) sensors estimate soil water level using the dielectric properties of soil, which are highly dependent on moisture content. The

Soil texture	Depletion in water holding capacity (kPa)		
	30%	50%	70%
Sand	20	30	60
Loamy Sand	25	40	67
Sandy Loam	28	50	80
Silt Loam	80	150	250

Table 2.
Soil matric potential (SMP) for 30%, 50%, and 70% of soil water depletion for different soil types [4].

Material	Dielectric permittivity
Air	1
Soil Minerals	3–7
Organic Matter	2–5
Ice	5
Water	80

Table 3.
Example of dielectric permittivity.

changes in the dielectric permittivity correlate with changes in the circuit frequency, which also correlates with soil moisture level. As soil moisture content increases, the dielectric permittivity increases. **Table 3** shows an example of dielectric permittivity for different materials.

Common FDR sensors are the EC-5 and 10-HS manufactured by Meter Group (Pullman, WA, USA), which provide the value as volumetric water content (cm^3/cm^3). This sensor comes with factory calibration, which should be validated with the site-specific soil condition. A previous study described the correction equations of soil moisture sensors for different types of soils to improve the accuracy of the reading [5]. Most of the cases, the sensors provide the trends of moisture level changes, which allows understanding of how the water flows in soils. This information is helpful to determine the field capacity of the field and evaluate the irrigation practice, such as under- or over-irrigation.

2.4.3 Soil moisture sensor: Placement and installation

There are several considerations when installing soil moisture sensors. Sensors should be installed between plants in a representative area within a row. For the drip irrigation system, ensure to install the sensors in the wetting zone, as shown in **Figure 2**. The figure clearly shows the wetting zone in the soil. For center pivot irrigation systems, avoid placing the soil moisture sensors close to a wheel track or lane edge. Additionally, sensor locations that are not representative of most of the field, such as the top of the hill, low areas, and the edge of the field should be avoided. The placement of the soil depth of the sensors is important. The user should be considered based on the crop's root depth (**Figure 3**). For example, the corn root

Figure 2.
Demonstrated wetting zone under the drip irrigation system using a blue dye (Benton Harbor, MI, USA).

Figure 3.
Installed FDR soil moisture sensors in a blueberry field (west olive, MI, USA).

system typically grows up to 36-inch soil depth. The recommended soil moisture placements are 6-, 18-, 24-, and 36-inch depths. Typical effective root zone moisture extraction depths for crops are described in **Table 4** [6]. It is important to monitor these effective moisture extraction soil depths to improve the accuracy and precision of irrigation scheduling. Soil moisture sensors measure only a small volume of the soil surrounding sensor. Therefore, the sensor installation technique is critical in obtaining accurate readings.

Crop	Effective root zone depth (inches)	Crop	Effective root zone depth (inches)
Alfalfa	36	Peppers	18
Asparagus	36	Potatoes	18
Apples	30	Pumpkins	24
Beets	18	Radish	6
Blueberries	18	Strawberries	6
Carrots	18	Sorghum	24
Corn	24	Soybean	24
Cucumber	18	Snap beans	18
Eggplant	18	Spinach	6
Grapes	36	Squash	24
Lettuce	6	Sweet Potatoes	18
Melons	24	Tomatoes	24
Onions – bunch	6	Watermelons	24
Peas	18	Wheat	24

Table 4.
Effective root zone water extraction depth in unrestricted soils [6].

2.5 Plant-based irrigation scheduling method

A common method of plant-based irrigation scheduling is using an sap flow sensor. Sap flow is the measurement of the water, nutrients, hormones, and anything else in the water that flows through the stem of a plant. The sensors use a heater and thermocouples to measure the amount of heat carried by the sap. This can then be converted to sap flow in units of grams per hour. Once the sensors are installed and the parameters are set, the system will record and calculate the sap flow, which can be downloaded from the system at any time. The sap flow sensor has been previously used in woody plants or other herbaceous plants [7, 8]. A common sap flow sensor is Dynagage Flow 32-1 K Sap Flow system, manufactured by Dynamax (Houston, TX). **Figure 4** shows the installed Flow 32-1 K Sap Flow system in a potato field (Lakeview, MI). An example of sap flow sensor data is shown in **Figure 5**. The result shows that the transpiration started at 9:30 am and stopped at 8:00 pm. Based on the total amount of water used by the plant, irrigation scheduling can be developed to maintain adequate soil moisture levels for plant growth. This approach is similar to weather-based irrigation scheduling methods, but uses directly measured values of water uptake from the plant.

2.6 IoT (internets of things) sensor technology

Agricultural technology industry is moving toward Agriculture 4.0, which includes the internet of things (IoT) and the utilization of big data to improve practices and efficiencies. Many microcontroller systems, such as Arduino and ESP 32, can be used

DOI: http://dx.doi.org/10.5772/intechopen.107386

Figure 4.
Installed Dynagage flow 32-1 K sap flow system in a potato field (Lakeview, MI, USA).

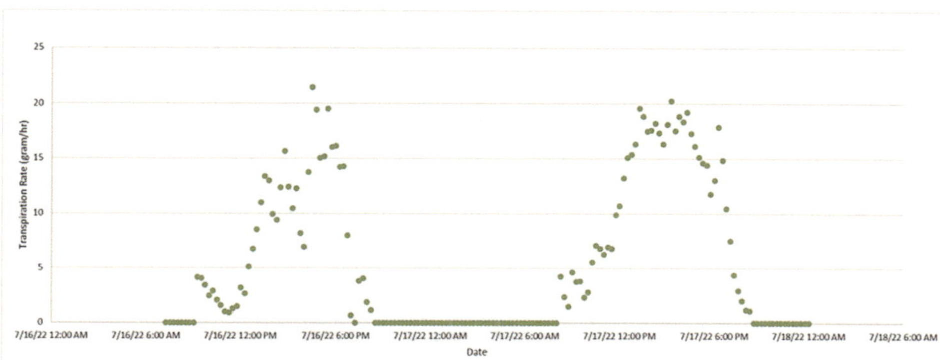

Figure 5.
SAP flow data result from a potato field (Lakeview, MI, USA).

in agricultural fields. Analog or digital soil moisture sensors can be connected to a microcontroller system to measure soil conditions. In addition to the soil conditions, other irrigation information, including water pressure, energy usage, irrigation system uniformity, and environmental conditions, can be measured using a microcontroller system. Many microcontroller systems allow sending data to a web server using Wi-Fi, cellular, or long range radio (LoRa) network system. The advantages of the remote monitoring system are that it allows monitoring the performance of data loggers and sensors and detecting any problems without having to visit the field. It also allows farmers to make timely farm management decisions. Michigan State University team has developed LOCOMOS (IoT-based Low-Cost Sensor Monitoring System), which measures soil and environmental conditions in irrigated fields, including multiple depths soil moisture levels, leaf wetness duration, temperature, humidity, soil temperature, and precipitation. The data is collected every 15 minutes and sent to the LOCOMOS IoT cloud web service. Then the data output is displayed on the IoT dashboard website and a smartphone app. IoT-based irrigation management has been very positive to most farmers as they can see soil moisture status in realtime without visiting the field.

2.7 Smartphone APP-based irrigation scheduling

Numbers of smartphone apps are available for irrigation scheduling. A large number of irrigation scheduling decision support tools have been developed in recent decades, many of them available via mobile apps. For example, water irrigation scheduling for efficient application (WISE) was developed by Colorado State University as an irrigation scheduling mobile app that uses evapotranspiration data and the water balance method [9]. WISE collects weather data from Colorado Agricultural Meteorological Network (CoAgMet) and Northern Colorado WATER Conservation District (NCWCD) weather stations. Cotton SmartIrrigation App (Cotton App), developed by the University of Georgia and the University of Florida, is an evapotranspiration-based irrigation scheduling tool that estimates root zone. Soil water deficits based upon weather data, soil parameters, crop phenology, crop coefficients, and irrigation rates [10]. In addition to cotton, SmartIrrigation offers separate apps for vegetables, soybean, turf, avocado, strawberry, citrus, and blueberry [11]. The apps obtain meteorological data from Florida Automated Weather Network (FAWN) and Georgia Automated Environmental Monitoring Network (GAEMN). While these and other similar systems improve the ability to operationally monitor crop water needs, they primarily weather data-based and do not take in situ soil moisture observations or data into account. Within growing season, observations of soil moisture can be extremely useful in monitoring crop water usage and plant available water in the rooting zone. LCOOMOS-APP is developed by Michigan State University's Irrigation Team, which is an easy-to-use smartphone app that accounts for in-field sensor data for determining irrigation management decisions. LOCOMOS-APP connects with a IoT cloud web server, which collects sensor data from each in-field LOCOMOS station. LOCOMOS-APP requires username and password to log-in. Once a user signs in, the app displays all registered devices to the specific user. When the user clicks a device, it will show available soil water (%) and recommended irrigation amount (in). Available soil water (%) value can help farmers when to irrigate. Recommended irrigation amount (in) value can help farmers how much to irrigate. Additionally, the app shows daily disease severity value (DSV) and cumulative DSV, and precipitation data. Cumulative DSV values can help farmers to apply fungicide in a timely manner. Overall, the advantage of the smartphone app-based irrigation scheduling tool is that it is an easy-to-use scheduling tool for farmers. This may increase the adoption of irrigation scheduling tools.

3. Conclusion

In conclusion, irrigation scheduling will be more important as the water resource is limited. Irrigation scheduling is critical to ensure optimum irrigation application. The benefits of applying the correct amount of water at the correct time are improved yield, reduced pumping costs, reduced nitrate leaching into groundwater or streams, improved soil health, and maximized return on investment. In the future, the incorporation of IoT sensor technology with AI Machine Learning techniques will be expected to increase the precision and accuracy of irrigation recommendations for different crop types.

Acknowledgements

Thank you to Michigan State University Irrigation Team, including Steve Miller, Lyndon Kelley, Brenden Kelley, Catherine Christenson, Allison Smith, and Hunter Hansen for their support on research and extension activities.

Conflict of interest

The authors declare no conflict of interest.

Author details

Younsuk Dong
Michigan State University, East Lansing, MI, USA

*Address all correspondence to: dongyoun@msu.edu

References

[1] Marurice V. Michigan Irrigation Guide. East Lansing, MI, USA: Michigan State University Cooperative Extension Service; 1977

[2] GardnerWH. Water content. In: Klute A, editor. Methods of Soil Analysis. Part 1. Physical and Mineralogical Methods. 2nd ed. Madison, WI, USA: American Society of Agronomy and Soil Science Society of America; 1986. pp. 493-544

[3] United States Department of Agriculture, National Resources Conservation Service. Soil Bulk Density/Moisture/Aeration. Washington DC, USA: United States Department of Agriculture, National Resources Conservation Service; 2011. Available from: https://www.nrcs.usda.gov/Internet/FSE_DOCUMENTS/nrcs142p2_053260.pdf

[4] Irmak S, Payero JO, VanDeWalle B, Rees J, Zoubek G. Principles and Operational Characteristics of Watermark Granular Matrix Sensor to Measure Soil Water Status and its Practical Applications for Irrigation Management in Various Soil Textures (EC783). Lincoln, NE, USA: University of Nebraska Extension; 2014

[5] Dong Y, Miller S, Kelley L. Performance evaluation of soil moisture sensors in coarse- and fine-textured Michigan agricultural soils. Agriculture. 2020; **10**:598. DOI: 10.3390/agriculture10120598

[6] United State of Department Agriculture Natural Resources Conservation Services. Crops. In: Ross E, Hardy L, editors. Irrigation Guide Part 652. National Engineering Handbook. Washington DC, USA: United State of Department Agriculture Natural Resources Conservation Services; 1997

[7] Ismanov M, Henry C, Espinoza L, Francis P. Soybean plant sap flow in high water demand and final growth stages. Soybean Research Studies. 2018;**2017**:12

[8] Liu H, Sun J, Duan A, Liu Z, Liang Y. Experiments on variation of tomato sap flow under drip irrigation conditions in greenhouse. Transactions of the Chinese Society of Agricultural Engineering. 2010;**26**(10):77-82

[9] Bartlett AC, Andales AA, Arabi M, Bauder TA. A smartphone app to extend use of a cloud-based irrigation scheduling tool. Computers and Electronics in Agriculture. 2015;**111**: 127-130. DOI: 10.1016/j.compag.2014.12.021

[10] Vellidis G, Liakos V, Andreis JH, Perry CD, Porter WM, Barnes EM, et al. Development and assessment of a smartphone application for irrigation scheduling in cotton. Computers and Electronics in Agriculture. 2016;**127**: 249-259. DOI: 10.1016/j.compag.2016.06.021

[11] Migliaccio KW, Morgan KT, Vellidis G, Zotarelli L, Fraisse C, Rowland DL, et al. Smartphone apps for irrigation scheduling. In: 2015 ASABE/IA Irrigation Symposium: Emerging Technologies for Sustainable Irrigation-a Tribute to the Career of Terry Howell, Sr. Conference Proceedings. MI, USA: American Society of Agricultural and Biological Engineers, St. Joseph; 2015. pp. 1-16

Chapter 9

Fundamentals of Irrigation Methods and Their Impact on Crop Production

Fawibe Oluwasegun Olamide, Bankole Abidemi Olalekan, Sokunbi Uthman Tobi, Mustafa Abdulwakiil Adeyemi, Joseph Oladipupo Julius and Fawibe Kehinde Oluwaseyi

Abstract

Water is the most precious resource on earth which is the sustenance of life. However, the competition for available water resources has intensified due to climate change and increase in global population. With a significant decrease in freshwater availability for crop production, agriculturists are open to innovation that could help save water and maximize crop production per unit drop of water. To ensure food security of a growing population, crop cultivation practices have continued to incorporate water-saving irrigation techniques to cope with water deficits, and increase crop production in an eco-friendly environment. This chapter discussed the different irrigation types based on driven-force and their specific advantages; fertigation; designing irrigation systems and scheduling of irrigation; water conservation through mulching; and water management for sustainable Integrated Pest Management (IPM). The introduction of water-saving techniques and their successful application has significantly reduced water loss through unproductive outflows and increase water and nutrients use efficiencies thereby promoting crop production. However, to achieve more success in the future, deliberate policy by government on irrigation and immense contributions from scientists would be required.

Keywords: irrigation, water productivity, mulching, crop production

1. Introduction

Water is the most precious resource on earth and is also the most abundant constituent of most organisms. This implies that most organisms including plants depend on water for their survival. Plants absorb a large quantity of water from the soil for physiological and biochemical processes that transform into growth and development of the plant. The importance of water to crop production made it an essential factor in ensuring food security. The trending issues of climate change resulting in erratic precipitation patterns, and increasing desert encroachment pose a threat to farmers in producing food that will meet the demand of the global population. To ensure food security in a changing world, additional water supply in the form of irrigation is necessary.

The artificial application of water to soil to meet the water need of crops and to maximize production is termed Irrigation. Irrigation systems are of two types based on their driven force (gravity-driven and pressure-driven). Whatever irrigation method is adopted, its purpose is to attain better water management and a higher yield.

2. Gravity-driven or surface irrigation

Gravity-driven irrigation is conventional and has been in use since time immemorial. This approach does not use pumps and relies on the ability of water to move through resistance. The irrigation system is efficient on plain topography for even distribution of water. It has three phases which are the advance, storage, and recession.

The advanced stage is the period of water introduction to the field. Water flows over the field to the end of the field with the help of gravity until the field becomes flooded. However, the storage period is the time frame for water to infiltrate the soil; whereas, the recession period begins after the source of the water is cut off. The water infiltrates the soil more and dries up as a result of evaporation and the closing of the water source. The success of surface irrigation depends on the water holding capacity of the soil, field slope, soil surface roughness, and the shape of the flow cross-section. Examples of surface irrigation include continuous flooding and furrow irrigation (**Figure 1**).

2.1 Continuous flooding

Continuous flooding is the process of artificially submerging a leveled land under water. It is a system predominantly used for rice cultivation in many regions of the world. Among continents, Asia is ranked the largest producer of rice, and it is responsible for 75% of the total global production. Rice is an aquatic plant but can survive under different soil conditions. However, the introduction of water-saving techniques and the release of drought-resistant varieties continuously prove that flooding is dispensable for rice production. In paddy, the field is irrigated until the water level reaches 5–6 cm above the ground level and is continuously maintained throughout the cropping season or drained two weeks before harvest when the rice plant is at physiological maturity. The soil condition under a continuous flooding system is anaerobic and the degree of anaerobicity depends on the level of water and oxygen availability [1].

Figure 1.
Gravity-driven irrigation system: continuous flooding system (A), furrow irrigation system (B).

Paddy fields account for about 40% of global irrigation [2] and it uses 2 to 3 times the volume of water required by other cereals such as wheat and maize to produce 1 kg of rice grains [3]. More than half of the water needed for irrigation in Asia is utilized in rice fields; however, most of this water is lost through unproductive water outflows such as evaporation, lateral seepage, deep percolation, and runoff. Apart from its excessive loss of irrigation water, the continuous flooding system is a major source of greenhouse gases such as methane and carbon dioxide thereby contributing negatively to the environment.

As a result of the decline in freshwater, more water-saving irrigation practices such as alternative wetting and drying (AWD), System of rice intensification (SRI), Ground cover production system (GCRPS), Drip irrigation with film mulch (DIP) had been introduced.

2.2 Furrow irrigation

This involves supplying water to the field along the furrow. The furrows are usually small parallel channels that serve as reservoirs of water on the field. The water gets to the plant root through lateral seepage.

The gravity-driven irrigation method requires minimal capital to construct and the energy required for it to work is obtained from free-flowing gravity. The system is easily controlled and does not require high technical know-how. Surface irrigation can be used on sloppy land. However, the irrigation method can affect plant growth and development due to the reduction in plant respiration caused by flooding. It could also increase the loss of water through deep percolation, runoff, infiltration, and evaporation.

3. Pressure-driven irrigation

The increase in global population resulting in rapid urbanization and industrialization have intensified competition for available water resources resulting in the decrease of fresh water available for crop production. Freshwater resources are becoming increasingly scarce and droughts are becoming more common as a result of climate change. Despite moderate rainfall in some regions of the world, over 50% of irrigation demands for crop production are met by pumping from underground aquifers, thereby depleting aquifers at an alarming rate. Therefore achieving food security requires high yields with efficient use of water resources [4]. Water-saving irrigation techniques that involve the use of pressure rather than gravity have been developed to help cope with water deficits and ensure maximum food production per unit drop of water.

A pressurized irrigation system involves the supply of water with the effort of pressure. This system is designed to achieve higher efficiency than the conventional method. The techniques help in quantifying the exact amount of water or nutrient to be supplied at a particular point in time. The choice of a pressurized irrigation system depends on the knowledge of the plant type, soil type, landscape characteristics, required flow rate, operating pressure, and cost. Examples of pressure-driven irrigation system are drip irrigation and sprinkler irrigation.

3.1 Drip irrigation system

A drip irrigation is an efficient irrigation system used for row cropping. In this system, water is directly supplied to the soil surrounding the root region with the of

Figure 2.
Drip irrigation set up: surface drip irrigation (A) and subsurface drip irrigation system (B).

the drip tubes laid on the soil surface (surface drip) or that are buried few centimeters below the ground level (subsurface) (**Figure 2**). The advantages and efficiency of drip irrigation has increase it acceptance and use by agriculturists around the globe, most especially in the arid and semi-arid regions where there is limited freshwater availability. The precise application of water to the root region of crop without wetting the entire farm plots makes drip irrigation an efficient water-saving technique compared with others. In a drip irrigation system, only a fraction (between 15% and 60%) of the soil surface is wet [5]. The drop by drop sequence of watering reduces surface runoff and percolation; hence, providing better disease management and salinity control [6]. Other benefits of drip irrigation include improved crop quality, efficient fertilizer and other chemical usages, limited weed growth, and improved agronomic practices [7].

In a drip irrigation system, emitter spacing is necessary to ensure precise delivery of irrigation water. This largely depends on the planting distance or vice versa. The effectiveness and efficiency of a drip emitter is an important factor that affects water distribution and the performance of a drip irrigation system. The rate of water delivery by emitters varies and their use is based on the soil types and the water-use efficiency of the crop. Emitter clogging is mostly related to the quality of irrigation water. The turbidity of water as a result of physical (sand particles), biological (bacteria), and chemical (inorganic fertilizer, salts) composition results in emitter clogging. The compounds gradually settle around the water passage until the clusters could not allow further passage of water. Clogging affects the productivity of the crops around the affected emitters and in turn reduces yield outcome. However, to prevent emitter clogging, water could be treated and made less turbid before application. The combination of strategies, such as installation of a filtration system, the use of sedimentation tank and tube settlers, frequent flushing of the irrigation system, and chlorination of the irrigation system, could mitigate emitter clogging.

3.1.1 Fertigation and its effect on crop production

Fertigation is the synchronous supply of nutrients or soluble fertilizer and water to the soil through drip irrigation system (**Figure 3**). The introduction of fertigation to crop production proffers a solution to the problem of flooding and overfertilization. Water and liquid fertilizer are harmoniously applied to the rhizosphere which makes

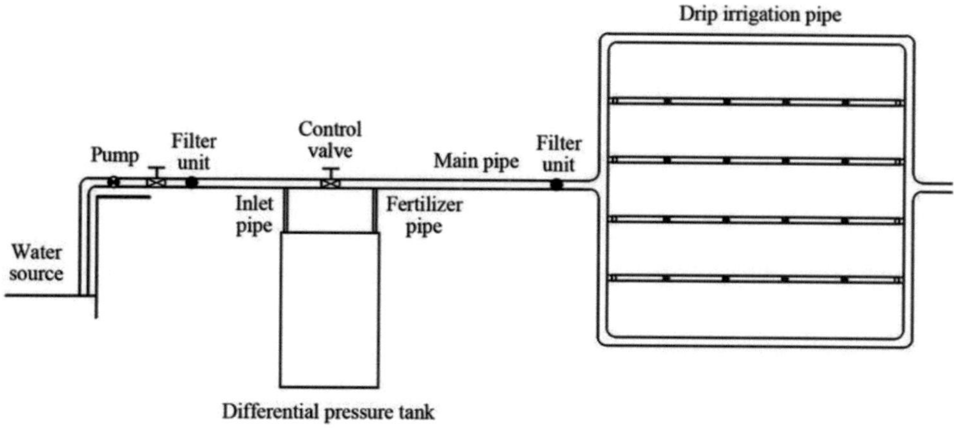

Figure 3.
Schematic illustration of fertigation system.

nutrients to be readily available for plant uptake. Fertigation results in increased crop yield and more efficient fertilizer usage [8]. Apart from increasing crop yield, fertigation reduces nutrient losses to the environment. Plants easily absorb soluble fertilizer thereby reducing nitrogen losses as nitrous oxide to the atmosphere. A well-designed fertigation system takes into consideration the appropriate rate of fertilizer and water, duration and frequency of supply to improve water, and nutrient uptake of the crop while at the same time reducing nutrient loss via leaching [9]. An appropriate liquid fertilizer applied through fertigation reduces leaf burn, stem scorching, and root death as mostly observed in the direct application of solid inorganic fertilizer close to the root zone of crops. Furthermore, fertigation reduces disease and pest infestation on crops, attributable to dryness of the plant shoot thereby creating a non-conducive environment for pathogens. The system was created to maximize the use of available water and mineral resources; thus, preventing runoff as it is not affected by wind.

To improve crop production through fertigation, the application of fertilizer should be done optimally to reduce acidification of the soil, and environmental degradation [10]. However, in case of overfertilization with the use of fertigation, continuous and frequent application of water regime should follow to reduce fertilizer concentration at the root region. Previous reports have documented that the use of fertigation increased both nutrient-use efficiency and water-use efficiency of crop. Nutrient use efficiency increased by 25%, and nitrogen and potassium application reduced by 20% as compared to the use of solid inorganic fertilizer [11]. Also, Ashrafi et al. [12] reported that the absorption rate of solid inorganic fertilizer was estimated to be 10–40%; whereas, the absorption rate of similar concentration on fertigated field was estimated to be 90%. Cotton yield increased by 50% on fertigated plots when compared to cotton supplied with surface irrigation with direct fertilizer application [13]. According to Hebbar et al. [14], drip fertigation enhanced tomato yield by 20–30% as compared to furrow irrigated tomatoes. The yield of Chili was also reported to increase by 52% and saved 40% and 50% of water and nitrogen, respectively through fertigation compared with a check-basin irrigation treatment [15]. Irrigation and nutrient management are the most effective methods for increasing agricultural output [16], and both management can be accomplished by fertigation. However, for successful use of fertigation, knowledge of soil fertility and crop nutrient uptake requirement is necessary.

Advantages of drip irrigation

- Despite its low operating cost as compared with the sprinkler, it is less affected by the speed of the wind.
- It increases the yield due to the efficient use of water and nutrients.
- It helps reduce the cost of weeding and herbicide use, especially when combined with film mulch.
- It is suitable to use in difficult topography.
- It helps to reduce environmental contamination and soil compaction when mineral nutrients are supplied through fertigation.
- In this system, there is little water contact with leaves thereby reducing the risk of plant diseases.

Disadvantages of drip irrigation

- The major disadvantage of a drip irrigation system is the initial installation cost.
- The cost of maintaining drip irrigation pipes might be a challenge to low-income farmers.
- It could be easily damaged by farm equipment, sunlight, rodents, wildlife, etc.
- Fertigation i.e. using the drip irrigation system for nutrient and fertilizer application may bring about the corrosiveness of the system and clogging of emitters.
- Drip irrigation needs to be replaced more frequently than other systems.

3.2 Sprinkler irrigation system

Sprinkler irrigation involves watering plants through a process that imitates natural rainfall. Water is sprinkled into the air through a series of pipes to form droplets before landing over leaves and areas within reach. The water is sprayed through a high-pressure sprinkler or guns. Though, sprinkler irrigation can be used on different land slopes; it is mostly used on flat ground such as lawns, golf courses, crops, landscapes, and flat terrains. There are different types of sprinkler irrigation systems, these include centre pivot system, rain gun system, side roll system, perforated pipe system and rotating head system (**Figure 4**). Each system is made up of the following components: pump unit, mainline, laterals, and sprinklers. The pump unit takes water from the source while the laterals distribute water from the pump unit to the sprinklers. To ensure efficient delivery of water, several sprinklers must be operated close together, ensuring an overlap of distribution patterns, since the heaviest water application is close to the sprinkler.

In a central pivot system, the machine moves in the shape of a circle, and water is sprayed on the crops beneath the circle. A rain gun system necessitates the use of a high-pressure machine that shoots water into the sky and dropped it on the farm in the form of rain. Side roll systems are made up of pipes attached to the middle of a

Figure 4.
Types of sprinkler irrigation system: Centre pivot system (A), rain gun system (B), side roll system (C), perforated pipe system (D) rotating head system (E).

wheel, which is perforated to drop water on the crops below as the wheel rolls across the field. In a perforated drain pipe system, a pipe is perforated to allow water to drain out of it; whereas, a rotating head system makes use of a pipe with spraying head nozzles to water the field. The application rates of sprinklers differ depending on their nozzle size, spray radii, and operating pressure.

Sprinkler irrigation is adaptable to most soil types but it is preferable for sandy soil with low water holding capacity. The water droplets wet both the soil and the crops and are accessible through uptake by the root and foliar penetration. However, sediment-free water is required to avoid blockage of the nozzle.

Advantages of sprinkler irrigation

- This system allows efficient use of water and reduces extra labour required for fertilizer, pesticide, and herbicides application.
- It is more efficient in irrigating plants with higher concentration per unit area of land such as cereals and vegetables.
- It is more effective and efficient for shallow-rooted plants, it is the only form of irrigation that could supply water to less than 1inch depth [17].
- It can last longer than drip irrigation.
- Applicable for agricultural, landscape, and nursery irrigation.

Disadvantages of sprinkler irrigation

- Initial cost of setting it up is high.
- It requires more pressure than drip irrigation, which increases the cost of energy to be used.
- Due to its complex structure, it requires high operating costs.
- Uneven distribution of water is possible due to the ability of wind to control the movement of water.
- Foliar application of water and nutrient could have a detrimental effect on leaves (leaf rot, senescence, and leaf burn) and fruit (fruit rot)
- It could bring about the inefficient delivery of water to understorey crops
- Rate of water evaporation is high in a sprinkler irrigation system.
- It has potential for runoff and erosion compared to drip irrigation.

4. Designing irrigation system

An irrigation design system is a way of determining the efficiency and effectiveness of water use, which involves management that affects the performance, yield, and quality of crops. One of the reasons for an irrigation system is the common phenomena of extreme weather events (e.g., floods and droughts). Recently, advancements in irrigation technologies are increasing. The use of robotics, smart controllers and remote sensing, and soil moisture sensors are gradually integrated into irrigation management [18]. However, the effectiveness of these technologies depends on the design of the irrigation system.

The quality of agricultural products could be improved by adopting this high-pressurized irrigation system, as the supply of water to crops is through piping. Designing effective irrigation systems and equipment will not only save money but will also conserve water, and results in improved agricultural production. The factors to be considered when designing irrigation systems and scheduling irrigation include:

4.1 Water source

The source of water is a determining factor in designing an efficient irrigation system. There are three main sources of water which include, groundwater, surface water and rainwater. Ground water is found under rocks, for example, spring water. Surface water includes water found on the surface of the earth examples are ocean, river, streams and lakes. Furthermore, rainwater from the atmosphere could be collected and used for irrigation. Depending on geographical location, the source and quantity of water for irrigation could differ which in turn could determine the type of irrigation system to be adopted. Also, the quality of available water needs to be considered.

4.2 Field characteristics

Field characteristics such as field size, topography, and soil types are determining factors in the choice of irrigation system, its design, crop type, and planting pattern.

- **Field size:** The bigger the field, the greater the number of crops that it will contain. Field size will determine the number and size of pumps that will sufficiently irrigate the farm. When designing an irrigation system, the field size will also predetermine the most appropriate source of water that could be effective. Moreover, the pressure for the water to travel to the desired destination will be based on the field size.

- **Land topography:** Land slope refers to the elevation of land over a specific distance. The flow of water depends on the topography of the land. Water flows from a high elevations to a lower elevations. There is a possibility of flooding at low elevation while there could be insufficient water availability at high elevation. Flat terrain land will allow even distribution of water; however, appropriate irrigation techniques will be required for a contoured or sloppy land.

- **Soil properties:** The water retention capacity of the soil is determined by its properties which includes structure, texture and organic content. Knowledge about the soil property will give agriculturist an edge in decision making. The soil properties will impact how well plant roots absorb nutrients from it. Nutrient mobility is faster in sandy soil due to its low water holding capacity, whereas it is slower in clay soil due to its high water holding capacity. With a limited water holding capacity, nutrients in the soil leak out, making nutrition uptake by the plant's roots difficult. Planning of irrigation system requires the knowledge of soil properties for optimum yield.

- **Plant type:** Plants require different irrigation systems, water application rates and application schedules. Some plants' are susceptible to insects and pests attack, and disease infestation when their foliar organs are exposed to excessive water or moist condition. Such plants are preferably irrigated through drip irrigation with plastic-film mulch; whereas, crops with high foliar tolerance to water or that require their leaves to be wet should be irrigated with a sprinkler. The duration of life of plants also determines their water requirement, for example, annual crops require less water than perennial plants [19]. The evapotranspiration rate of plants varies, and this influences irrigation scheduling. Plants with high evapotranspiration rates require more frequent irrigation than plants with low evapotranspiration rates. Furthermore, the market value of a crop could influence the introduction of an irrigation system to sustain its production.

5. Improving irrigation system

Irrigation system improvement must take into account agricultural output as well as saving water. The introduction of new crops to an irrigated farm requires technicalities and acquaintance with the crop to the farming system. In the present day, artificial intelligence is an example of cutting-edge technological innovations in irrigation systems. A Photovoltaic (PV) irrigation system is an example of renewable energy resources for improving irrigation systems. The effective management of water for various irrigation uses on a farm depends on the architectural structure of the farm and the mode of operation of the farm operators. Therefore, the system of agriculture may vary from one country to the other. Hence, the need for government to coordinate a sustainable irrigation system in agriculture that suits their Nation [20]. The characterization of the improvement of irrigation systems is a deliberate issue and needs a wealth of knowledge from the technical programming aspect. The deprivation of farmers' provisional assets may lead to the failure of policy made by government parastatals. This may also result in low farmer turnout in a country or state.

There are factors toward implementing a particular irrigation system and crop cultivation in particular [21]. The integrating components include the irrigation pipe, water pump/tank, valve, emitters, and pressure gauge regulator for a typical drip irrigation system [21]. The capital for implementing the newly adopted irrigation system, the human resources in the coordination, the management of infrastructure and water supply, and the strategy in operating the resources for good crop production are all important to be considered. In summary, the improvement of the agricultural system including the irrigation system needs deliberate policy by government and immense contributions from the scientific community [22].

6. Irrigation scheduling

There are variations in water requirements by plants at different growth stages; hence a need for irrigation scheduling that could supply optimum water required by plants at the appropriate time. Irrigation scheduling considers when and how much water should be applied to plants [23, 24]. These could be predetermined by monitoring the soil water status and the crop water requirements. Soil moisture-based, evaporation-based, and plant-based measurements are the most common methods for scheduling irrigation to aid effective use of water and promote crop productivity.

Soil moisture content can be used to determine an irrigation schedule. The moisture content of the soil is measured with the aid of instruments, these include FDR soil moisture meter (DIK-321A, Daiki Rika Kogyo Co. Ltd., Kounosu, Japan) [25, 26] and when soil moisture goes below a critical level, irrigation commences. The soil moisture-based irrigation schedule takes into consideration the type of soil and its composition to determine the availability of water in the soil. Sandy, loam, and clay have a low, medium, and high availability of water, respectively.

The evapotranspiration schedule takes into consideration soil evaporation and plants transpiration rates. The amount of water required by a plant is determined by balancing the amount of water input into the soil and the amount of water loss. Evapotranspiration data allow us to better understand when to irrigate an actively growing plant.

Also, the plant observation method could be used to determine the irrigation schedule. This method takes into consideration the changes in plant characteristics to determine when to irrigate the plant. There are common morphological symptoms of plants under stress or low water deficit. These visible changes including chlorosis, dried leaves, curling of leaves, and stunted growth were used to assess the timing of irrigation. To determine chlorosis and water stress for irrigation scheduling, a chlorophyll meter (SPAD 502 PLUS, Minolta corporation, Ltd., Japan) and a chlorophyll fluorometer (PAM-2000, Walz Co., Ltd., Effeltrich, Germany) [25] are used.

7. Mulching and its impact on crop production

Mulching is a strategy for enhancing soil conditions that involve covering the soil surface with various materials. It involves covering the soil around the plants' root zone to protect the roots from an adverse effect of the micro-climate. Mulching has become a popular agricultural practice not just for its immediate economic benefits, such as improved yields, earlier harvests, better fruit quality, and less water usage, but also for its increased soil microbial performance. Mulching creates an environment for the plant to perform at its best as it improves soil temperature, conserves soil moisture, reduces weed pressure and certain insect pests, and makes more efficient use of soil nutrients, among other benefits [27]. The use of mulch reduces the impact of raindrops on the soil surface [28, 29]; thereby, improving the hydrothermal regime of the soil and soil physical properties such as texture, porosity, and infiltration rate.

7.1 Types of mulching

The mulching materials can either be organic or synthetic (**Figure 5**).

Figure 5.
Types of mulch: Organic mulch, for example, rice straw (A) and plastic-film mulch (B).

- **Organic mulches**: These are mulches that can be easily degraded. They are usually available on the farm. The gradual decomposition of the organic mulches increases organic matter and soil fertility (**Figure 5A**). They host and nurture various beneficial soil organisms such as bacteria, fungi, insects, and worms and their remains in the soil do not create post utilization disposal problems. For example, leaves, paddy straw, grasses, sawdust, sugarcane trash, etc.

- **Paddy straw:** It has a unique property of not absorbing water and makes water available to plants. Among all the organic mulches, paddy straw has the longest life span. It also serves as a nutrient reservoir and gradually releases them into the soil.

- **Sawdust:** It is a small granular chip wood that is obtained as the finished product in the sawmills. Easy to apply and inexpensive with high C/N ratio. It retains moisture for longer periods.

- **Sugarcane trash:** It is the residue gotten from sugar cane after the removal of its juice. Helps conserve moisture and reduces weed growth and should be avoided in an area where there is an incidence of termites.

- **Synthetic mulches**: These are mulches that cannot be easily degraded (**Figure 5B**). They synthesized materials that need prior work before using in the field. They are referred to as non-biodegradable as a result of the natural decomposition of organic mulches. They are available in different colors and thicknesses. Much expensive when compared to organic mulches and should be disposed of at the end of the growing season, for example, plastic films

- **White plastic mulch:** It is good for establishing crops under hot summer conditions and has little effect on soil temperature. It repels some insects and reflects more light on the plant as compared to black mulch.

- **Black plastic mulch:** It is the most predominant colored mulch and acts as an opaque black body absorber and radiator. It does not allow sunlight to reach the soil; therefore, it suppresses the growth of the weed. It helps to increase soil temperature and improves mineralization and nutrient absorption. It encourages plant growth through the warming of the soil during the winter season.

- **Transparent plastic mulch:** It is also known as clear plastic mulch. It absorbs little solar radiation with a transmission of 85% - 95%. It raises soil temperature drastically and affects the plant's growth adversely.

- **Degradable plastic mulch:** This can either be biodegradable or photo-degradable. It can be degraded by microorganisms or by sunlight.

- **Biodegradable plastic mulch:** It is made from plant starches such as corn, wheat, and potatoes which can be broken down by microbes. It can be easily plowed into the ground after harvest.

- **Photo degradable plastic mulch:** It is formulated to break down after a certain period of exposure to sunlight. It has similar qualities to black or clear plastic film. Examples are plastigone and biolane.

7.2 Benefits of mulching

7.2.1 Conservation of soil moisture

Several abiotic variables could be responsible for the loss of moisture from the soil. These variables include high winds, elevated temperatures, harsh climatic conditions, etc. Mulching helps to reduce weed infestation and water loss through evaporation. Straw mulch has been shown to minimize evaporation by up to 35%. However, mulching reduces direct soil water evaporation, making more water available for transpiration. This way of water conservation helps the plants to maintain water balance, especially in regions with little precipitation per annum. Also, mulching reduces erosion and nutrient loss by protecting the soil surface.

7.2.2 Minimize soil compaction and erosion

Compaction caused by heavy equipment or machinery is becoming a serious problem on many agricultural lands [30]. The addition of organic mulch materials can help to ease the problem of compaction. These materials prevent compaction due to heavy implements or machinery and from wind and water erosion. It can also reduce the compaction of soil, which can negatively affect the roots of crops, thereby reducing their growth and development. Some grasses and legumes have been used as organic mulch, which serves as the best example of living mulch on the slopes and reduces soil erosion by aggregating the soil particles by binding them into a complex unit.

7.2.3 Regulation of soil temperature

Mulching helps to maintain soil temperature stability, which is beneficial to crop growth and development. Studies have shown that mulch can keep the soil cool during extremely hot weather as well as during normal or warm temperatures. Extreme temperatures have a negative impact on newly emerging plant roots, limiting nutrition and water intake. Plants may be stressed as a result of the extreme temperature conditions under which they grow, and newly established roots may be unable to absorb the proper amount of water and essential plant nutrients [30]. Various types of mulch have different effects on soil temperature. Some mulches increase the soil temperature as compared to bare soil due to the absorption of solar radiation. Moreover, it has been observed that plastic-film mulch and organic mulch materials are better at maintaining a favorable soil temperature compared to other mulch materials.

7.2.4 Reduces infiltration rate

Organic mulch helps to retain water at the soil surface allowing water to slowly penetrates thereby minimizes surface runoff [31]. As reported by Abu-Awwad [32], covering the soil surface reduced the amount of irrigation water required by pepper and onion crops by 14–29% and 70%, respectively.

7.2.5 Reduced fertilizer leaching

Fertilizer loss due to leaching is reduced as excessive rainfall is drained around the root zone, especially in sandy soil. Also, the use of organic mulch increases soil organic carbon which improves the water and nutrient holding capacity of the soil. Mulching with coconut fronds increased leaf N, P, and K content in chili [33]. Findings have shown faster plant growth, early fruiting, reduced P, and increased N concentration in leaves and fruits of crops when mulch is used.

7.2.6 Reduces weed infestation

Mulching reduces the germination and nourishment of many weeds by providing a physical barrier between the soil and the atmosphere. The mulching operation promotes the reduction of weed seed germination and weed growth and keeps weeds under control. Weed seed germination can be prevented or physically suppressed by covering or mulching the soil surface. Weed control can be achieved with materials like rice and wheat straws. Covering the soil surface can prevent weed seed germination or physically suppress seedling emergence. Organic mulch such as rice straw and sugarcane bark can provide effective weed control.

7.2.7 Organic matter improvement

Mulches decompose and restore organic matter and plant nutrients to the soil. Improving the physicochemical and biological properties of the soil which in turn increases crop productivity. Organic mulches do not only help to maintain soil moisture, but they also greatly enhance soil nutrients by adding organic matter. Lal et al. [34] reported a decrease in bulk density under straw mulch (1.42 g cm-3) compared to bare soil (1.50 g cm-3). Khurshid et al. [31] concluded that organic matter was significantly higher when more mulch was applied.

7.2.8 Reduces harvesting period

Vegetables such as cucumbers, muskmelons, watermelons, eggplants, and peppers usually respond well to mulching in terms of early maturity and higher yields. In comparison to control, organic mulches cause earlier blooming, resulting in fewer days between fruit set and harvest in tomato crops [35]. Polyethylene used as mulch reduced the growth season and increased the earliness and productivity of various vegetable crops [36, 37].

7.2.9 Improves quality and yield

Mulch keeps fruits clean from touching the ground and reduces soil rot, fruit cracking, and blossom end rot in many circumstances. Fruits are smoother and have fewer scars. Plastic mulch when properly laid prevents dirt from splashing onto the plants during rainfall, reducing grading time. Moreover, straw mulch can also improve the yield and quality of early potatoes, cabbage, and other vegetables.

7.2.10 Reduction of diseases

Mulch can reduce the force of irrigation water or the beating motion of raindrops, which can convey disease spores. These spores attach themselves to vulnerable plants' leaves and branches. Mulches provide food for a variety of beneficial soil organisms that compete with entering harmful spores or emit compounds that suppress diseases. They minimize the possibilities of illness occurring in plants. Many soil bacteria are inhibited by organic mulches, which compete with or digest pathogenic organisms through a variety of enzymatic processes. Mulches play a crucial role in integrated pest management (IPM).

7.2.11 Remediation of heavy metals

Heavy metals are harmful to the health of both animals and humans. Mulches are an excellent source for removing heavy metals from soils. *Eucalyptus* leaves are commonly used to remove heavy metals from soil solutions [38]. In forest environments, woodchips and compost can form complexes with copper metal, converting them to a non-toxic form for crop plant growth [39].

7.3. Negative impacts of mulching on crop production

7.3.1 Competition for resources

Mulches most especially organic compete with the main crop for resources such as water, nutrients, oxygen, carbon dioxide, and space. The inter and intra-specific rivalry for the resources could be fierce. Both types of competition are harmful to the growth and development of the main crop.

7.3.2 Allelopathic effects

Allelopathy is the term used to describe the limitation of seed germination and plant growth caused by the release of allelochemicals by some plants or organic mulches. Allelochemicals inhibit weeds in crop plants; however, previous studies have shown that when plants like eucalyptus, acacia, and pine were mulched, they lowered or completely suppressed the growth of numerous weed species, demonstrating their allelopathic actions. Narrow-leaved plants, such as grasses, are not as badly damaged as broad-leaved plants or dicot species [40].

7.3.3 Weed infestation

The partially decomposed organic mulch materials could act as carriers of various weed seeds. Incorporation of mulch to a deeper depth could help to mitigate the problem of weed seed because at a deep depth the growth of weeds is suppressed before getting to the surface. Organic mulch can inhibit the growth of weeds by depleting air and resources necessary for their growth, thereby promoting healthy plants and soil. Previous study has shown that weed suppression is directly related to the depth of mulch [30]. Organic mulches that are applied at a higher depth can reduce weed species as compared to those applied at shallow depths.

7.3.4 Nitrogen insufficiency

Organic mulching results in nitrogen deficit in the soil. They require nitrogen to decompose since they contain high structural carbohydrates such as lignin, cellulose, and hemicellulose and reduced non-structural carbohydrates. As a result, they compete with the crop for nitrogen, lowering their C/N ratio. Though the accumulated nitrogen will be released to the soil after decomposition, the crop may suffered nitrogen deficiency at critical stages of plant growth, resulting in chlorosis.

8. Water management for sustainable integrated pest management

Integrated pest management is the method of controlling pests, especially insect pests that invade farmland [41]. Crop production on farmland becomes more vulnerable to pest invasion if it is not closely monitored, which can result in crop damage and significant financial loss for farmers. The objectives of integrated pest management can be classified into three categories.

- Sustainable ecosystem maintenance and the reduction of pesticide negative impacts.

- Cost-effective farm production.

- Maintaining and putting human and animal health in check.

However, the role of farmers and stakeholders in the management of crop production and agricultural farmland is imperative. The success of the control tactics must be measured using indicators based on monitoring of harmful and beneficial organisms, pesticide use, and their impact on the environment [41]. Greenhouse gas emissions as a result of pesticides and fertilizer application on agricultural land contribute largely to global warming potential [1]. The connection between measuring CO2 gas fluxes emission and pesticides applied through irrigation is critical in examining the impact of the chemical–water ratio applied to the soil rhizosphere and mineral nutrients available to the crop [42, 43].

Irrigation is an essential agricultural practice for food, pasture, and fiber production in semiarid and arid areas [42]. Fertigation allows flexibility in the application timing when injections can be made virtually any time during the season from the point of the seedling establishment until harvest. The intensive use of water in the irrigation system is inherent in the cumulative effect of the modeling concept of the irrigation system. The inclusive role of the model and pesticide application is important in an integrated pest management system. However, the use of water in a non-essential way may lead to a high pest in-breeding rate. Hence, there is a need to plan the amount of water pumped into the irrigation system and also calculate the relative drip chemigation used [42].

To ensure a sustainable integrated pest control system, the connecting pipes and the osmothermal capacity of the pipe used for surface irrigation must be regularly examined for an effective irrigation system. The internet of things (IoT) and big data collection are new advancements in the application of addressing irrigation system defects while also monitoring integrated pest management activities [42, 44]. Relative data used in IoT is the collection of previous crop performance and farm production activities. This will enable the farmer to predict the future production of the crop. A large amount of data must be collected to amuse the net profit on crop production.

The cost of controlling pests on farmland must be reduced and be effective to make it a success. Combining the relative evaluation of the integrated pest management innovation system and the new event on irrigation system on crop output is not excessively expensive as compared to the success it will bring to farmland. Pesticide application through fertigation is a common example of combining insect pest management with a drip irrigation system. As a result, there exist methods for managing crop water availability as well as applying chemical pesticides and liquid fertilizer efficiently for agricultural production management. Furthermore, the building of irrigation system components that suit the topography of the soil and planting pattern is critical for successful crop production.

Though there is no specific way of designing an irrigation system for sustainable integrated pest management, what matters is making sure it is designed in a way that will reduce water usage without hampering the efficacy of the pesticide and also protecting the environment. The use of chemigation systems is an advancement that could help in achieving the goals for integrated pest management.

Author details

Fawibe Oluwasegun Olamide[1*], Bankole Abidemi Olalekan[1], Sokunbi Uthman Tobi[1], Mustafa Abdulwakiil Adeyemi[1], Joseph Oladipupo Julius[1] and Fawibe Kehinde Oluwaseyi[2]

1 Federal University of Agriculture, Abeokuta, Nigeria

2 Chiba University, Matsudo, Japan

*Address all correspondence to: fawibeoo@funaab.edu.ng

References

[1] Fawibe OO, Honda K, Taguchi Y, Park S, Isoda A. Greenhouse gas emissions from rice field cultivation with drip irrigation and plastic film mulch. Nutrient Cycling in Agroecosystems. 2019;**113**(1):51-62. DOI: 10.1007/s10705-018-9961-3

[2] Tuong TP, Bouman BAM, Mortimer M. More rice, less water - integrated approaches for increasing water productivity in irrigated rice-based systems in Asia. Plant Production Science. 2005;**8**(3):231-241. DOI: 10.1626/pps.8.231

[3] Bouwman BAM, Hengsdijk H, Hardy B, Bindraban PS, Tuong TP, Ladha JK. Water-wise rice production. In: Proceedings of the International Workshop on Water-Wise Rice Production. Philippines: International Rice Research Institute; 2002. p. 356

[4] McClung AM, Rohila JS, Henry CG, Lorence A. Response of U.S. rice cultivars grown under non-flooded irrigation management. Agronomy. 2020;**10**:55

[5] Fanish SA, Muthukrishnan P, Santhi P. Effect drip fertigation on field crops. Agricultural Review. 2011;**32**(1):14-25

[6] Hanson BR, May DM. The effect of drip line placement on yield and quality of drip-irrigated processing tomatoes. Irrigation Drainage Systems. 2007;**21**:109-118

[7] Wang P. Rice Cultivation under Field Conditions with Drip Irrigation and Mulch in Arid Areas of China. Unpublished working paper. China: Urumqi Agricultural and Environmental Institute for Arid Areas in Central Asia; 2012

[8] Hagin J, Lowengart A. Fertigation for minimizing environmental pollution by fertilizers. Fertilizer Research. 1996;**43**:5-7

[9] Gardenas AI, Hopmans JW, Hanson BR, et al. Two-dimensional modelling of nitrate leaching for various fertigation scenarios under micro-irrigation. Agricultural Water Management. 2005;**74**(3):219-242

[10] Rahman KMA, Zhang D. Effects of fertilizer broadcasting on the excessive use of inorganic fertilizers and environmental sustainability. Sustainability. 2018;**10**(3):1-15. DOI: 10.3390/su10030759

[11] Mattos D, Kadyampakeni DM, Oliver AQ, Boaretto RM, Morgan KT, Quaggio JA. Soil and nutrition interactions. In: The Genus Citrus. Brazil: Elsevier Inc.; 2020. pp. 311-331. DOI: 10.1016/B978-0-12-812163-4.00015-2

[12] Ashrafi RM, Raj M, Shamim S, Lal K, Kumar G. Effect of fertigation on crop productivity and nutrient use efficiency. Journal of Pharmacognosy and Phytochemistry. 2020;**9**(5):2937-2942

[13] Janat M, Somi G. Performance of cotton crop grown under surface irrigation and drip fertigation, II: Field water-use efficiency and dry matter distribution. Communications in Soil Science and Plant Analysis. 2001;**32**(19 and 20):3063-3076

[14] Hebbar SS, Ramachandrappa BK, Nanjappa HV, Prabhakar M. Studies on NPK drip fertigation in field-grown tomato (Lycopersicon esculentum mill.). European Journal of Agronomy. 2004;**21**:17-127

[15] Singh AK, Singh RB. Effect of mulches on nutrient uptake of

Albiziaproceraand subsequent nutrient enrichment of coal mine overburden. Journal of Tropical Science. 1999;**11**:345-355

[16] Yousaf M, Li J, Lu J, Ren T, Cong R, Fahad S, et al. Effects of fertilization on crop production and nutrient-supplying capacity under rice-oilseed rape rotation system. Scientific Reports. 2017;**7**(1):1-9. DOI: 10.1038/s41598-017-01412-0

[17] Chauhdary JN, Bakhsh A, Arshad M, Maqsood M. Effect of different irrigation and fertigation strategies on corn production under drip irrigation. Pakistan Journal of Agricultural Sciences. 2017;**54**(04):855-863. DOI: 10.21162/pakjas/17.5726

[18] Dukes MD. Water conservation potential of landscape irrigation smart controllers. Transactions of the ASABE. 2012;**55**(2):563-569

[19] Vico G, Brunsell NA. Tradeoffs between water requirements and yield stability in annual vs, perennial crops. Advances in Water Resources. 2018;**112**:189-202

[20] Arif SS, Pradipta AG, Subekti E, Prabowo A, Sidharti TS, Soekarno I, et al. Toward modernization of irrigation from concept to implementations: Indonesia case. IOP Conference Series: Earth and Environmental Science. 2019;**355**:12-24

[21] Sarker KK, Hassan A, Murad KF, Biswas SK, Akter F, Rannu RP, et al. Development and evaluation of an emitter with a low-pressure drip irrigation system for sustainable eggplant production. Agricultural Engineering. 2019;**1**:376-390

[22] Simons GWH, Bastiaanssen WGM, Chuma MJM, Ahmad B, Immerzeel WW. A novel method to quantify consumed fractions and non-consumptive use of irrigation water: Application to the Indus basin irrigation system of Pakistan. Agricultural Water Management. 2020;**236**:1-19

[23] Davis SL, Dukes MD. Irrigation scheduling performance by evapotranspiration-based controllers. Agricultural Water Management. 2010;**98**:19-28

[24] Kisekka I, Migliaccio KW, Dukes MD, Schaffer B, Crane JH, Bayabil HK, et al. Evapotranspiration-based Irrigation for Agriculture: Sources of Evapotranspiration data for Irrigation Scheduling in Florida. United State of America: University of Florida Institute IFAS extension; 2019. pp. 1-4

[25] Fawibe OO, Hiramatsu M, Taguchi Y, Wang J, Isoda A. Grain yield, water-use efficiency, and physiological characteristics of rice cultivars under drip irrigation with plastic-film-mulch. Journal of Crop Improvement. 2020;**34**(3):414-436. DOI: 10.1080/15427528.2020.1725701

[26] Park S, Nishikoji H, Takahashi S, Fawibe OO, Wang P, Isoda A. Rice cultivation under drip irrigation with plastic film mulch in Kanto area of Japan. Universal Journal of Agricultural Research. 2021;**9**(4):101-110

[27] Zhang DQ, Liao YC, Jia ZK. Research advances and prospects of film mulching in arid and semi-arid areas. Agricultural Research in the Arid Areas. 2005;**23**:208-213

[28] Farooq M, Siddique KHM, Rehman H, Aziz T, Lee DJ, Wahid A. Rice direct seeding: Experiences, challenges, and opportunities. Soil and Tillage Research. 2011;**111**:87-98

[29] Seguy L, Husson O, Charpentier H, Bouzinac S, Michellon R, Chabanne A,

et al. Principles, Functioning, and Management of Ecosystems Cultivated under Direct Seeding Mulch-Based Cropping Systems (DMC). France: CIRAD; 2012. Available from: http:// agroecologie.cirad.fr

[30] Chalker-Scott L. Impact of mulches on landscape plants and the environment - a review. Journal of Environmental Horticulture. 2007;**25**:239-249

[31] Khurshid K, Iqbal M, Arif MS, Nawaz A. Effect of tillage and mulch on soil physical propertiesand growth of maize. International Journal of Agriculture& Biology. 2006;**8**:593-559

[32] Abu-Awwad AM. Irrigation water management for efficiency water use in mulched onion. Journal of Agronomy and Crop Science. 1999;**183**:1-7

[33] Hassan SA, Ramlan ZA, Inon S. Influence of K and mulching on growth and yield of chilli. Acta Horticulture. 1994;**369**:311-317

[34] Lal H, Rathore SVS, Kumar P. Influence ofirrigation and mixtalol spray on consumptive use of water, water use efficiency and moisture extraction pattern of coriander. Indian Journal of Soil Conservation. 1996;**24**:62-67

[35] Ravinderkumar, Srivatsava BK. Influence ofdifferent mulches on flowering an fruit setting of wintertomato. Crop Research. 1998;**12**:174-176

[36] Goreta S, Perica S, Dumicic G, Bucan L, Zanic K. Growth and yield of watermelon on polyethylenemulch with different spacing and nitrogen rates. HortScience. 2005;**40**:366-369

[37] McCann I, Kee E, Adkins J, Ernest E, Ernest J. Effect of irrigation rate on yield of drip-irrigated seedless watermelon in humid region. Science Horticulture. 2007;**113**:155-161

[38] Salim R, El-Halawa RA. Efficiency of dry plant leaves (mulch) for removal of lead, cadmium and copper from aqueous solutions. Process Safety and Environmental Protection. 2002;**80**:270-276

[39] Kiikkila O, Derome J, Brugger T, Uhlig C, Fritze H. Copper mobility and toxicity of soil percolation water to bacteria in metal polluted forest soil. Journal of Plant Soil. 2002;**238**:273-280

[40] Schumann AW, Little KM, Eccles NS. Suppression of seed germination and early seedling growth by plantation harvest residues. South African Journal of Plant and Soil. 1995;**12**:170-172

[41] El-Shafie. Integrated Insect Pest Management. London, UK: IntechOpen; 2018. DOI: 10.2307/1939250

[42] Koech R, Langat P. Improving irrigation water use efficiency: A review of advances, challenges and opportunities in the Australian context. In Water. 2018;**10**(1771):1-17. DOI: 10.3390/w10121771

[43] Liu JL, Ren W, Zhao WZ, Li FR. Cropping systems alter the biodiversity of ground- and soil-dwelling herbivorous and predatory arthropods in a desert agroecosystem: Implications for pest biocontrol. Agriculture, Ecosystems and Environment. 2018;**266**:109-121. DOI: 10.1016/j.agee.2018.07.023

[44] Pramanik M, Khanna M, Singh M, Singh DK, Sudhishru B, Ranjan A, et al. Automation of soil moisture sensor-based basin irrigation system. Smart Agricultural Technology. 2022;**2**:100032

Chapter 10

Effect of Irrigation Depths and Salinity Levels on the Growth and Production of Forage Palm Orelha de Elefante Mexicana

Mariana de Oliveira Pereira, Jailton Garcia Ramos, Carlos Alberto Vieira de Azevedo, André Alisson Rodrigues da Silva, Geovani Soares de Lima, Luciano Marcelo Fallé Saboya, Patrícia Ferreira da Silva and Gustavo Bastos Lyra

Abstract

Although the adaptation of forage palm to the Brazilian semi-arid, it may be influenced by soil and climatic conditions of this region, irregular rainy periods, high annual evapotranspiration and soils with a low water retention capacity. These factors may reduce crop production during dry seasons, including forage. The present research aimed at analyzing the effect of irrigation with different water depths and levels of salinity on Orelha de Elefante Mexicana cultivar. The study was carried out in pots in the Federal University of Campina Grande, from September 2017 to December 2018. Experimental design was randomized blocks in a factorial scheme 4 x 4, with 4 replications. Four irrigation water depths were applied (25, 50, 75, and 100%), as a function of water retention capacity of soil and four levels of electrical conductivity: 0.60; 3.00; 5.40 and 7.80 dS m^{-1}. Morphometric and production variables were evaluated. Plant growth was not affected by irrigation water depth and levels of salinity, except the thickness of secondary cladode. Primary cladodes showed the greatest average values (4.03 cladodes) for 376.00 mm depth. The other variables evaluated did not present significant effects under treatments. Saline water did not affect the total production of the cultivar.

Keywords: *Opuntia stricta* (Haworth) Haworth, water availability, water salinity

1. Introduction

Brazilian semi-arid is characterized by irregular rainy periods, high annual evapotranspiration and soils with a low water retention capacity, limiting livestock activities in this region [1]. These conditions affect the production during dry seasons, including the reduction of forage for animal feed. In this context, Orelha de Elefante

Mexicana (Haworth) Haworth) is an important forage palm that can mitigate the effects of low performance of the livestock. Thus, the efficiency of soil water use by *Opuntia* species is around 100 and 150 liters of water for each kilogram of dry matter produced, while grasses need 250 and 350 liters to produce the same quantity of dry matter [2].

Forage palm species are cactus with a great exploitation potential in the Brazilian northeast, constituting an important resource during periods of drought, due to its high potential of phytomass production in semi-arid region [3]. Despite forage palm adaptation to the region, local meteorological conditions influence plant development, since hydric deficit may cause a reduction of water content and hydric potential, resulting in loss of turgescence, closure of stomata and reduction of growth, which, consequently, promote a decrease in the final production. Thus, irrigation practice is very important to the production system [4].

The usage of poor water quality has been an alternative for producers in the northeast region to minimalize water scarcity in plants. However, it is important to highlight that the available water in several Brazilian semi-arid regions has high soluble salt contents. In this context, palm water needs may modify, changing water absorption process and evapotranspiration due to salt accumulations in soil, contributing to its degradation [5]. According to Ribeiro, Moreira, Seabra Filho and Menezes [6], salinity is one of the abiotic stresses limiting agricultural production the most, since it presents negative effects on vegetal development.

Crops that are sensible to saline water show the need of studies that aim at analyzing viable technologies to producers, in order to minimize salt effects on plants.

Thus, the present research aimed to evaluate the effect of water depths and levels of saline water on the growth and production of forage palm Orelha de Elefante Mexicana (*O. stricta* (Haworth) Haworth).

2. Material and methods

2.1 Localization and characterization of the experimental area

The study was conducted open to sky at the experimental area of the Federal University of Campina Grande (UFCG), in the municipality of Campina Grande (7° 12′52,56"S; 35°54′22,26"O and 532 m of altitude), state of Paraíba, from September 26, 2017 to December 11, 2018, totalizing 442 days. According to Köppen climate classification, the region has a mesometric, sub-humid, Csa climate, dry season (4 to 5 months) and rainy season (autumn to winter).

During the experiment, climate conditions were monitored by the automatic weather station at Brazilian National Institute of Meteorology (INMET), (7.22°S; 35.90°O and 546 m of altitude), located approximately 1200 m of distance (horizontal line) from the experimental area (**Figure 1**).

2.2 Experimental design and treatments

Experimental design was randomized blocks in a factorial scheme 4x4, with 4 replications, totalizing 64 experimental parcels. Four irrigation water depths were applied (L1 = 25%, L2 = 50%, L3 = 75% and L4 = 100%) as a function of soil water depletion taking into consideration the value of soil water retention capacity and four levels of electrical conductivity: S1 = 0.60 dS m^{-1}; S2 = 3.00 dS m^{-1}; S3 = 5.40 dS m^{-1} and S4 = 7.80 dS m^{-1}; applied on forage palm Orelha de Elefante (*O. stricta* (Haworth) Haworth).

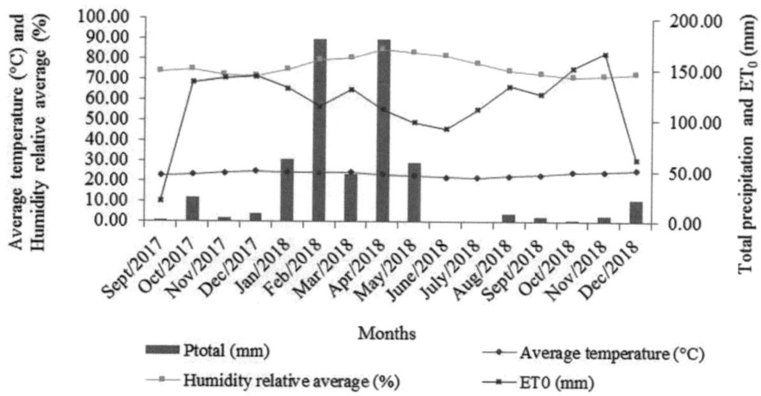

Figure 1.
Mean air temperature conditions (mean temperature), relative humidity of air (mean temperature), total precipitation (Ptotal) and reference evapotranspiration (ET_o) of the region in analysis during the research.

Grain size			Texture classification	Soil density	Particles density	Porosity	Humidity (% dry soil basis)	
% Sand	Silt	Clay	—		g cm^{-3}	%	0.10 atm	15.00 atm
82.79	11.86	5.35	Loam Sandy	1.58	2.72	41.91	7.50	2.85

Table 1.
Physical characterization of soil.

Ca	Mg	Na	K	SB	H	Al	CEC	O.M.	Available P	pH H2O (1:2.5)
		meq 100 g^{-1} of soil						%	mg 100 g^{-1}	—
1.07	2.41	0.20	0.13	3.81	0.56	0.0	4.37	1.15	0.57	6.75

SB—sum of basis; CEC—cation exchange capacity; O.M—organic matter.

Table 2.
Chemical characterization of soil (fertility).

The research was carried out in 120 L pots open to the sky, with a space of 1.30 m between rows and 1.00 m between plants and one plant per pot. The pots were used as drainage lysimeters.

A layer of crushed stone (2.40 kg) was put at the bottom of each pot, covered with a texture fabric, a layer of coarse sand (2.10 kg) and 170 kg of soil (0.268 m^3). Soil analysis was performed in the Irrigation and Salinity Laboratory (LIS) of The Federal University of Campina Grande. Physical characterization (**Table 1**), Fertility (**Table 2**) and Salinity (**Tables 3** and **4**) were evaluated.

2.3 Plant material and fertilization

The cladodes of forage palm (Orelha de Elefante Mexicana) evaluated in the experiment were obtained from the Experimental station Lagoa Bonita, National Institute of Semi-arid (INSA), located in the countryside of the municipality of

CE	Chloride	Carbonate	Bicarbonate	Sulphate	Ca	Mg	K	Na
mmhos cm^{-1}				meq L^{-1}				
0.53	5.00	0.00	0.22	—	1.50	3.50	3.76	2.56

Table 3.
Chemical characterization of soil (salinity).

SP	SAR	ESP	Salinity	Soil classification
%	—	—	—	—
22.17	1.62	4.57	Non-saline	Normal

SP—saturation percentage; SAR—sodium absorption ratio; ESP—exchangeable sodium percentage.

Table 4.
Initial chemical characterization of soil (salinity).

Campina Grande – PB. Secondary cladodes with a homogeneous height were used. After cutting cladodes, they remained during 15 days in shadow to shed moisture and heal injures. Cladodes were treated with bordeaux mixture 48 hours before sowing, in order to prevent fungi and bacteria [7, 8]. Planting was performed at a 45° angle to avoid the fall of the rackets, considering the wind factor in east-west.

Fertilizations followed the recommendations of Novais, Neves and Barros [9] for experiments using urea, potassium chloride (KCl) and monoammonium phosphate (MAP), keeping equal doses in every pot. Fertilization was divided in 30% as basal dressing and the difference was divided and applied every month during the experiment.

2.4 Irrigation

Treatments were carried out at 108 days after sowing and concluded at the end of the cycle (442 days after sowing), totalizing 334 days of application, period necessary for establish forage palm.

Water depths were determined based on water retention capacity of soil (D1 = 25%, D2 = 50%, D3 = 75% and D4 = 100% of WRC), with a variable irrigation frequency and determined by the depletion of soil water content that corresponded to the water depths to be replaced. Water retention capacity was determined by water availability in soil, according to Salassier, Soares & Mantovani [10]:

$$AWC = (FC\text{–}WP) \text{ x } \rho \quad (1)$$

$$WRC = AWC \text{ x } Z \quad (2)$$

Where, AWC: Available water capacity in soil (mm cm^{-1} of soil); FC – field capacity (% of weight); WP: wilting point (% of weight); ρ_b – bulk density (g cm^{-3}); WRC – water retention capacity (mm) and Z – effective root system depth (mm).

Variation of water storage was determined based on Lopes et al. [11] at 0.15 m:

$$\Delta TWS = (\theta_2\text{–}\theta_1) \text{ x } z \quad (3)$$

Where, ΔTWS – water storage variation (mm d^{-1}); θ_2 – average humidity at the end time (cm^3 cm^{-3}) corresponding to the day; θ_1 – average humidity at the initial time (cm^3 cm^{-3}) corresponding to humidity of the previous day; z – depth for balancing.

For the evaluation of the reduction in moisture as the function of water retention capacity perceptual, daily collections of soil samples were performed at 15 and 30 cm (corresponding to the radicular zone of great distribution of forage palm *O. stricta* (Haworth) Haworth), in order to determine moisture using electric oven [12].

Based on the physical analysis of soil, soil-water characteristic curve was determined and adjusted by van Genuchten model using the computer program *Soil Water Retention Curve fit* (SWRT fit), considering granulometry (%) and density particles values (g cm^{-3}) [13]. Thus, water content in soil (humidity in a given volume cm^3 cm^{-3}) was determined as a function of humidity, obtained by gravimetry method, using an electric oven in the radicular system depth.

Water was prepared by adding commercial sodium chloride (without iodine), calcium chloride and magnesium chloride in the proportions 7:2:1, respectively, in order to increase electrical conductivity of water, according to the methodology proposed by Richards [14]. The dilutions were performed in four 500 L polyethylene pots, in which, every pot corresponded to a different saline level. The water used came from the Water and Sewerage Company of Paraíba (CAGEPA). Chemical water analyses were performed in the Irrigation and Salinity Laboratory (LIS/UFCG) (**Tables 5** and **6**).

2.5 Growth variables

Growth variables were analyzed at the end of the cycle, at 334 days after treatments application or 442 days after sowing, according to the methodology proposed by Borges et al. [15] and consisted of: length of primary cladode (PCL, cm) and secondary cladode (SCL, cm); width of primary cladode (PCW, cm) and secondary cladode (SCW, cm); perimeter of primary cladode (PCP, cm) and secondary cladode (SCP) and thickness of primary cladode (PCT, mm) and secondary cladode (SCT, mm). A measure type was used for height, width, length and perimeter; and a digital caliper for thickness, 0.05 mm precision.

Primary (PCA cm^2) and secondary (ACS, cm^2) cladode areas were estimated considering cladode width and length, following the methodology proposed by Santos et al. [16] for forage palm (*Opuntia*):

$$CA = CL \text{ x } CW \text{ x } 0.693 \tag{4}$$

CA – cladode area (cm^2); CL – cladode length (cm); CW – cladode width (cm) and; 0.693 – correction factor as a function of cladode.

Ph	EC	Ca	Mg	Na	K
—	dS m^{-1}		mg L^{-1}		
6.70	0.03	1.20	2.40	0.70	0.0
Chloride	**Bicarbonate**	**Carbonate**	**Sulphate**	**Hardness**	**Alkalinity**
		mg L^{-1}			
0.0	20.13	0.0	0.0	13.20	16.50

Table 5.
Chemical parameters of rainy water used in the experiment.

Ph	EC	Ca	Mg	Na	K
—	dS m^{-1}		mg L^{-1}		
7.40	0.37	16.00	8.00	19.00	32.70
Chloride	**Bicarbonate**	**Carbonate**	**Sulphate**	**Hardness**	**Alkalinity**
		mg L^{-1}			
46.79	75.64	0.0	35.04	73.12	62.00

Table 6.
Chemical parameters of tap water used in the experiment.

2.6 Production variables

At 442 days after sowing, or 334 days after treatments application, production evaluations were performed. Number of primary cladodes (NPC), secondary cladodes (NSC) and total number of cladodes (TNC) per plant were obtained by direct counting, according to Borges et al. [15].

2.7 Statistical analyses

Growth and production data were submitted to the distribution normality test (Shapiro–Wilk test) at 5% probability. The variables that did not demonstrate distribution normality were altered using quadratic root. After normality test, variance analysis by F test at 1 and 5% probability was used for isolate irrigation depths factors and for the interaction depths versus salinity.

Data that presented significant effects were adjusted by polynomial, linear and quadratic regression. Statistical analyses (Shapiro–Wilk and F test) were performed using Sisvar software, 5.6 version [17].

3. Results and discussion

3.1 Irrigation

The results concerning average frequency of irrigation as a function of depths and salinity levels are shown in **Table 7**.

Treatment	Frequency (days)	Treatment	Frequency (days)	Treatment	Frequency (days)	Treatment	Frequency (days)
L1S1	2	L2S1	5	L3S1	9	L4S1	10
L1S2	2	L2S2	6	L3S2	10	L4S2	10
L1S3	2	L2S3	5	L3S3	9	L4S3	13
L1S4	3	L2S4	6	L3S4	8	L4S4	12

L1S1, L1S2, L1S3 and L1S4 (25% water retention capacity - WRC and 0.60; 3.00; 5.40 and; 7.80 dS m^{-1}, respectively); L2S1, L2S2, L2S3 and L2S4 (50% of WRC and 0.60; 3.00; 5.40 and; 7.80 dS m^{-1}, respectively); L3S1, L3S2, L3S3 and L3S4 (75% of WRC and 0.60; 3.00; 5.40 and; 7.80 dS m^{-1}, respectively) and; L4S1, L4S2, L4S3 and L4S4 (100% of WRC and 0.60; 3.00; 5.40 and; 7.80 dS m^{-1}, respectively).

Table 7.
Average frequency of irrigation as a function of depths and salinity levels.

According to the results, average frequencies of irrigation varied between 2 and 13 days, considering treatments. In general, the lowest frequencies were obtained under the greatest saline levels (S3 = 5.40 dS m^{-1} and S4 = 7.80 dS m^{-1}).

The lowest frequencies of irrigation using the highest saline waters may have occurred due to the salt effects on soil, altering its physical–chemical properties.

When comparing the same saline level to the other depths, it was observed a progressive reduction of irrigation frequencies, this occurred because irrigation was determined by water decay in soil as a function of water retention capacity. Thus, with the increase of water depths, it was necessary a longer period of time (lower frequency of irrigation) in order to promote the water decay in soil and the application of irrigation.

The total value irrigated per treatment (depth x salinity) are shown in **Table 8**.

In general, the greatest irrigation frequencies resulted in the highest values of accumulated irrigation depths (treatments with 25 and 50% of water retention capacity – WRC). Furthermore, the influence of salt in irrigation water was verified in the crop hydric consumption, demonstrated by treatments with 5.40 and 7.80 m^{-1} that presented the lowest irrigation depths.

According Souza et al. [18], the accumulation of salts in soil is related to irrigation that compromises chemical, physical and biological properties of soil. Especially in semi-arid regions, where there is a predominance of evaporation over precipitation. Salts present in soils reduce the osmotic potential of its solution and may decrease water availability in plants.

3.2 Growth

3.2.1 Primary cladodes

The results of variance analysis (F test at 1 and 5% probability) for the variables of length, width, area, perimeter and thickness of primary cladodes at 334 days after treatment applications, corresponding to a total cycle of 442 days, are shown in **Table 9**.

According to the data obtained (**Table 9**), there were not significant statistical differences ($p > 0.05$) for any growth variable evaluated. Thus, treatments did not influence the growth parameters of primary cladodes.

Average values were observed for length (PCL), width (PCW), area (PCA), perimeter (PCP) and thickness (PCT), as the following: 30.02 cm; 24.27 cm; 551.86 cm^2; 79.25 cm and; 15.56 mm, respectively.

Treatment	Irrigation (mm)	Treatment	Irrigation (mm)	Treatment	Irrigation (mm)	Treatment	Irrigation (mm)
L1S1	434.88	L2S1	375.42	L3S1	310.32	L4S1	352.76
L1S2	505.37	L2S2	344.51	L3S2	309.64	L4S2	385.57
L1S3	507.63	L2S3	415.31	L3S3	326.78	L4S3	299.22
L1S4	380.61	L2S4	347.39	L3S4	337.42	L4S4	317.95
Mean	457.12	L2S4	370.65	L3S4	321.04	L4S4	338.87

Table 8.
Total accumulated irrigation at the end of the cycle as a function of water depths and salinity levels.

Variation source	DF	Quadratic Average (QA)				
		PCL[1]	PCW	PCA[1]	PCP	PCT
Depth	3	66.51^{ns}	6.46^{ns}	44623.12^{ns}	48.01^{ns}	1.45^{ns}
Salinity	3	41.13^{ns}	5.83^{ns}	19373.73^{ns}	51.82^{ns}	3.67^{ns}
Depth x Salinity	9	50.73^{ns}	1.71^{ns}	17987.23^{ns}	18.07^{ns}	2.56^{ns}
Block	3	71.47^{ns}	8.04^{ns}	53399.78^{ns}	103.92^{ns}	0.45^{ns}
Error	45	55.33	4.43	29940.33	49.86	2.08
CV (%)	—	9.80	8.68	12.83	8.91	9.28
General average	—	30.02	24.27	551.86	79.25	15.56

^{ns}*Non-significant by F test ($p > 0.05$).*
[1]*Data transformed by quadratic root.*

Table 9.
Summary of variance analysis of primary cladode length (PCL), primary cladode width (PCW), primary cladode area (PCA), primary cladode perimeter (PCP) and primary cladode thickness (PCT) of forage palm Orelha de Elefante at 442 days of cycle.

Donato et al. [19] state that morphometric characteristics of forage palm are rarely influenced by management, which was also verified in the present study. The results differed from two other authors, as Silva et al. [20] and Lima et al. [21]. These authors verified that the adoption of management practices on forage palm results in significant effects on the plant growth, by irrigating with different sowing spacing or by supplying fertilization.

Silva et al. [20] evaluated the growth of forage palm clones in semi-arid region conditions and its relations with meteorological variables and verified statistical differences for the average length of primary cladode with 27.73 cm for Orelha de Elefante Mexicana. This value is close to the one observed in this study for the same cultivar.

Lima et al. [21] studied morphological and productive characteristics of forage palm Gigante irrigated with saline water (5.25 dS m^{-1}) and submitted to intensive cutting, and obtained average length of 37.87 cm for the second cycle. Significant differences were not observed for this variable. The authors only verified significant effects for average width (20.95 cm), thickness (18.62 cm) and area (583.46 cm^2). The aforementioned authors concluded that the differences observed for width and thickness of cladode may have resulted from the best efficiency of physiological and biochemical process of plant, as photosynthesis, respiration and transpiration, being influenced by management practices.

Pereira et al. [22] evaluated the growth of forage palm clones (Orelha de Elefante Mexicana, IPA Sertânia and Miúda), in the municipality of Serra Talhada in the state of Pernambuco, under drip irrigation with a permanent depth (7.50 mm) and three intervals of water application (7, 14 and 28 days). The authors concluded that irrigation promoted the best biometric increments for the evaluated clones. For Orelha de Elefante Mexicana were observed the following average values in the primary cladodes, in absolute terms: 23.80 cm of cladode length and 11.80 mm of cladode thickness. In relative terms, for the frequencies of 7, 14 and 28 days, these values were the following: 18.70 cm (7 days), 15.40 cm (14 days) and 16.20 cm (28 days) of cladode width; 68.60 cm (7 days), 59.00 cm (28 days) of cladode perimeter; and 285.00 cm^2 (7 days), 310.00 cm^2 (14 days) and 377.00 cm^2 (28 days) of cladode area. The total water received by the crop was equivalent to 558.00 mm (7 days), 475.00 mm (14 days) and 438.00 mm (28 days). Even with different irrigation frequencies performed in this research, these depth values are close to those verified by Pereira et al. [22], which did not present significant effects for primary cladodes as well as the present study.

Sarmento et al. [23] evaluated the influence of different irrigation frequencies on the growth and production of forage palm Orelha de Elefante Mexicana (*O. stricta* (Haworth) Haworth) submitted to different frequencies of irrigation (0, 7, 14 and 21 days) and verified an increasing linear effect in width and length of primary cladodes due to the increase in irrigation frequency. This did not occur to the thickness of primary cladode. Means observed for primary cladode length were the following: 31.16 cm (without irrigation); 32.83 cm (21 days); 34.97 cm (14 days) and; 36.07 (7 days). For average width: 23.86 (without irrigation); 26.46 cm (21 days); 26.73 (14 days) and; 28.90 (7 days). Thickness: 19.14 mm (0 days); 17.60 mm (21 days); 18.36 mm (14 days) and; 17.32 mm (7 days). And perimeter: 92.88 mm, 91.80 mm, 91.15 mm and 92.66 mm (0, 21, 14 and 7 days), respectively. The results corroborate with the values obtained in this study, showing that the growth of primary cladodes were not affected, even with different irrigation depths contributing to different frequencies.

3.2.2 Secondary cladodes

The summary of variance analysis using F test at 1 and 5% probability for length, width, area, perimeter and thickness of secondary cladode at 334 days after treatment applications are shown in **Table 10**.

According to the results of variance analyses by F test (**Table 10**), as for primary cladodes, treatments did not present effects ($P > 0.05$) on the variables evaluated: length, width, area and perimeter.

The variables of secondary cladode present average values that did not depend on the applied treatment: 21.32 cm of length, 18.66 cm of width, 287.93 cm^2 of area and

Source of variation	**DF**	**Quadratic Average (QA)**				
		SCL	**SCW**	**SCA**	**SCP**	**SCT[1]**
Depth	3	3.68^{ns}	2.09^{ns}	2573.55^{ns}	15.42^{ns}	4.63^{ns}
Salinity	3	7.18^{ns}	2.79^{ns}	4025.10^{ns}	36.87^{ns}	5.53^{*}
Depth x Salinity	9	4.00^{ns}	4.11^{ns}	3657.90^{ns}	34.93^{ns}	1.85^{ns}
Block	3	7.11^{ns}	4.02^{ns}	5040.10^{ns}	37.58^{ns}	1.22^{ns}
Error	45	2.39	2.04	1946.91	21.93	1.76
CV (%)	—	7.26	7.65	15.32	8.02	6.53
General average	—	21.32	18.66	287.93	58.38	9.39

**Significant ($p \leq 0.05$).*
^{ns}Non-significant ($p > 0.05$) by F test.
[1]Data transformed by quadratic root.

Table 10.
Summary of variance analysis of secondary cladode length (SCL), secondary cladode width (SCW), secondary cladode area (SCA), secondary cladode perimeter (SCP) and secondary cladode thickness (SCT) of forage palm Orelha de Elefante at 442 days of cycle.

58.38 cm of cladode perimeter. These results were inferior to the results of primary cladodes due to plant morphometric characteristics.

Sarmento et al. [23] obtained secondary cladode length in different frequencies of irrigation: 27.77 cm (without irrigation) 27.28 cm (21 days); 28.23 cm (14 days) and; 29.26 cm (7 days). Regarding cladode width, from the lowest frequencies to the highest frequencies, the average values were the following: 21.84 cm; 24.33 cm; 24.69 cm and; 25.52 cm. And the perimeters: 87.73 cm, 83.24 cm, 85.38 cm and 86.96 cm from the lowest to the highest. In spite of the absence of significant effects, frequent irrigation contributed to the growth of forage palm Orelha de Elefante Mexicana. The results observed by the authors for the species *O. stricta* (Haworth) Haworth were superior to the ones obtained in the present study. The variables evaluated in this work may have been influenced by saline water reducing the growth in relation to the conditions of plant cultivation using non-saline water, even without statistical significance for treatments.

Borges et al. [15] studied three different forage palm (Orelha de Elefante Mexicana, Miúda and Baiana) submitted to nitrogen fertilization via fertigation and observed average cladode length varying between 25.40 to 27.52 cm in relation to the error pattern for Orelha de Elefante Mexicana. Thus, this difference between length obtained in the present study and the values verified by the authors is due to the average values between primary and secondary cladodes.

Sales et al. [24] evaluated the vegetative growth of forage palm Gigante under different densities of cultivation in Curimataú (river located in the states of Paraiba and Rio Grande do Norte) in the state of Paraiba. It was verified that the average value between densities of cultivation for cladode width was 18.98 cm and cladode length 33.89 cm at 710 days after sowing. On the other hand, the variable cladode area was affected by treatments showing values of 440.12 cm^2, 397.95 cm^2 and 383.05 cm^2. The authors did not perform regular intervals of irrigation during the experiment; the water depth applied was made through precipitation. Thus, the present study as well as the research carried out by the authors did not obtain significant statistical differences in relation to the application of treatments on the growth variables analyzed, except for plant height.

Pereira et al. [22] observed average values for secondary cladodes: 10.40 cm of length, 12.60 of width, 27.30 cm of perimeter and 74.60 cm^2 of area. The results for these variables were inferior to the ones of the present study. Statistically, in this research, the results were not significant ($P > 0.05$) in relation to the treatments applied. However, the salt in water may have influenced plant growth as salinity may have contributed to elevate forage palm evapotranspiration due to sodium (Na^+) and to stomatal adjustment that fomented plant development. Moreover, according to Campos [25], water availability through irrigation increases real evapotranspiration of forage palm plant in comparison to rainfed cultivations. In this context, the plant increases its transpiration due to greater water availability.

This hypothesis can be affirmed by Fonseca et al. [26] that concluded that higher average values of morphometric characteristics of forage palm under conditions of hydric availability indicates that even with use of saline water, irrigation provides better conditions of crop development due to the increment of photosynthetic taxes.

There were only significant statistical differences ($P \leq 0.05$) for secondary cladode thickness in relation to salinity levels in the irrigation water applied. Effects of water depths and its interaction with salinity were not observed in forage palm thickness. Data showed low dispersion in the coefficient of variation (CV) of 6.53%. The other variables did not show significant effect ($P > 0.05$) for any factor evaluated.

There was an increment for cladode thickness up to 9.79 mm under a salinity level of 1.22 dS m^{-1}, verified by the graphic adjustment equation. On the other hand, with a higher salinity level, cladode thickness decreased. The value of R^2 in the adjustment equation was 0.70 (**Figure 2**).

Forage palm thickness is related to the accumulation of water in its cladodes. This stored water, according to Nobel [27], may favor palm gas exchange. This content may be an indicative of stress tolerance caused by saline water.

The effects of treatments ($P < 0.05$) on secondary cladode thickness may denote the development of tolerance mechanisms to salinity through juiciness. According to Willadino and Camara [28], sodium tends to be transported via xylem and accumulate in plant shoot system. Thus, plant may have developed juiciness in order to mitigate the effects of accumulated salts on secondary cladodes or even due to the osmotic adjustment, the responsible for causing a potential difference and induce water movement into guard-cell.

Although, Freire [29] concluded that higher salinity levels of 3.60 dS m^{-1} or more, result in the decrease of forage palm juiciness. This maximum salinity level on thickness was 60% lower than the one verified by the author.

It can be inferred that osmotic adjustment contributed to prevent restrictions to plant stomatal opening, promoting the increase of transpiration and the reduction of water content in cell and, consequently, its thickness. This justifies the reduction of thickness of secondary cladode due to the increase of salt in irrigation water.

Cladode thickness is one of the species characteristics directly correlated to plant turgidity. Thus, the higher the thickness, the greater water quantity in cells, which is one of the main attributes of CAM plants [22].

Pereira et al. [22] obtained average values of cladode thickness in OEM, IPA and Miúda cultivars: 11.80; 18.50 and 14.10 mm, respectively. The thickness observed by the authors was approximately 17% superior.

Sarmento et al. [23] verified a significant effect of cladode thickness as a function of irrigation frequencies, which presented a decreasing quadratic effect, showing values of 12.31 (0 days); 13.58 (21 days); 11.44 (14 days); 11.30 (7 days). On the other hand, it did not occur for primary cladode.

When comparing the results of the thickness of primary cladode to the results of Pereira et al. [22], Sales et al. [24] and Sarmento et al. [23], it was verified that thickness was lower than the ones verified by the other authors. However, the

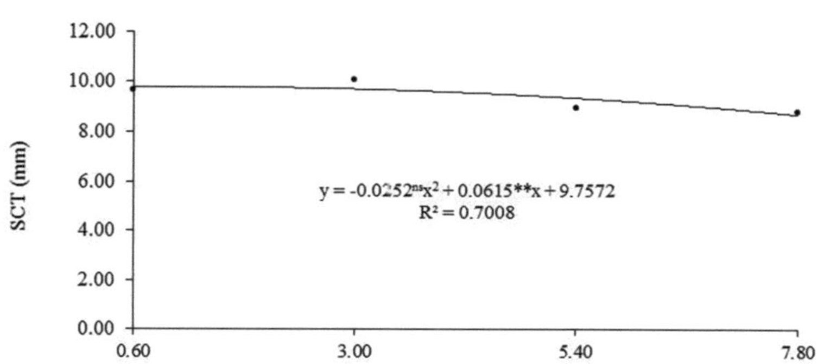

Figure 2.
Secondary cladode thickness (SCT) of forage palm (Orelha de Elefante) as a function of salt in irrigation water (EC_w) at 442 days of the cycle.

aforementioned works studied the species *Opuntia fícus-indica* Mill while the present research analyzed the species *O. stricta* (Haworth) Haworth.

According to Rocha, Voltolini, and Gava [30], cladode thickness is of great importance for the photosynthetic capacity and for water storage in plant.

Orelha de Elefante clone has a great potential to adapt to conditions of low water availability in soil, presenting a greater capacity of water storage in its cladodes [22]. Thus, this capacity also contributes to plant tolerance in relation to salts.

3.3 Production

3.3.1 Number of primary and secondary cladodes

The results of variance analysis by F test at 1 and 5% probability showed a significant effect ($P \leq 0.05$) only for the number of primary cladodes in relation to water depths. Regarding the other variables, the number of secondary cladodes and the total number of cladodes did not present significant statistical differences (**Table 11**).

The number of primary cladodes showed a quadratic tendency with R^2 equal to 0.8643 (**Figure 3**). The greatest average was presented by the number of primary cladodes, based on the graphical adjustment equation, for the water depth of 376.00 mm with 4.03 cladodes. The same depth is the closest to L2 (370.66 mm), which corresponds to 50% of water retention capacity of soil – WRC. The other depths presented average values of 3.43 cladodes (L3–75% of WRC equal to 321.04 mm), 3.76 cladodes (L4–100% of WRC equal to 338.87 mm) and 2.72 cladodes (L1–25% of WRC equal to 457.12 mm), also obtained using the adjustment equation.

Thus, it was verified that the number of primary cladodes increased due to the water depth increase up to 376.00 mm and, for values higher than that, it decreased.

The other variables did not present significant effect ($P > 0.05$) with average values of 11.48 of secondary cladodes (NSC) and 17.20 of total number of cladodes (TNC).

Cavalcante, Leite, Pereira and Lucena [31] evaluated forage palm Orelha de Elefante Mexicana with and without cure of cladodes and did not verify statistical

Source of variation	DF	Quadratic Average (QA)		
		NPC[1]	NSC	TNC
Depth	3	4.89*	3.43^{ns}	1.64^{ns}
Salinity	3	2.89^{ns}	12.56^{ns}	9.18^{ns}
Depth x Salinity	9	1.78^{ns}	14.25^{ns}	14.52^{ns}
Block	3	1.02^{ns}	20.31^{ns}	15.10^{ns}
Error	45	1.73	19.35	16.66
CV (%)	—	11.79	21.19	12.06
General average	—	5.66	11.48	17.20

**Significant ($p \leq 0.05$).*
^{ns}Non-significant by F test.
[1]Data transformed by quadratic root.

Table 11.
Summary of variance analysis for the number of primary cladodes (NPC), number of secondary cladodes (NSC) and total number of cladodes (TNC) of forage palm Orelha de Elefante Mexicana at 442 days of the cycle.

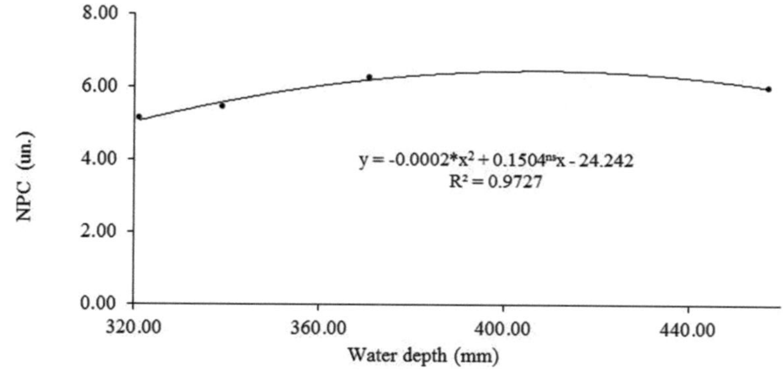

Figure 3.
Number of primary cladodes of forage palm Orelha de Elefante Mexicana as a function of water depths at 442 days of the cycle.

differences ($P > 0.05$) in the number of primary cladodes, number of secondary cladodes and total number of cladodes. But, the average values were 2.35 for primary cladodes, 5.37 for secondary cladodes and 12.31 for total number of cladodes. However, tertiary cladodes were considered by the authors when counting the number of total cladodes, and, in the present research, tertiary cladodes were not observed in forage palm. The values obtained in this study were superior to the ones by Cavalcante, Leite, Pereira and Lucena [31] probably due to the application of water depths, promoting the production of the cultivar.

Lima et al. [21] did not verify significant interaction ($P > 0.05$) in the number of primary cladodes, but obtained significant effect in the total number of cladodes with a maximum value of 20.60 of cladodes per plant. The authors also used saline water on forage palm, but productive characteristics were evaluated in forage palm Gigante. Lima et al. [21] concluded that the great number of cladodes tent to increase CO_2 capture, which results in a higher photosynthetic tax, contributing to maximize the production.

Sarmento et al. [23] obtained values that corroborate with the present research. The authors verified that Orelha de Elefante Mexicana, under different irrigation frequencies (0, 7, 14 and 21 days), presented number of cladodes equal to: 6.06 (0 days), 4.92 (21 days), 4.85 (14 days) and 4.82 (7 days) of primary cladodes; 12.40 (0 days), 11.91 (21 days), 11.91 (14 days) and 10.50 (7 days) of secondary cladodes; 18.60 (0 days), 17.20 (21 days), 16.94 (14 days) and 15.44 (7 days) of total number of cladodes.

According to Borges et al. [15], the greatest number of cladodes in forage palm plants reflects in a greater production. The authors verified that Orelha de Elefante Mexicana under fertigation presented 13.00 cladodes per plant, which is equivalent to the total number of cladodes.

Pereira et al. [22] verified that hydric availability conditions did not affect the number of primary cladodes and the total number of cladodes of forage palm *O. stricta* (Haworth) Haworth. However, there was significant effect for the number of secondary cladodes. The authors observed the following average values: 8.11 of primary cladodes and 13.50 of total number of cladodes. The number of secondary cladodes showed the following values, respectively: 7.67 units (7 days frequency with a total water depth of 558.00 mm); 1.56 units (14 days frequency with water depth of 475.00 mm) and; 4.22 units (frequency of 28 days with water depth of 438.00 mm).

In the present study, the number of secondary cladodes presented average values of 11.48 units for water depths that varied between 321.04 mm and 457.12 mm. Comparing these results to the ones obtained by Pereira et al. [22], there was a better use of water irrigation by plant in the production of cladodes, since palm produced a quantity of 33% higher under water depths inferior than the ones applied by the authors.

4. Conclusions

The research analyzed the effect of irrigation with different water depths and levels of salinity on Orelha de Elefante Mexicana cultivar. This cultivar can produce even under saline conditions and irrigation with saline water is an alternative strategie for forage production in semiarid regions.

The growth of forage palm Orelha de Elefante Mexiacana was not affected by water depths and by the salt in water irrigation, except thickness of secondary cladodes, which was positively affected by water electrical conductivity.

The production of primary cladodes was affected by water availability for forage palm Orelha de Elefante Mexicana, which was maximized by water depth corresponding to 50% of water retention capacity of soil presenting 4.03 cladodes.

Regarding the conditions to carry out the research, saline water did not affect the total production of Orelha de Elefante Mexicana cultivar.

Author details

Mariana de Oliveira Pereira[1*], Jailton Garcia Ramos[1], Carlos Alberto Vieira de Azevedo[1], André Alisson Rodrigues da Silva[1], Geovani Soares de Lima[1], Luciano Marcelo Fallé Saboya[1], Patrícia Ferreira da Silva[1] and Gustavo Bastos Lyra[2]

1 Academic Unit of Agricultural Engineering, Federal University of Campina Grande, Campina Grande, Paraíba, Brazil

2 Department of Environmental Sciences, Forest Institute, Seropédica, Federal Rural University of Rio de Janeiro, Rio de Janeiro, Brazil

*Address all correspondence to: marianapereira.agri@gmail.com

References

[1] Silva LM, Fagundes JL, Viegas PAA, Muniz EN, Rangel JHA, Moreira AL, et al. Produtividade da palma forrageira cultivada em diferentes densidades de plantio. Ciência Rural. 2014;**44**:2064-2071. DOI: 10.1590/0103-8478cr20131305

[2] Larcher W. Utilização de carbono e produção de matéria seca. In: Larcher W, editor. Ecologia vegetal. São Paulo: EPU; 1986. pp. 74-160

[3] Campos FS, Gois GC, Vicente SLA, Macedo A, Matias AGS. Alternativa de forragem para caprinos e ovinos criados no semiárido. Nutritime Revista Eletrônica. 2017;**14**:5004-5013

[4] Silva TGF, Santos JEO, Primo JTA, Moura MSB, Silva SMSE, Silva MC, et al. Dinâmica de água no solo em uma área de cultivo com clones de palma forrageira. Anais do Congresso Brasileiro de Meteorologia. Gramado, RS: Sociedade Brasileira de Meteorologia; 2012. p. 205

[5] Toledo JV, Zolnier S, Silva TGF, Boehringer D, Steidle Neto AJ. Alterações na evapotranspiração de cultivares de cana-de-açúcar sob distintos níveis de salinidade aplicados na fertirrigação. Revista Engenharia Agrícola. 2017;**37**(5):940-952. DOI: 10.1590/1809-4430-Eng.Agric. v37n5p940-952/2017t

[6] Ribeiro AA, Moreira FJC, Seabra Filho M, Menezes AS. Emergência do maracujazeiro-amarelo sob estresse salino em diferentes substratos. Revista Brasileira de Engenharia de Biossistemas. 2016;**10**:27-36. DOI: 10.18011/ bioeng2016v10n1p27-36

[7] Gava CAT, Lopes EB. Produção de mudas de palma forrageira utilizando fragmentos de cladódios. Petrolina: Embrapa Semiárido; 2012

[8] Rocha JES. Palma forrageira no Nordeste do Brasil o estado da arte. 21st ed. Sobral: Embrapa Caprinos e Ovinos; 2012

[9] Novais RF, Neves JCL, Barros NF. Ensaio em ambiente controlado. In: Oliveira AJ, editor. Métodos de pesquisa em fertilidade do solo. Brasília: Embrapa SEA; 1991. pp. 189-125

[10] Salassier B, Soares AA, Mantovani EC. Manual de irrigação. 8th ed. Viçosa: Ed. UFV; 2008

[11] Lopes AS, Pavani LC, Corá JE, Zanini JR, Miranda HA. Manejo da irrigação (tensiometria e balanço hídrico climatológico) para a cultura do feijoeiro em sistemas de cultivo direto e Convencional. Engenharia Agrícola. 2004;**24**:89-100. DOI: 10.1590/ S0100-69162004000100011

[12] Matos RM, Dantas Neto J, Lima AS, Silva PF, Borges VE, Galvão Sobrinho T. Teor de umidade por diferentes métodos em Neossolo do semiárido brasileiro. Revista Brasileira de Agricultura Irrigada. 2017;**11**(4):1588-1597. DOI: 10.7127/rbai.v11n400622

[13] Seki K. SWRC fit - a nonlinear fitting program with a water retention curve for soils having unimodal and bimodal pore structure. Hydrol. Earth System Science Discussion. 2007;**4**:407-437. DOI: 10.5194/hessd-4-407-2007

[14] Richards LA. Diagnosis and Improvement of Saline and Alkali Soils. Washington: US Department of Agriculture; 1954

[15] Borges VE, Matos RM, Silva PF, Dantas Neto J, Sobrino TG. Análise de crescimento de três cultivares de palma forrageira. Educação Agrícola Superior. 2015;**30**:7-14

[16] dos SDC, Farias I, Lira, de M, A, Dos SMVF, De AGP, RSB C, et al. Manejo e utilização da palma forrageira (*Opuntia* e *Napolea*) em Pernambuco. Recife: IPA; 2006. p. 48

[17] Ferreira DF. SISVAR - a computer statistical analysis system. Ciência e Agrotecnologia. 2011;**35**:1039-1042. DOI: 10.1590/S1413-70542011000600001

[18] Souza MVP, Sousa GG, Sales JRS, Freire MHC, Silva GL, Viana TVA. Água salina e biofertilizantes de esterco bovino e caprino na salinidade do solo, crescimento e fisiologia da fava. Revista Brasileira de Ciências Agrárias. 2019; **14**(3):5672. DOI: 10.5039/agraria.v14i3a5672

[19] Donato PER, Pires AJV, Donato SLR, Bonomo P, Silva JA, Aquino AA. Morfometria e rendimento da palma forrageira 'Gigante' sob diferentes espaçamentos e doses de adubação orgânica. Revista Brasileira de Ciências Agrárias. 2014;**9**:151-158. DOI: 10.5039/agraria.v9i1a3252

[20] Silva PF, Matos RM, Borges VE, Melo JAP, Dantas Neto J. Características morfológicas de três cultivares de palma forrageira sob fertilização mineral em Campina Grande-PB. Revista Enciclopédia Biosfera. 2015;**11**:385-397

[21] Lima GFC, Rego M, Teles M, Dantas FDG, Lôbo RNB, Silva JGMD, et al. Morphological characteristics and forage productivity of irrigated cactus pear under different cutting intensities. Revista Caatinga. 2016;**29**:481-488. DOI: 10.1590/1983-21252016v29n226rc

[22] Pereira PC, Silva TGF, Zolnier S, Morais JEF, Santos DC. Morfogênese da palma forrageira irrigada por gotejamento. Revista Caatinga. 2015; **28**(3):184-195. DOI: 10.1590/1983-2125 2015v28n321rc

[23] Sarmento WGC, Oliveira JFF, Cardoso DB, Santos EA, Mendonça CGS, Andrade AP, et al. Crescimento e produtividade da palma forrageira cv. orelha de elefante mexicana (Opuntia stricta Haw) submetida a diferentes frequências de irrigação. In: Anais da XXVII Semana de Zootecnia da UFRPE. Recife, PE: Universidade Federal Rural de Pernambuco; 2019. pp. 1-3

[24] Sales AT, Leite MLMV, Alves AQ, Ramos JPF, Nascimento JP. Crescimento vegetativo de palma forrageira em diferentes densidades de plantio no Curimataú Paraibano. Tecnologia & Ciência Agropecuária. 2013;7:19-24

[25] Campos ARF. Manejo de irrigação na palma forrageira: definição de critérios com base no potencial matricial da água no solo. Cruz das Almas, BA, Basil: Tese de doutorado Universidade Federal do Recôncavo Baiano; 2018

[26] Fonseca VA, Santos MR, Silva JA, Donato SLR, Rodrigues CS, Brito CFB. Morpho-physiology, yield, and water-use efficiency of Opuntia ficus-indica irrigated with saline water. Acta Scientiarum Agronomy. 2019;**41**:1-11. DOI: 10.4025/actasciagron.v41i1.42631

[27] Nobel PS. Achievable productivities of CAM plants: Basis for high values compared with C3 and C4 plants. The New Phytologist. 1991;**119**:183-205. DOI: 10.1111/j.1469-8137.1991.tb01022.x

[28] Willadino L, Camara TR. Tolerância das plantas à salinidade: aspectos fisiológicos e bioquímicos. Enciclopédia Biosfera. 2010;**6**(11):1-23

[29] Freire JL. Avaliação de clones de palma forrageira (*Opuntia* e *Nopalea*) sob irrigação e salinidade. Tese de

doutorado: Universidade Federal Rural de Pernambuco, Recife, PE, Brasil; 2012

[30] Rocha RS, Voltolini TV, Gava CAT. Características produtivas e estruturais de genótipos de palma forrageira irrigada em diferentes intervalos de corte. Archivos de Zootecnia. 2017;**66**:363-371

[31] Cavalcante AB, Leite MLMV, Pereira JS, Lucena LRR. Crescimento de palma forrageira em função da cura de segmentos dos cladódios. Tecnologia & Ciência Agropecuária. 2017;**11**:15-20

Chapter 11

Anaerobic Filters: Alternative Solution for the Treatment of Domestic Wastewater for Reuse in Vegetable Irrigation

Valdemiro Pitoro, Rodrigo Sánchez-Román, João Queluz, Tamires Da Silva, Sérgio Jane and Kevim Muniz

Abstract

This chapter is the result of a research conducted in the experimental area of the Department of Rural Engineering, School of Agronomic Sciences, São Paulo State University, Botucatu, São Paulo, Brazil, for 165 days, between the months of February and August 2018, with the objective to evaluate the performance of a wastewater treatment plant (WWTP) at a pilot-scale composed of six anaerobic filters (tank-in-series), filled with gravel #1 in the treatment of domestic wastewater (DW) for agricultural reuse. The parameters were monitored was: pH, electrical conductivity, total suspended solids, biochemical oxygen demand, chemical oxygen demand, total nitrogen, total phosphorus, and potassium. The results indicate that the WWTP performed satisfactorily and provided treated wastewater (TWW) with acceptable quality for agricultural reuse in irrigation of vegetable crops. It was observed that the mean concentrations of the pollutants decreased as wastewater advanced through the filters stage, presenting high removal efficiency at the 6th filter (TWW 6) than the 3rd filter (TWW 3), with statistical analysis corroborating that there are significant differences between the quality of TWW 3 and TWW 6 for most of the parameters evaluated, suggesting that the increase of the number of filters in this treatment system proposed improves the treated wastewater quality.

Keywords: agricultural irrigation, agricultural wastewater reuse, wastewater reuse, wastewater treatment, water scarcity

1. Introduction

The reuse of domestic wastewater (DW) in agricultural irrigation is seen as a fundamental alternative to alleviate water scarcity in the world [1], besides contributing to the reduction of environmental impacts, costs in treatment, and the discharge of DW to natural water bodies [2].

The availability of wastewater (WW) throughout the year is indicated as one of the most important aspects of water reuse in agriculture, which will reduce the dependence on precipitation, especially in regions with arid or semi-arid climates [3]. The recycling of nutrients available in the WW is an important economic alternative, as in many cases, it allows farmers to reduce or even eliminate the application of conventional fertilizers in their production fields [2]. In the literature, there are numerous experimental researchers in which WW has been used successfully for irrigation of agricultural vegetable crops [2, 4].

Despite the numerous benefits provided by effluent reuse in irrigation, it is important to emphasize that improper application can be harmful to plants, animals, producers, and consumers, as well as to the soil [5], because unlike drinking water, reuse water may contain high concentrations of bacteria, viruses, salts, and heavy metals, depending on its source and treatments [6].

The direct discharge of UTWW into soil and water bodies has negative impacts on the various components of the production process. Therefore, the combination of effluent reuse with appropriate treatment and recycling methods could be vital to provide a quality effluent for reuse, especially one of acceptable quality for agricultural irrigation. There are currently several alternatives or wastewater treatment technologies for reuse in irrigation, ranging from expensive and complex, to low cost and simple in structure, implementation, and maintenance. The selection of the effluent treatment technology to be adopted should respect issues related to the economic, social, and environmental conditions of the beneficiary community, as well as the quality recommended for the purpose it is intended to apply [7, 8].

The wastewater treatment technologies used by the sanitation companies are not feasible for low-income rural communities, due to the high cost of implementation, operation, and maintenance [9], as well as the large dispersion of the population in rural areas. Thus, it is important to develop decentralized, low-cost, and easy-to-operate technologies for the treatment of DW effluents [10], and to provide optimum water supply for non-potable purposes, such as irrigation of agricultural crops.

There are several technologies that can be used for the treatment of DW for reuse in vegetable irrigation [11], such as: anaerobic filters (ANF), aerobic filters (AEF), septic tanks (ST), constructed wetlands (CWs), filter membranes, chlorination, sand filter, UV disinfection, among others. ST is the simplest and oldest low-cost system widely used as decentralized treatment technology, but, post-treatment is usually required because of the high soluble organic matter and pathogens content that remains in the effluent [12]. ST alone contributes to total suspend solids (TSS) (62%), chemical oxygen demand (COD) (31%) and fecal coliforms (FC) (31%) removal [13]. Similar results were presented by Del Castillo et al. [9], which found removal efficiency of 35% for COD, 73% for TSS, and 33% for biochemical oxygen demand (BOD). According to Bouted and Ratanatamskul [14], the aerobic treatment system has been recognized as a highly efficient system, but they have relatively high running costs in terms of energy. CWs are attracting interest as potential low-cost treatment solutions [15] and are practiced for primary and secondary treatment of DW [16]. However, when used alone to treat DW, CWs might not be able to meet quality guidelines for agricultural reuse [9].

In the ANF, the treatment process occurs in the absence of oxygen, and the highlight compared to other treatment systems according to da Silva et al. [17], is its high efficiency in the degradation of solid organic waste, converting organic matter into biogas, which can be used for thermal, electrical or mechanical energy generation, biofertilizer or substrate that can be used to improve the physicochemical and biological properties of the soil. In general, the ANF removes total dissolved solids (TDS) and TSS through close contact with anaerobic bacteria attached to the filter media. Refers. [18] reported that ANF can present removal efficiencies higher than 70% for TSS and BOD, and 90% for fecal coliforms (FC).

Every treatment technology has different benefits, limitations, cost requirement and land area, payback period, and removal efficiency. Among all above mentioned technologies, ANF stands out because is a low-cost and sustainable treatment technology [18]. Tripathi et al. [19] also highlight that the operational simplicity is one of the main advantages of using ANF in wastewater treatment, specifically because they can be operated without the need for electricity, which makes them suitable for developing countries and rural communities or regions isolated from large urban centers. However, despite the diversity of their application and benefits provided, similarly to CWs, when ANF are used as a standalone technology, the effluent often does not meet the quality guidelines for agriculture reuse [9].

Abegunrin et al. [20] report the limitation of the application of ANF in the treatment of WW with higher concentrations of suspended solids, which is why it is commonly used for post-treatment. Tonon et al. [21] observed that the BOD and COD removal efficiency decrease with the increase of the hydraulic loading rate. According to de Oliveira Cruz et al. [18], nitrogen compounds and phosphorus concentration do not change during anaerobic treatment, thus, requiring an additional aerobic step or CWs to increase the quality of the final treated effluent. The improvement of the anaerobic (filter) system to become more efficient in treating DW is challenging, compared to the traditional aerobic treatment systems [22]. To deal with the limitations of the anaerobic (filter) DW treatment system, researchers are studying the performance of ANF in different environments [14], plant operational conditions [9], and alternative filled mediums [14] to identify the best way to improve their efficiency.

It is evident the need to study and develop low-cost technologies for WW treatment, which can contribute to the rational replacement of potable water, for treated DW in certain activities. It is also evident, the importance of discussing the need for the reuse of lower quality water in less sensitive activities concerning water quality, especially in irrigated agriculture. In this context, this research aims to evaluate the performance of ANF filled with gravel #1 (as inert material) in the treatment of DW for agricultural reuse.

2. Materials and methods

2.1 Site description

The research was developed in the experimental area of the Department of Rural Engineering, School of Agronomic Sciences (FCA), São Paulo State University (UNESP), Botucatu, São Paulo, Brazil, at coordinates 22° 50′ 48″ S, 48° 26′ 06″ W and altitude of 817.74 m. The climate of the region is defined as type Cfa (Koppen): humid subtropical climate (mesothermal) with rainy summer and dry winter, the average temperature of the warmest month is above 22°C and the average annual precipitation around 1,501.4 mm.

2.2 Design and operation of the wastewater pilot-scale plant

The research comprised two phases; the first (Phase I) [23] the wastewater treatment plant (WWTP) had two beds, one filled with gravel #1 and other filled gravel #4. Each bed was composed of three vertical filters connected in series and operated for 105 days. Based on the results of Phase I, in the second phase (Phase II), the WWTP comprised a bed composed of six filters (**Figure 1**) made in 200 l plastic barrels (with dimensions of 0.90 m high and 0.50 m diameter), filled only with gravel #1, the water level was maintained at 10 cm from the barrel surface (to prevent overflow), providing an average porosity of 48% about the inert material, corresponding to a daily application rate of 95 liters and a hydraulic detention time (HDT) of 5.4 days.

The choice of the plastic barrel for making the filter was due to its low price and capacity to resist the weight of the inert material, weather conditions and the change in the structure of the WWTP in Phase II aimed to improve the quality of the TWW, because, the values of the BOD of the TWW obtained in the Phase I were above the range established for irrigation of food and non-food crops according to US Environmental Protection Agency (EPA) [8]; therefore, the increase in the number of filters aimed to increase the HDT and improve the performance of the WWTP, essentially for BOD and the other parameters associated with it.

The WWTP also had a 1,000-liter water tank (to receive and store the WW) and two 150-liter water tanks, one before (influent storage) and after (final TWW storage) the 6th filter, the second tank also served to store the TWW used in the irrigation system. The design of the system was proposed (**Figure 2**) in order to reduce the space requirement.

DW (**Table 1**) used in this research came from SABESP´s Domestic WWTP in Botucatu city (secondary effluent). This WWTP presents a mixed treatment system, composed of equalization, Up-flow Anaerobic Sludge Blanket and activated sludge. The WWTP is responsible for the WW treatment of the urban area of the municipality of Botucatu, which is predominantly DW, with little contribution of industrial discharges.

2.3 Water quality sampling

In Phase II, the research was conducted for 165 days. Samples were collected from the outlet of the influent (IF) storage, TWW released from the filter 3 (TWW 3) and

(a) (b)

Figure 1.
Frontal (a) and lateral (b) view of the WWTP set up at the FCA, UNESP, Botucatu.

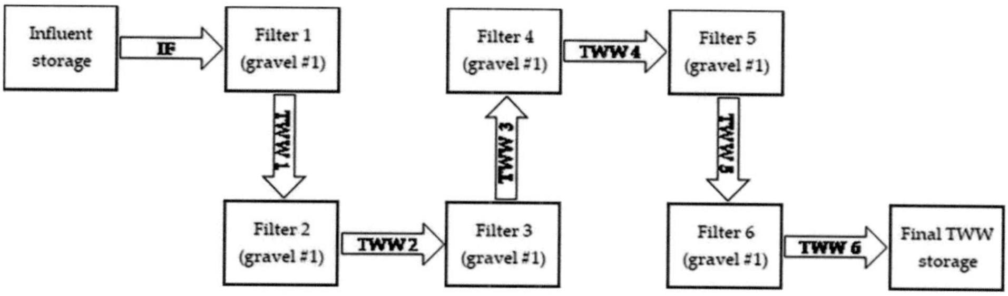

Figure 2.
General scheme of the WWTP (pilot-scale) set up at the FCA, UNESP, Botucatu.

	41 days	70 days	100 days	130 days	165 days	Average (SD)	Stand. for Agric. Reuse*
pH	7.87	8.17	7.5	6.9	7.03	7.49 (±0.62)	6.5–8.4[a]
EC	577	752	693	719.5	717.67	691.83 (±0.06)	≤700[a,d]
TS	389.67	808	1270.83	1085.67	1291.67	969.17 (±377.62)	
TSS	44.33	512.33	920.33	737.33	931	629.07 (±341.19)	<500[a]
TDS	345.33	295.67	350.5	348.33	360.67	340.1 (±25.50)	<450[a]; <500[c]
BOD	48	54	91	87	74	70.81 (±17.82	<10[a]; <30[a]
COD	600.8	445	1251	706.33	1000.8	800.8 (±299.4)	
TN	76.03	65.6	96.7	77.8	84.03	80.03 (±11.13)	<10[a]
TP	10.56	13.76	7.92	13.18	12.68	11.62 (±3.01)	
K	80	71.5	76.9	97.6	84	82.00 (±9.65)	
TC	$2{,}06 \times 10^9$	$2{,}19 \times 10^{10}$	$1{,}99 \times 10^7$	$1{,}35 \times 10^7$	$5{,}17 \times 10^6$	*	
FC	$7{,}42 \times 10^8$	Absent	$1{,}37 \times 10^6$	$2{,}36 \times 10^6$	$3{,}36 \times 10^5$	*	≤1000[b, c]

*EC—electrical conductivity ($\mu S\ cm^{-1}$), TS—total solids ($mg\ l^{-1}$), TSS—total suspended solids ($mg\ l^{-1}$), TDS—total dissolved solids ($mg\ l^{-1}$), BOD—biochemical oxygen demand ($mg\ l^{-1}$), COD - chemical oxygen demand ($mg\ l^{-1}$), TN—total nitrogen ($mg\ l^{-1}$), TP - total phosphorus ($mg\ l^{-1}$), K - potassium ($mg\ l^{-1}$), TC—total coliform (MPN 100 ml^{-1}), FC—fecal coliforms (MPN 100 ml^{-1}). SD: standard deviation. *Standard water quality for agriculture reuse.*
[a]Ref. [8].
[b]Ref. [7].
[c]Ref. [24].
[d]Ref. [25].

Table 1.
Characteristics of wastewater quality from SABESP´s domestic WWTP (influent in this research).

filter 6 (TWW 6). The first sampling was performed 41 days after the beginning of the operation of the WWTP and the other samplings (4) were performed at 30-day intervals.

The WW quality analyses were performed in the Laboratory of Water Quality of the Department of Rural Engineering located at the School of Agronomic Sciences, UNESP, Botucatu.

All WW quality parameters such as total solids (TS), TSS, BOD, COD, total nitrogen (TN), total phosphorus (TP) and microbial population were determined according to the Standard Methods [26]. pH and EC values were measured in situ by digital pH and conductivity meter.

The microbiological indicator adopted for the fecal organisms was the concentration of total coliforms (TC) and *Escherichia coli* (*E. coli*), and its determination was performed using IDEXX Colilert Quanti-Tray System as per the manufacture´s procedures (https: //www.idexx.com/files/colilert-procedure-en. pdf, accessed on 13 June 2021). The most probable number (MPN) was used to find the concentration of TC and *E. coli* in the samples [27], expressed in MPN 100 ml^{-1}.

2.4 Statistical analysis

The results of the performance evaluation of the WWTP were submitted to analysis of variance to verify whether or not the change in the structure of the WWTP in Phase II had significant effects on the quality of the TWW, specifically in the parameters: pH, EC, TSS, BOD, COD, TN, TP and K. The statistical analysis considered two treatments, which comprised sample collection points, identified by: TWW 3 and TWW 6, with three repetitions per sampling, an entirely randomized design with repetitions in time. The variables that showed significant differences were subjected to Tukey's test, at the 5% probability level, with the aid of the Software Sisvar version 5.6 [28].

3. Results and discussion

As shown in the **Table 1**, the quality of the influent used in this research does not meet the standard water quality recommended for reuse in agricultural irrigation.

The performance evaluation of the WWTP was essentially based on the comparison of the physical, chemical, and microbiological characteristics of TWW 3 and TWW 6, with IF. The main findings are presented and discussed below.

During the experiment, the maximum, minimum, and average temperatures were: 26, 16, 20°C, respectively; and the effluent operated in the range of 18–25°C. Thus, it can be stated that the filters operated between temperature ranges considered psychrophilic and mesophilic for microorganisms; the latter is considered by Chernicharo [29] as the one that anaerobic digesters have been commonly designed for because it presents the best results. Regarding the occurrence of rain, during the experimental period, a total of 230.3 mm was recorded, and it is believed that this did not influence the results because it did not coincide with the periods of sample collection. However, it can be recommended to place covers on the filters to prevent the entry of rainwater, and consequently compromise the microbial activity in the beds, which comprise the main agents of decomposition of organic matter.

Table 2 shows the average and standard deviation of the measurements corresponding to each sampling point and provides an overview of the overall performance

	41 days	70 days	100 days	130 days	165 days	Average (SD)
	pH					
IF	7.87	8.17	7.5	6.9	7.03	7.49 (±0.62)
TWW 3	8.1	8.07	7.7	7.7	7.1	7.73 (±0.38)
TWW 6	8.13	8.3	7.8	7.7	7.5	7.89 (±0.30)
	Electrical conductivity (μS cm^{-1})					
IF	577	752	693	719.5	717.67	691.83 (±0.06)
TWW 3	592	721.67	677	711	684.67	677.27 (±0.05)
TWW 6	556.67	674.33	630	595.5	611	613.5 (±0.04)
	Total solids (mg l^{-1})					
IF	389.67	808	1270.83	1085.67	1291.67	969.17 (±377.62)
TWW 3	357.5	377.33	368.67	350.33	345.67	359.9 (±13.04)
TWW 6	346.66	348.66	285	320.33	304.33	321 (±27.38)
	Total suspended solids (mg l^{-1})					
IF	44.33	512.33	920.33	737.33	931	629.07 (±341.19)
TWW 3	7	2.33	9.67	9	2.67	6.13 (±3.23)
TWW 6	4.4	0.33	3.67	1.67	0.33	2.1 (±2.21)
	Total dissolved solids (mg l^{-1})					
IF	345.33	295.67	350.5	348.33	360.67	340.1 (±25.50)
TWW 3	350.5	375	359	341.33	343	353.77 (13.78)
TWW 6	342.17	348.33	281.33	318.67	304	318.9 (27.57)

IF: is influent; TWW: 3 is treated wastewater collected in filter 3; TWW: 6 is treated wastewater collected in filter 6; SD: standard deviation.

Table 2.
Average values of water quality parameters measured at each sampling point.

system for the parameters pH, EC, TS, TSS and TDS. It can be observed that mean concentrations of EC, TS, TSS and TDS decrease as wastewater advances through the sampling points. The results of pH indicate that there was a tendency of increasing as WW advanced through the sampling points, however, they remained within the range considered adequate (pH between 6 and 8) for the anaerobic digestion [29]; they also remained within the adequate range (pH between 6.5 and 8.5) for reuse in crop irrigation [25].

According to the parameters proposed by Ayers and Westcot [25], TWW with EC less than 700 μS cm^{-1} is not restricted from use in irrigation, even for salinity-sensitive crops. Thus, according to these parameters, the EC of the present research presents satisfactory quality for reuse in irrigation without restrictions.

The results of TS, TSS, and TDS, indicate that the WWTP performed satisfactorily in the reduction of solids. For example, an average reduction of 62.9% of ST and 99% of TSS was registered in TWW 3, 67% of TS, and 100% of TSS in TWW 6, concerning the concentration of these in the IF.

The levels of removal of TS and TSS obtained in this research were higher than those observed by Reinaldo et al. [30] evaluating a sewage treatment system consisting of a decant digester with a biological filter followed by constructed wetland and solar reactor, which obtained average efficiencies of removal of TS and TSS of 61 and 99% respectively; and [31] studying treatment system in ANF followed by two constructed wetlands, observed average efficiencies of removal of TS and TSS of 29 and 74% respectively. Fia et al. [31] associated the low efficiency of TS and TSS removal to low temperatures (average of 17.4 ± 2.2°C) recorded during the experimental period, which according to them, kept the fluid viscosity high, which resulted in lower sedimentation velocity of the biomass produced.

Low concentration of TSS observed in this research can be beneficial in rural areas for reuse in agriculture, since the chances of clogging irrigation equipment, such as emitter, would be lower. According to Lamm et al. [32], when the TSS concentration is less than 50 mg l^{-1} there is a lower risk of dripper clogging.

3.1 Chemical and biochemical oxygen demand

The efficiency of the WWTP in the reduction of organic matter can be considered satisfactory; because, an average reduction of BOD and COD was observed in the order of 33.3 and 96.2% in TWW 3, 62.9, and 96.7% in TWW 6 respectively; and registering average values of BOD and COD in TWW 6 of 26.24 mg l^{-1} and 26.78 mg l^{-1} respectively (**Figure 3**).

According to EPA [8], BOD values of irrigation water for crops consumed cooked and raw should not exceed 30 and 10 mg l^{-1} respectively. Since kale is a vegetable that is commonly consumed cooked, it is observed that only TWW 6 meets the required quality standard, with a mean value of 26.24 mg l^{-1}. These results show that the WWTP was not only a filter-in-series but a biological treatment unit as well. One way of demonstrating that the treatment provided by WWTP proposed in this research was not restricted to physical removal by filtration, is the evaluation of the BOD concentrations. de Oliveira Cruz et al. [18] observed COD removal of 41% (lower than our results), and assumed that their anaerobic filter was not restricted to mechanical actions such as fixation, interception, and adsorption, given the fact that the degradation and consumption of soluble material also occurred, by the action of the biofilm present in the interstices of the filled medium, this same behavior was observed by Tonon et al. [21].

Tonetti et al. [33] obtained BOD reduction values higher than 98% studying an alternative WW treatment system consisting of an ANF (with bamboo support material) followed by a sand filter. In this system, they observed that the ANF provided a BOD reduction of 47%. The same authors mention that although ANF usually present good removal efficiency, between 10 and 30% of organic matter is not degraded, which limits that their WW meets the quality standard required by the Brazilian legislation, and this need posttreatment.

Hashem and Qi [34] states that the COD/BOD ratio is an indication of the amount of non-biodegradable organic matter. The higher this ratio is, it is an indication of low organic matter degradation that may be related to the failure of the biological treatment system adopted or the quality of the sewage. The ratio (COD/BOD) with a value of 1.02 recorded in this research is lower than that obtained by Colares and Sandri [35] who obtained 2.08 and observed BOD removal efficiencies in the order of 79.01% evaluating the performance of ST followed by constructed wetlands in the treatment of domestic sewage. Therefore, the low reduction of BOD observed in this research compared to the results found in the literature can be associated with the quality

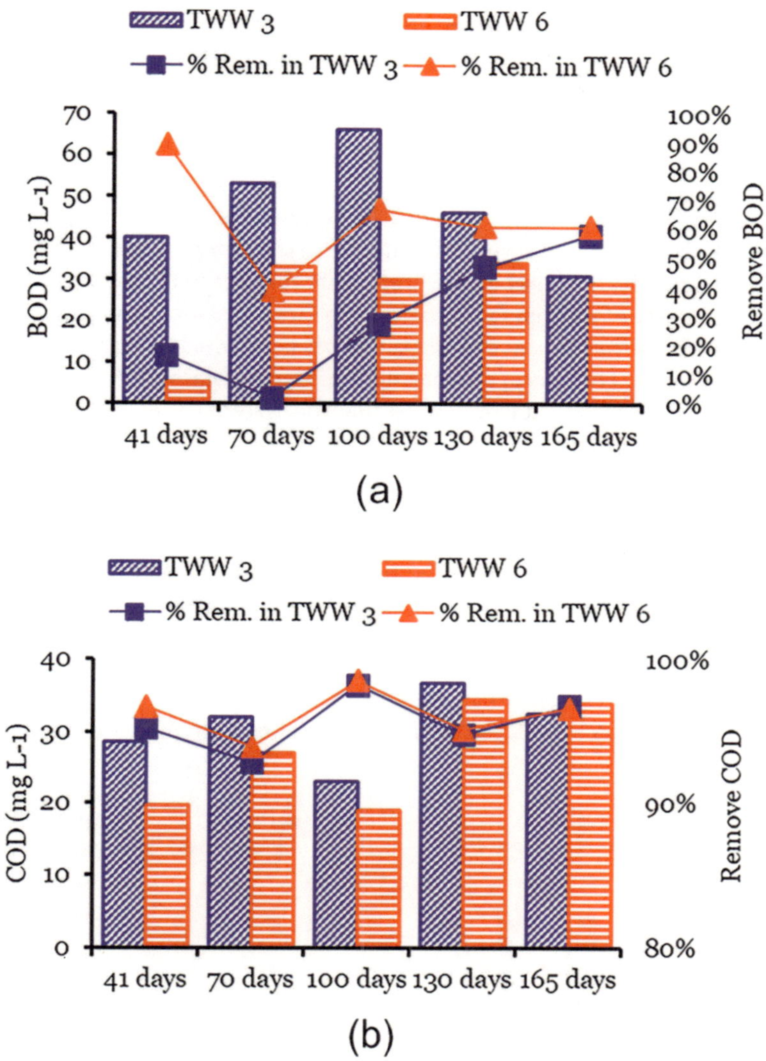

Figure 3.
BOD (a) and COD (b) concentration and remove efficiency comparison during the operation of WWTP.

of the sewage and not necessarily the capacity of organic matter degradation of the WWTP of this research.

Reduction efficiencies of BOD and COD lower than our results were observed by Fia et al. [31], obtained 55 and 35% removal of COD and BOD respectively, operating a treatment system in downflow ANF followed by constructed wetlands systems. These authors, in common, indicated the quality of the raw sewage (low organic load) as the main reason for obtaining unsatisfactory results.

Although the BOD reduction was relatively low compared to what is found in the literature, it is important to mention that on average, the results obtained to meet the [8] requirements for irrigation of cooked consumed crops and the probable range of pollutant removal (represented by BOD) established in NBR 13.969 for ANF [36].

3.2 Total nitrogen, total phosphorus, and potassium

The performance of the WWTP regarding the reduction of TN, TP, and K indicated a removal efficiency of 70,2, 46.5, and 80%, resulting respectively in mean concentrations in the TWW6 of 23.83 mg l^{-1} of total nitrogen (**Figure 4a**), 6.22 mg l^{-1} of total phosphorus (**Figure 4b**) and 16.40 mg l^{-1} of potassium (**Figure 4c**).

Bueno et al. [37] refers that fixed bed ANF or operated under static conditions and without recirculation are commonly unable to remove phosphate and nitrogen compounds efficiently. Bastos et al. [38] refer that the fact that anaerobic treatment occurs in the absence of oxygen is also a limiting factor for nitrogen reduction because nitrification requires the presence of dissolved oxygen, which is only possible in an aerobic environment. Therefore, comparing the results obtained in this research with those found in the literature, it can be seen that the system generated a satisfactory reduction of these elements.

Ucker et al. [39] refer that the removal of nutrients in effluents is more effective in treatment systems that involve the cultivation of plants. These authors, in their research evaluating a CW's system using vetiver grass, they observed efficiencies of phosphorus and ammonia nitrogen reduction around 80.35 and 83.3% in modules with plants and, 44.45 and 42.55% in modules without plants respectively; the removal of phosphorus in modules without plants, was close to that obtained in the present research (46.5%).

The mean value of TP concentration observed during the research is within the range indicated by Bastos et al. [38], which varies between 4 and 12 mg l^{-1}; and TN

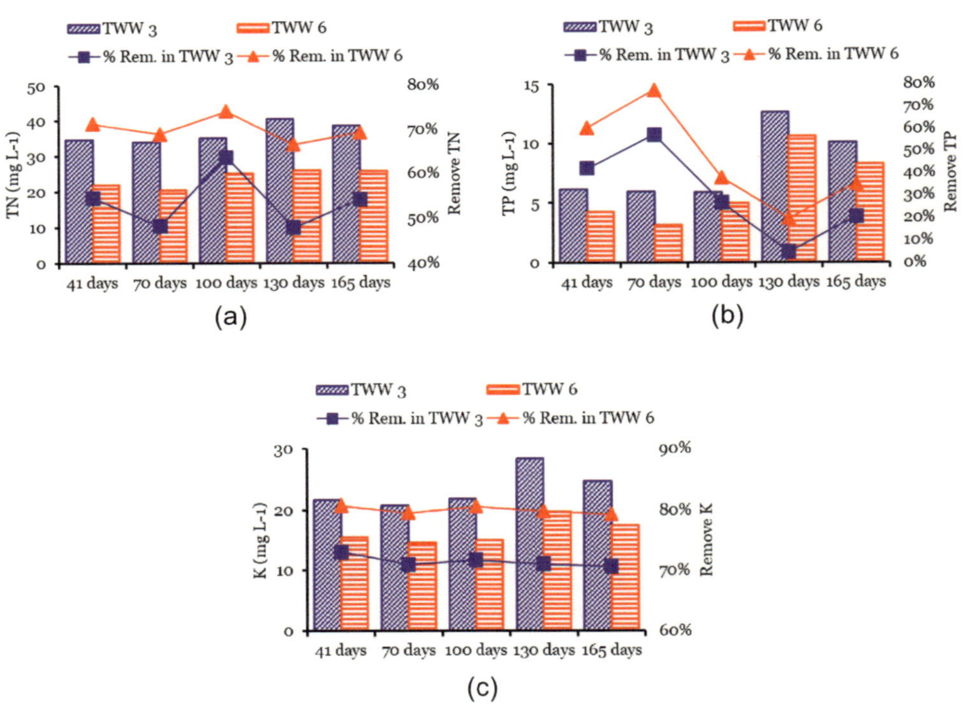

Figure 4.
TN (a), TP (b) and K (c) concentration and remove efficiency comparison during the operation of WWTP.

is slightly above the range indicated by the same authors (35–60 mg l^{-1}), for sewage considered domestic. Comparing the characteristics of TWW 6 to the standards established by Brazilian and international guidelines, it was found that the average concentration of TN is in the range considered of moderate restriction for use in irrigation. Phosphorus exceeded the limits established by CONAMA Resolution 357/05 [24]; and potassium was not the target of comparisons because it is not included in any of the guidelines adopted for this purpose, although it is indicated by them as an element to be considered in analyses to determine the quality of water for irrigation.

3.3 Overall evaluation of the treatment system

The overall evaluation of the performance of the WWTP based on the results of the statistical analysis (**Table 3**) indicates there are significant differences in the quality of TWW 3 and TWW 6 for most of the parameters evaluated, demonstrating that changing the structure of the WWTP, specifically increasing the number of bed filters and consequently increasing the HDT in Phase II had a positive effect on the reduction of pollutants present in the IF. The positive effect of increasing the HDT on the quality of the TWW was also observed by Bouted and Ratanatamskul [14], in which operating isolated ANF in the treatment of influent under different temperatures and HDT (9, 18, and 27 h) obtained greater reductions of TSS, COD, PT and NT in the treatment system with the highest HDT (**Table 4**).

Another fact that should be highlighted is concerning BOD because, in average terms, only TWW 6 with an average BOD of 26.24 mg l^{-1} presented quality suitable for irrigation of food and non-food crops (BOD less than 30 mg l^{-1}) according to EPA [8].

The effect of the changes introduced in the WWTP layout in Phase II is best evidenced by the microbiological characteristics of the TWW; for, although removal efficiencies of TC and *E. coli* around 100% have been recorded in TWW 3 and TWW

Parameters (Units)	*P*-value (*F* test)	MSD	Water quality per filter stage		CV (%)
			TWW 3	TWW 6	
pH (μS cm^{-1})	0.2308 ns	0.16	7.73a	7.89a	4.39
EC (mg l^{-1})	0.0006**	9.09	677.3a	613.5b	6.96
TSS (mg l^{-1})	0.0004**	0.91	6.13a	2.10b	67.2
BOD (mg l^{-1})	0.0000**	1.09	47.16a	26.24b	31.85
COD (mg l^{-1})	0.1091 ns	4.64	30.82a	26.78a	23.25
TN (mg l^{-1})	0.0000**	3.27	36.23a	23.83b	14.75
TP (mg l^{-1})	0.1076 ns	1.97	8.09a	6.22a	43.07
K (mg l^{-1})	0.0000**	1.28	23.43a	16.40b	16.02

*TWW 3: is treated wastewater collected in filter 3; TWW: 6 is treated wastewater collected in filter 6. MSD: minimum significant differences; CV: coefficient of variation. Means that do not share the same lower-case letter in the row are significantly different, by the Tukey's test; ns: not significant at 5% ($P < 0.05$).**Significant at 1% ($P < 0.01$).*

Table 3.
Average values and statistical analysis results of physical and chemical water quality.

	41 days	70 days	100 days	130 days	165 days
Total coliforms (MPN 100 ml^{-1})					
IF	2.06×10^9	2.19×10^{10}	1.99×10^7	1.35×10^7	5.17×10^6
TWW 3	2.16×10^5	3.26×10^6	6.65×10^5	7.82×10^5	1.08×10^6
TWW 6	1.45×10^4	2.76×10^5	1.24×10^3	7.70×10^3	4.35×10^3
E. coli (MPN 100 ml^{-1})					
IF	7.42×10^8	Absent	1.37×10^6	2.36×10^6	3.36×10^5
TWW 3	Absent	1.99×10^5	4.10×10^3	7.40×10^4	Absent
TWW 6	5200	Absent	21	328	97

IF is influent; TWW 3: treated wastewater collected in filter 3; TWW 6: treated wastewater collected in filter 6.

Table 4.
Total coliforms and E. coli concentration comparison during the operation of WWTP.

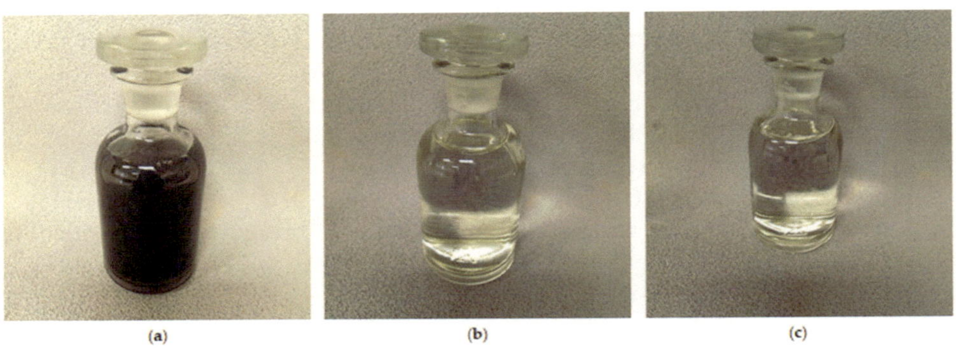

Figure 5.
Photos of water aesthetics (clarity) from different points of samples collected. (a) Sample of influent; (b) Sample of TWW 3; (c) Sample of TWW 6.

6 concerning the IF, only in TWW 6 did it reach acceptable quality standards for reuse in the irrigation of vegetables consumed raw or fruits that develop in contact with the soil, and in their consumption, the outer skin is not removed.

Based on the results presented in **Table 4**, it is observed that the quality standards established for reuse of TWW in irrigation according to World Health Organization (WHO) [40] and CONAMA Resolution 357/05 [24] are only met in the effluent collected in TWW 6, which from the third collection (100 days after the start of operation of the WWTP) showed a population of fecal coliforms below 1000 MPN 100 ml^{-1}; while for TWW 3, the *E. coli* population remained above 1000 MPN 100 ml^{-1} in all samplings performed.

Figure 5 shows a visual indicator of the level of treatment obtained in the ANF WWTP implemented in the experimental area of the Department of Rural Engineering, School of Agronomic Sciences, Botucatu Campus. The figure shows that the samples of TWW 3 already indicated satisfactory levels of solids and organic load reduction.

Good water aesthetics can be a result of low turbidity and TSS concentration, which is important when the aim is reuse. According to de Oliveira Cruz et al. [18], these characteristics favor a more aesthetically pleasing TWW with fewer suspended solids to shelter pathogenic microorganisms, thus, decreasing the amount of chemical to clean it.

4. Conclusions

The WWTP with filter-in-series showed satisfactory performance for most of the parameters evaluated and provided final TWW with acceptable quality for irrigation of cooked consumed vegetables.

The results of the statistical analysis showed that there are significant differences in the quality of TWW 3 and TWW 6, and a lower concentration of salts, solid particles, organic load, and nutrients were observed in TWW 6.

The mean concentrations of the pollutants decrease as wastewater advances through the filters stage, indicating that the increase in the hydraulic retention time improves the performance of the WWTP.

TWW 6 presented good water aesthetics as a result of the low turbidity and TSS concentration, an aspect that is important when the aim is reuse.

The change in the structure of the WWTP, increasing the number of filters from three to six filled with gravel #1, improved the capacity of the proposed system to reduce the concentration of pollutants present in the IF.

It is recommended that the TWW 6 be used for irrigation 100 days after the start of operation of the WWTP proposed in this research because it was only after the third sampling that the final TWW meets the water quality standards for irrigation established by Brazilian and international guidelines.

Acknowledgements

The authors acknowledge the scholarship provided by the Ministry of Science, Technology and Higher Technical Professional Education (Ministério da Ciência e Tecnologia Ensino Superior Técnico Profissional-MCTESTP, Mozambique) and funding provided by the Coordination for the Improvement of Higher Education Personnel (Coordenação de Aperfeiçoamento de Pessoal de Nível Superior—CAPES, Brazil).

Author contributions

V.S.J.P. designed and conducted the experiments and prepared the original manuscript draft. R.M.S.-R. and J.G.T.Q. reviewed the manuscript. V.S.J.P. and T.L.S. collected and analyzed treated wastewater samples. S.A.J. and V.S.J.P. were involved in data preparation and statistical analysis.

Funding

This research was funded by Ministry of Science, Technology and Higher Technical Professional Education (Ministério da Ciência e Tecnologia Ensino

Superior Técnico Profissional-MCTESTP, Mozambique) and Coordination for the Improvement of Higher Education Personnel (Coordenação de Aperfeiçoamento de Pessoal de Nível Superior- CAPES, Brazil).

Conflicts of interest

There are no conflicts to declare.

Author details

Valdemiro Pitoro[1*], Rodrigo Sánchez-Román[2], João Queluz[3], Tamires Da Silva[2], Sérgio Jane[4] and Kevim Muniz[2]

1 Faculty of Agricultural Sciences, Department of Rural Engineering, Lúrio University, Sanga, Mozambique

2 School of Agriculture, São Paulo State University (Unesp), Botucatu, Brazil

3 Environmental Studies Center, São Paulo State University (Unesp), Rio Claro, Brazil

4 Faculty of Agricultural Sciences, Department of Plant Production and Protection, Lúrio University, Sanga, Mozambique

*Address all correspondence to: vpitoro@gmail.com

References

[1] Libutti A, Gatta G, Gagliardi A, Vergine P, Pollice A, Beneduce L, et al. Agro-industrial wastewater reuse for irrigation of a vegetable crop succession under Mediterranean conditions. Agricultural Water Management. 2018;**196**:1-14

[2] Urbano VR, Mendonça TG, Bastos RG, Souza CF. Effects of treated wastewater irrigation on soil properties and lettuce yield. Agricultural Water Management. 2017;**181**:108-115

[3] Ungureanu N, Voicu G, Vlăduț V, Voicu G. Water scarcity and wastewater reuse in crop irrigation. Sustainability (Switzerland). Multidisciplinary Digital Publishing Institute. 2020;**12**:1-19

[4] Cirelli GL, Consoli S, Licciardello F, Aiello R, Giuffrida F, Leonardi C. Treated municipal wastewater reuse in vegetable production. Agricultural Water Management. 2012;**104**:163-170

[5] Balkhair KS. Microbial contamination of vegetable crop and soil profile in arid regions under controlled application of domestic wastewater. Saudi Journal of Biological Sciences. 2016;**23**:S83-S92

[6] Wang Z, Li J, Li Y. Using reclaimed water for agricultural and landscape irrigation in China: A review. Irrigation and drainage. Wiley Online Library. 2017;**66**:672-686

[7] WHO. Guidelines for the Safe Use of Wastewater, Excreta and Greywater. World Health Organization; 2006. p. 114

[8] US Environmental Protection Agency. Guidelines for water reuse. Development. 2012

[9] Del Castillo AF, Garibay MV, Senés-Guerrero C, Yebra-Montes C, de Anda J, Gradilla-Hernández MS, et al. Mathematical modeling of a domestic wastewater treatment system combining a septic tank, an up flow anaerobic filter, and a constructed wetland. Water. 2020;**12**:1-20

[10] Tonetti AL, Coraucci Filho B, Guimarães JR, de Cruz LM, Nakamura MS. Avaliação da partida e operação de filtros anaeróbios tendo bambu como material de recheio (Start up Evaluation and Operation of Anaerobic Filters Filled up with Bamboo). Engenharia Sanitaria e Ambiental. SciELO Brasil. 2011;**16**:11-16

[11] Baettker EC, de Carvalho KQ, Freire FB, Passig FH, de Morais JL. Alternative materials as support medium of anaerobic filters for the treatment of synthetic sanitary sewage. Engenharia Sanitaria e Ambiental. SciELO Brasil. 2018;**23**:1091-1102

[12] Mattos L, Cruz DO, Graziella B, Araujo L, Tonetti AL. Using coconut husks in a full-scale decentralized wastewater treatment system: The influence of an anaerobic filter on maintenance and operational conditions of a sand filter. 2019;**127**:454-459

[13] Lohani SP, Khanal SN, Bakke R. A simple anaerobic and filtration combined system for domestic wastewater treatment. Water. 2020;**3**:41-45

[14] Bouted C, Ratanatamskul C. Effects of temperature and HRT on performance of a novel insulated anaerobic filter (IAF) system incorporated with the waste heat input for building wastewater treatment. Journal of Environmental Management. 2018;**206**:698-706

[15] Wu H, Zhang J, Ngo HH, Guo W, Hu Z, Liang S, et al. A review on the

sustainability of constructed wetlands for wastewater treatment: Design and operation. Bioresource Technology. 2015;**175**:594-601

[16] Parde D, Patwa A, Shukla A, Vijay R, Killedar DJ, Kumar R, et al. A review of constructed wetland on type, treatment and technology of wastewater. Environmental Technology & Innovation. 2021;**21**:101261

[17] da Silva JAR, Carvalho Terra AB, de Assis C, Florentino LA, Putti FF. Treatment of wastes in Brazil: A comparison bet ween compost techniques and anaerobic biodigesters. Revista em Agronegocio e Meio Ambiente. 2020;**13**:797-817

[18] de Oliveira Cruz LM, Gomes BGLA, Tonetti AL, Figueiredo ICS. Using coconut husks in a full-scale decentralized wastewater treatment system: The influence of an anaerobic filter on maintenance and operational conditions of a sand filter. Ecological Engineering. 2019;**127**:454-459

[19] Tripathi VK, Rajput TBS, Patel N, Kumar P. Effects on growth and yield of eggplant (Solanum Melongema l.) under placement of drip laterals and using municipal wastewater. Irrigation and Drainage. 2016;**65**:480-490

[20] Abegunrin TP, Awe GO, Idowu DO, Adejumobi MA. Impact of wastewater irrigation on soil physico-chemical properties, growth and water use pattern of two indigenous vegetables in southwest Nigeria. Catena. 2016;**139**:167-178

[21] Tonon D, Tonetti AL, Filho BC, Augusto D, Bueno C. Wastewater treatment by anaerobic filter and sand filter: Hydraulic loading rates for removing organic matter, phosphorus, pathogens and nitrogen in tropical countries. Ecological Engineering. 2015;**82**:583-589

[22] Liu W, Song X, Huda N, Xie M, Li G, Luo W. Comparison between aerobic and anaerobic membrane bioreactors for trace organic contaminant removal in wastewater treatment. Environmental Technology & Innovation. 2020;**17**:100564

[23] Pitoro VSJ, Lima TDS, Román RMS, Queluz JGT. Wastewater treatment in vertical anaerobic filters for application in drip irrigation. Irriga. 2018;**1**:15-28

[24] AMBIENTE-CONAMA CNDOM. Resolução CONAMA n°357/2005, de 17 de março de 2005: Dispõe sobre a classificação dos corpos de água e diretrizes ambientais para o seu enquadramento, bem como estabelece as condições e padrões de lançamento de efluentes, e dá outras providências. Diário Oficial da União (DOU). 2005

[25] Ayers RS, Westcot DW. Water Quality for Agriculture. Rome, Italy: Food and Agriculture Organization of the United Nations Rome; 1985

[26] APHA, AWWA, WEF. Standard Methods for Examination of Water and Wastewater. 22nd ed. Washington: American Public Health Association. Standard Methods; 2012

[27] EPA MPN Calculator. 2021. Available from: https://mostprobable numbercalculator.epa.gov/mpnForm

[28] Ferreira DF. Sisvar: A computer analysis system to fixed effects split plot type designs. Revista Brasileira de Biometria. 2019;**37**:529-535

[29] de Chernicharo L. Reatores anaeróbios. Vol. 5. Belo Horizonte: Departamento de Engenharia Sanitária e Ambiental—UFMG; 1997. p. 380

[30] Reinaldo GPB, Batista RO, Silva PCM, Lemos Filho LC, Ferreira Neto M, Santos DB. Desempenho de sistema decanto-digestor com filtro biologico seguido por alagado construido e reator solar no tratamento de esgoto domestico (Performance of digester decant system with biological filter followed by constructed wetland and solar reactor). An Interdisciplinary Journal of Applied Science. 2012;7:62-74

[31] Fia R, De Matos AT, Lambert TF, Fia FRL, De Matos MP. Tratamento das águas do processamento dos frutos do cafeeiro em filtro anaeróbio seguido por sistema alagado construído: II—remoção de nutrientes e compostos fenólicos (Wastewater treatment of coffee fruit processing in anaerobic filter system followed). Engenharia Agricola. Associação Brasileira de Engenharia Agrícola. 2010;**30**:1203-1213

[32] Lamm FR, Ayars JE, Nakayama FS. Microirrigation for Crop Production: Design, Operation, and Management. Amsterdam: Elsevier; 2006

[33] Tonetti AL, Coraucci Filho B, Stefanutti R, Figueiredo RF. O emprego do filtro de areia no pós-tratamento de efluente de filtro anaeróbio (The use of sand filter in the post- treatment of anaerobic filter effluent). Sanare Revista Técnica da Sanepar, Curitiba. 2004;**21**:42-52

[34] Hashem MS, Qi X. Treated Wastewater Irrigation—A Review. Vol. 13. Multidisciplinary Digital Publishing Institute; 2021. p. 1527

[35] Colares CJG, Sandri D. Eficiência do tratamento de esgoto com tanques sépticos seguidos de leitos cultivados com diferentes meios de suporte (Efficiency of sewage treatment with septic tanks followed by constructed wetlands with different support materials). Revista Ambiente e Agua. 2013;**8**:172-185

[36] ABNT. NBR 13969: Tanques sépticos—Unidades de tratamento complementar e disposição final dos efluentes líquidos—Projeto, construção e operação (Septic Tanks—Complementary Treatment Units and Final Disposal of Liquid Effluents – Design, Construction and Operation)1997. p. 60

[37] Bueno RF, Piveli RP, Campos F, Sobrinho PA. Simultaneous nitrification and denitrification in the activated sludge systems of continuous flow. Environmental Technology. 2018;**39**:2641-2652

[38] Bastos FS, von Sperling M, Haandel AC Van, Machado LCGT, Monteggia LO, Volschan I, et al. Nutrientes de esgoto sanitário: utilização e remoção (Nutrients from sanitary sewage: Utilization and removal). Projeto PROSAB. 2009;1:428

[39] Ucker FE, Almeida RA, Kemerich PDC. Remoção de nitrogênio e fósforo do esgoto sanitário em um sistema de alagados construídos utilizando o capim vetiver (Removal of nitrogen and phosphorus from wastewater in a constructed wetland system using vetiver grass). An Interdisciplinary Journal of Applied Science. 2012;7:87-98

[40] World Health Organization. Safe Use of Wastewater, Excreta and Greywater Guidelines for the Safe Use of. 2006

Chapter 12

Theoretical Approaches to Water Use Optimization for Rice Irrigation Systems in the Lower Kuban

Alina Buber, Yuri Dobrachev, Alexander Buber and Evgenii Ratkovich

Abstract

The object of research is irrigation systems that are hydraulically connected to a large specialized agricultural complex for rice production, located in the lower part of the Kuban River basin and experiencing an acute shortage of water resources. In the last decade, rice irrigation area decreased by 50%. Calculation technology is proposed for water resources management of the Lower Kuban basin basing on integrated use of six models developed in Russia: hydrological and meteorological forecasts; simulation models of yield formation to calculate crops water demand; water balance models to develop water use schedules for irrigation systems; hydrodynamic models to calculate water use scenarios; optimization models to choose a trade-off Pareto management options; and statistical models to calculate yield losses depending on water resources deficit. This technology allows adopting optimal trade-off solutions online with the Lower Kuban water resources management in interests of agriculture.

Keywords: water resources management, rice irrigation systems, water scarcity, simulation modeling, the Kuban River

1. Introduction

Over the long period of operation of the country's largest Nizhnekubansky water management complex, the operational characteristics of a significant part of its hydraulic structures (HS) have changed significantly in comparison with the design ones, and some of them have lost their management functions. The need to consider the current technical HS condition is due to the fact that they are the main structural and functional elements of an integral system of the reclamation and water management complex designed to provide water resources to 215 thousand hectares of irrigated rice lands. The reclamation water management complex of the Lower Kuban includes, in addition to the Krasnodarsky one, the Kryukovskoye, Varnavinskoye, and Shapsugskoye reservoirs, the Fedorovsky retaining and Tikhovsky water distribution waterworks, more than 100 functioning pumping stations and over 550 HS, more than 3 thousand km of irrigation and discharge channels, up to 156 thousand hectares of natural fishery reservoirs [1].

Irrigation systems and hydraulic structures of the agricultural complex of the Lower Kuban are shown in **Figure 1**.

The Krasnodarskoye reservoir is the main source of irrigation and performs a number of other important functions: population protection from floods, high water control with discharges up to 1200 m^3/s. The main function of the reservoir is to provide water for 12 rice irrigation systems, the needs of fisheries, and navigation conditions for shipping. As a result of siltation, the capacity of the Krasnodarskoye reservoir decreased, and the value of the temporary operation level dropped below the design by 90 cm. Failures in the technical condition are also noted at other reservoirs, waterworks and HS, including pumping stations, and water supply and drainage channels, which negatively affect the water supply of rice systems. Reduction of the total volume of flood water in reservoirs, which previously, according to design parameters, made it possible to avoid a shortage of irrigation water at the most stressful time in terms of agrometeorological conditions. The low efficiency of channels, pumping stations, and other HS as a result of their unsatisfactory technical condition are also the cause of water scarcity [2].

In recent years, the area of rice crops in the Krasnodar Territory has grown to 130 thousand hectares. The average yield in the region is about 7 t/ha, with a total water intake for rice irrigation of more than 2.5 km^3 [3]. An increase in air temperature and a decrease in precipitation in the rice-growing area during its growing season, associated with climate change in the Kuban, lead to an increase in irrevocable water consumption. High irrigation rates of rice, reaching an average of 20 thousand m^3/ha or more, on saline soils, additionally cause a shortage of irrigation water.

The current state of the water management complex, characterized by transformation and degradation of operational characteristics, is mainly due to the processes of wear of engineering systems and their destruction under the influence of natural

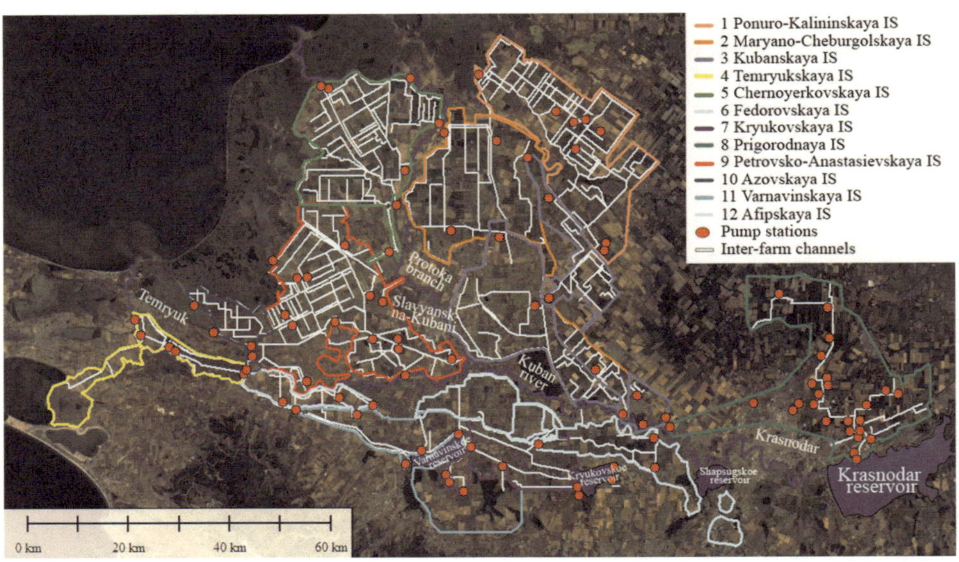

Figure 1.
Irrigation systems and hydraulic structures of the Lower Kuban.

and industrial factors, it is impossible to return to the previous design parameters. In the current situation, one of the productive ways to reduce risks and damages from systematically observed water scarcity is not only an obvious direct opposition to degradation factors, repair and replacement of HS, reduction of rice acreage, new rice varieties resistant to water scarcity, and low-volume irrigation methods, but also a flexible management system for resource distribution and use. Adaptive properties of the rice agro complex, as an integral result of the agricultural and agro-industrial (agricultural producers) complexes' water users operational and production activities can be achieved with the management system organically combining strategic and tactical goals. First of all, these are combined into single regional and local levels of monitoring the elements state of the rice agro complex and the distribution of water resources, static long-term, and real-time dynamic operation modes of water management and hydro-reclamation systems elements.

1. The strategic level involves the implementation of seasonal planning for the distribution of water resources in form of the most probable management scenarios, using information on rice acreage and their spatial distribution, on the technical state and characteristics of the HS for water intake and transportation to consumers and forecast data on the water content of surface and groundwater sources. Thus, already at the planning stage, water management activities are closely connected with the plans for rice-growing farms. The formation, selection, and analysis of the most suitable scenarios for water management for the upcoming cycle of agricultural production are implemented using hydrodynamic modeling of water flows in the river basin area within the boundaries of the agro-reclamation water management complex [4].

2. The tactical approach provides optimization of water use during the irrigation period basing on forecast hydrological and meteorological models in order to maintain optimal water consumption in conditions of limited water resources. The flexibility of management is achieved through the use of calculations in the operational mode of water distribution scenarios based on a hydrodynamic model, considering the requirements of water users and the available volume of water (inflow and water supply); simulation models of crop yield formation, allowing to calculate the needs of crops in irrigation water, to forecast yields considering restrictions on water resources; models of water balance of irrigation system fields for the formation of irrigation schedules and water distribution for irrigation of crop rotations within the farm; multifactorial statistical models for assessing production losses of gross yield in low-water years.

Calculations done periodically using this tool allow you to quickly respond to the supply of agricultural producers with water, adjusting the irrigation regime during the vegetation period, as well as through local redistribution and adjustment of water use. The analysis of the results obtained using the above approach at the end of the irrigation season following the harvest results will make it possible to form a water use schedule for the next year adapted to the expected environmental conditions, making the management system trainable.

The need for such an analysis arose due to the fact that, on the one hand, water demands are formed by agricultural producers, and their adjustment and implementation are carried out by the Federal State Budgetary Institution "Kubanmeliovodkhoz" management department, while the indicators of the operation service are rice yields and gross output from the entire territory of the reclamation and water management complex, forcing to control water consumption. On the other hand, water demands by agricultural producers are based on the principle of "from what has been achieved," i.e., the more, the easier it is to carry out the technological process of cultivation, which, on the contrary, creates conditions for increasing the risk of developing water scarcity.

2. Materials and methods

The following set of initial data was used to analyze the water supply and water use of agricultural producers of the Kuban reclamation agro complex based on the water balance of rice irrigation systems lands:

1. Production data of the Federal State Budgetary Institution "Kubanmeliovodkhoz" management department on water resources use (Form 1-BX) with information on water intake from the Kuban River basin, including reuse, on water supply for irrigation of rice and non-rice fields, on watering of rivers and reservoirs, efficiency of canals, on acreage of rice and other crops, and on the volume of drainage runoff.

2. Retrospective, current, and forecast hydrometeorological data of the Kuban River basin.

3. Maps of land use, soils, and groundwater levels, as well as schemes of water supply and disposal.

4. Satellite images, digital terrain models, and topographic maps.

5. Field observations and data from literary sources.

In addition, the following computer programs and models were used:

The river inflow forecast for the estimated period was carried out according to a special model using standard data from the Hydrometeorological Center. A retrospective series of hydro and meteorological elements is selected from the database, for which the recorded inflow is close to the current and forecast values of the Hydrometeorological Center. The selected data series are adjusted by normalizing coefficients and then used as input information for calculating daily forecast values on the same model. The algorithm for forming a meteorological forecast is similar.

Crop yield models for calculating the daily parameters of water exchange of agricultural lands are similar to those presented in [5]. Based on the results of numerical experiments, water consumption schedules for rice, corn, alfalfa, and wheat during irrigation and on rainfed lands were formed. Evaporation from the water surface of rice paddy fields and water objects was calculated according to the Ivanov formula [6].

The dynamics of the water balance of a structurally heterogeneous agricultural landscape were calculated with a monthly time step according to a simplified scheme for individual landscape elements and soil layers using traditional hydrological methods. The entire territory of the considered agricultural landscape was divided into separate elements by land category, type of land use (irrigated, rainfed), and water objects (canals). For each element of the agricultural landscape, the monthly values of the water balance components were calculated according to formula 1.

$$\Delta S = P + M + \Phi + p - E - Д_p + (\overline{\Pi} - \overline{O}) + (\underline{\Pi} - \underline{O}) \tag{1}$$

where ΔS is the change in moisture reserves for the $(S_2 - S_1)$ period; P is the meteorological precipitation; M is the irrigation rate; Φ is the filtration losses; p is the water exchange of subsoil waters and underground water basin; E is the evapotranspiration; $Д_p$ is the drainage flow; $(\overline{\Pi} - \overline{O})$ is the surface inflow and outflow of water; and $(\underline{\Pi} - \underline{O})$ is the inflow and outflow of groundwater.

Meteorological data used to calculate the water balance components are obtained from the database and forecast calculations. The water balance of each particular element of the agricultural landscape was adjusted in accordance with the interaction with neighboring elements using an assessment of the lateral inflow of groundwater.

The efficiency of irrigation water transportation from the intake to the rice fields through inter-farm and intra-farm channels was calculated considering their design, length, and efficiency. In addition, the efficiency of the channels was assessed based on the calculation of the channels' water balance as a water object. The channels' water balance considered evaporation from the water surface and vegetation cover of the slopes, as well as filtration losses of lateral outflow to neighboring landscape elements adjacent to the channels.

Both in case of the implementation of the optimal (calculated) version of water resources distribution between water users plan, and in accordance with the demands submitted by water users, it is necessary to have a number of hydrological and water management conditions that ensure the possibility of the waterworks functioning and the required volume of accumulated and incoming water. Production reports on water resources use of the Lower Kuban formed the basis of a water use scheme that displays the level and discharge conditions at the water intake points from the river Kuban and its branches (Kuban and Protoka) which are necessary for the functioning of rice irrigation systems (**Figure 2**).

For this scheme, many possible water distribution scenarios were formed with priorities and requirements for various water users and rice irrigation systems, including unscheduled releases from the Krasnodarskoye reservoir. Simulation of scenarios was performed on the hydrodynamic model of the Kuban River in the MIKE 11 software using the "Control structure" module [4].

With the help of a hydrodynamic model, water use scenarios were used to calculate water levels and discharges in controlled section lines and assess possible water scarcity for irrigation systems and reservoirs' drawdown. A matrix of water distribution strategies was constructed from the solutions obtained.

The matrix of solutions obtained in this way was analyzed in the Pareto Front Viewer software [7, 8] to form a set of optimal solutions in the sense of Edgeworth-Pareto by the beginning of negotiations between the decision-maker and water users. The choice of a particular solution is achieved by a compromise method as a result of discussing possible options.

The economic assessment of the chosen solution was carried out on a statistical model of the regional level (branch), designed to calculate crop losses depending on the acreage, weather conditions, crops' vegetation phase, and water scarcity. Production data for the last 10 years were used to develop a statistical model.

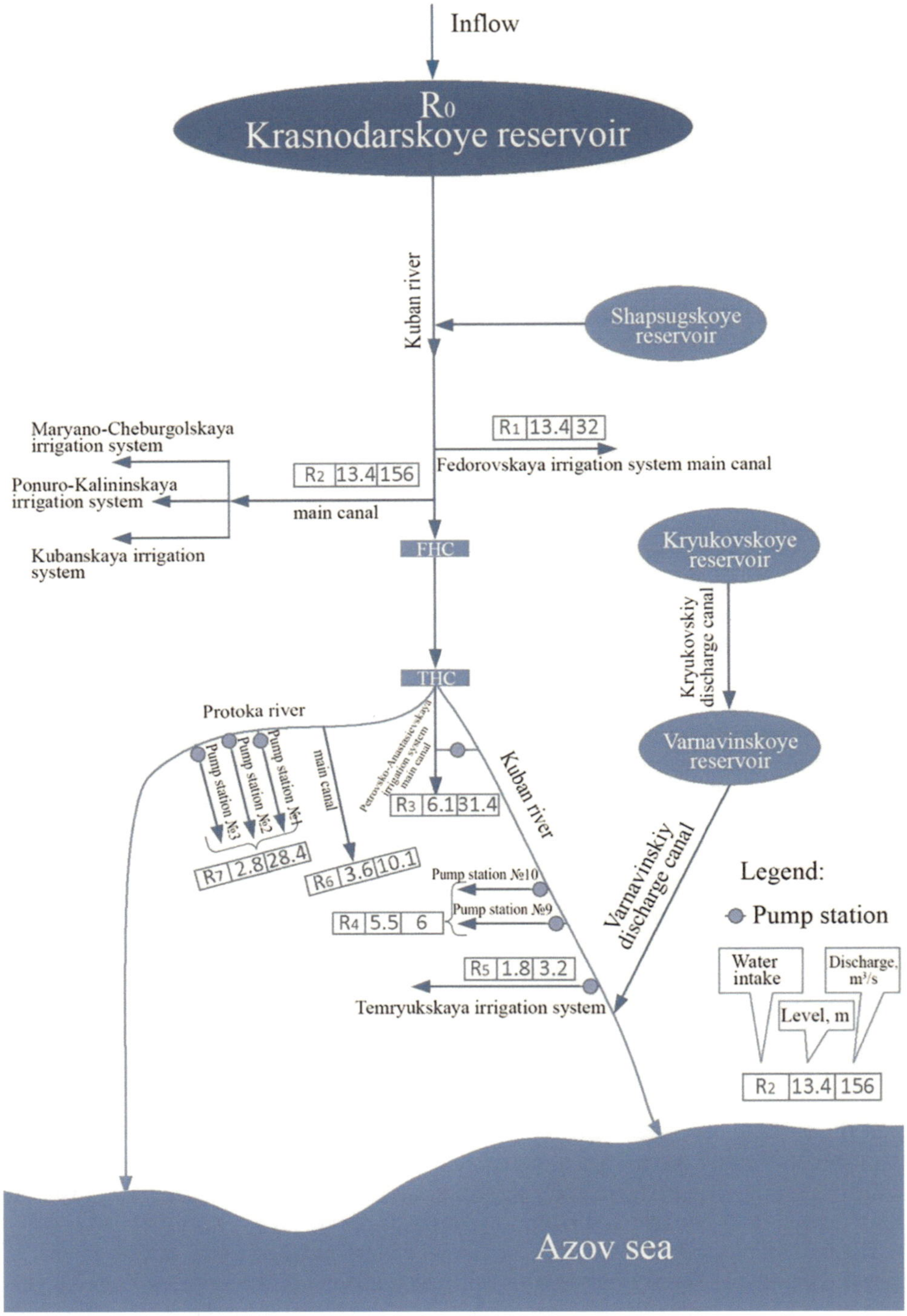

Figure 2.

Water use scheme of the lower Kuban. R0 – Krasnodarskoye reservoir; R1 – Fedorovskaya irrigation system; R2 – Kubanskaya, Ponuro-Kalininskaya and Maryano-Cheburgolskaya irrigation systems; R3 – Petrovsko-Anastasievskaya irrigation system; R4 – Pump stations No. 9 and No. 10 of Petrovsko-Anastasievskaya irrigation system; R5 – Temryukskaya irrigation system; R6 – Main canal of Chernoyerkovskaya irrigation system; R7 – Pump stations No. 1, No. 2, and No. 3 of Chernoyerkovskaya irrigation system.

3. Results and discussion

As a test site, three irrigation systems of the Kalininsky branch were selected, the total area of the agricultural landscape of which is about 150 thousand hectares. Rice irrigation systems' area is 16 thousand hectares, irrigated lands on non-rice crops have an area of more than 17 thousand hectares, and the array of rainfed agriculture area is 75 thousand hectares. The total volume of water supply to the agricultural landscape of the Kalininsky branch in 2018 is more than 500 million m^3.

The rice paddy fields of the branch, located on the territory of three irrigation systems (Ponuro-Kalininskaya, Kubanskaya, and Mariano-Cheburgolskaya), were divided into four groups, depending on their affiliation to the water outlet. Water supply from the Kuban River comes to the water outlet through main channels. The characteristics of the irrigation systems' outlets of the Kalininsky branch according to 2018 data are presented in **Table 1**.

Drainage flow disposal is provided by 4 pumping stations to the Kirpilsky Estuary and further to the Sea of Azov.

The water capacity of rice paddy fields at the beginning of the growing season was estimated by the relief, hydrophysical parameters of soils of different categories, the groundwater level, and the amount of precipitation accumulated during the winter period. The calculated values of evaporation from the rice paddy field surface and filtration were adjusted based on the results of hourly field observations conducted during the expedition of young scientists to the Kalininskoye LLC. Meteorological and soil parameters were monitored using the Davis Vantage Pro 2—automatic weather station—and the Veles-VP—soil humidity and temperature station. The map of the surface slopes of the test area was formed according to SRTM data.

A qualitative assessment of the water balance calculations result for the vegetation months was carried out by checking the fulfillment of Eq. (2) applied to the agricultural landscape as a closed system that is part of the catchment area. The values of the agricultural landscape components balance are given in **Table 2**.

$$P - Qs - Qu - E = N \tag{2}$$

where P is the water intake + precipitation; Qs is the drainage outflow; Qu + E is the filtration + evapotranspiration; and N is the groundwater and soils moisture capacity.

The high correlation, as well as the permissible variation of the residual error in fulfillment of the equality condition, indicates the applicability of the models and calculation methods used, as well as the correctness of the display of hydrological

No.	Irrigation system	Outlet	Rice sowing area, thousand ha	Water supply volume, million m^3
1	Ponuro-Kalininskaya	МПК	7	270
2	Kubanskaya	A-2	1,5	20
3	Mariano-Cheburgolskaya	P-3	2	90
4		P-4-3	1,3	18

Table 1.
The characteristics of the irrigation systems' outlets of the Kalininsky branch.

Month	P, mm	Qs, mm	Qu, mm	E, mm	N, mm
April	360	1022.2	63.4	81.1	391.2
May	2656	737.4	1561.2	133.8	402.5
June	2573	740.7	1444.5	178.1	333.3
July	2386	430.2	1609.0	215.7	356.2
August	2080	129.7	1451.0	145.2	306.7
September	145	251.4	260.3	27.4	321.7

Table 2.
Monthly values of the main agricultural landscape water balance components of the Kalininsky branch.

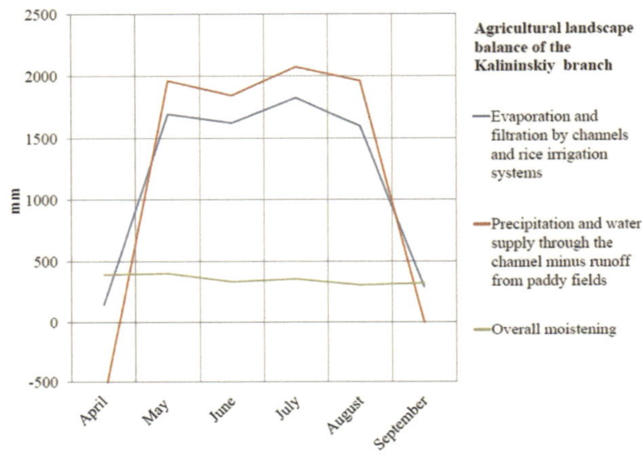

Figure 3.
Dynamics of the agricultural landscape water balance components of the Kalininsky branch for the vegetation period of 2018.

processes within the boundaries of the considered territory when solving the task (**Figure 3**). The results of water balance calculations were used to analyze and evaluate the effectiveness of water resources management in the region.

The calculated data compared with actual water supply volume data for the vegetation period are presented in **Table 3**.

Water supply failure was observed in July, August, and September, which, according to our calculations, could lead to a decrease in rice yield from the maximum by 7% (5.8 c/ha), which is lower than the average annual value.

The demand for water intake from the Kuban River formed on the basis of water balance calculations is a reasonable requirement of the water user. However, in addition to these requirements for the volume and discharge of water, gravity water intakes also need to meet the conditions on the water level in the water source. Such a requirement for the Kalininsky branch (one of the R2 water users) is the water level in the upstream of Fedorovsky waterwork, maintained at 13.4 m mark.

For the subsequent simulation of water distribution scenarios in the MIKE11 software, using the "Control structure" module, a hydrodynamic model of the Kuban River was formed, considering lateral tributaries, reservoirs, and waterworks. The model was calibrated for 7 stream gauges, the error was up to 5 cm, which indicates

Month	Irrigation rate, m³/ha		Wetting demand, million m³	Actual supply, million m³	Water supply failure, million m³
	Rice crops	Non-rice crops			
May	1200	309	24.6	79.8	
June	4300	444	83.1	101.1	
July	5200	457	99.4	72.6	26.8
August	5500	841	114.6	112.9	1.7
September	3700	406	65.5	59.1	6.4
Total	19900	2457	387.2	425.5	

Table 3.
Water supply schedule for Kalininsky branch on 2018.

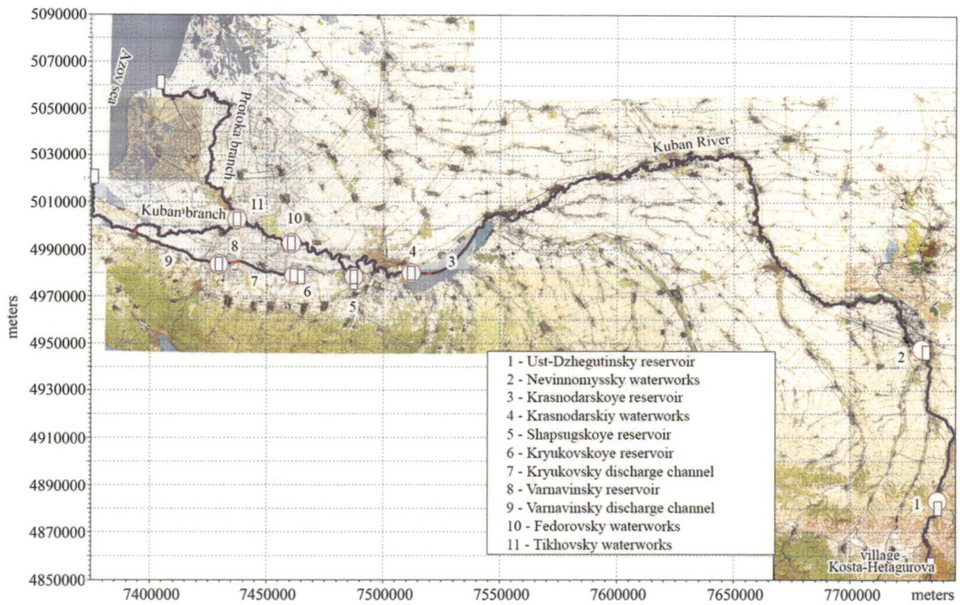

Figure 4.
Scheme of river network for hydrodynamic model.

adequate operation and the possibility of its use for scenario calculations [4]. The inflow to the Krasnodar reservoir was set according to the calculation on the hydrological forecast model. The initial data and boundary conditions for water intakes and hydraulic structures are established in accordance with the developed scenarios of water use. The scheme of the river network calculated on the hydrodynamic model in the MIKE11 is shown in **Figure 4**.

River network plotted from village Kosta Khetagurov to the Azov Sea:

- river network on river bed of the Kuban River and Kuban branch —778 km;
- river network on river bed of the Protoka branch —135 km;

- Kryukovskoye discharge canal —22 km;
- Varnavinskoye discharge canal —37 km;

Hydrodynamic model includes 195 cross sections with interval ≈ 5 km;

Reservoirs were defined by the bathygraphic function;

Boundary conditions: releases from reservoirs and waterworks, water intake by pump stations and canals, water level of the Azov Sea, lateral inflow from right and left banks.

In MIKE11, the control strategy describes the function of the gate level depending on the water level or discharge value at the controlled point. Using the "if" operator for the selected gates, it is possible to form a given management strategy, depending on water consumption, time, requirements of water users, etc. When selecting a management option, you can configure MIKE11 so that the choice is possible only between options representing different strategies. These management strategies are set using a list of "if" operators that allow you to implement a hierarchy of priorities and requirements of water users.

To form a set of alternative plans (solutions), a list of different options should be formed, usually associated with lexicographic ordering (groups of water users with the highest priority, normal priority, less significant, insignificant, etc.). The control command assigned to the first execution corresponds to the first "if" operator with a "true" condition. Therefore, it is important for the user to determine which "if" operator will be evaluated first, second, and third in the specified hierarchy. The "Control Structure" module also provides for the operation mode of the hydraulic unit according to the dispatcher schedule.

For each water user, on the basis of demands (irrigation schedules) during the vegetation period, the volumes and modes of water intake are determined based on their water requirements (nominal value and permissible "cut" or minimum level of water intake operation) in accordance with the current volume of water consumption and considering the long-term plan. The requirements are set by the level and expenditure functions from time for the particular section lines of the Kuban riverbed, branches Kuban and Protoka, and Krasnodar reservoir.

For the entire vegetation period (April-September of the current year), the water use schedule is formed in accordance with the data of the Hydrometeorological Center and the Construction Norms and Specifications-33-101-2003 daily hydrological series of flow, based on water management calculations.

The hierarchy of priorities is set in the hydrodynamic model in the "Control structure" module, which, by iteration based on the PID algorithm, allows you to select control at waterworks (Krasnodarskoye reservoir, Fedorovsky waterworks, and Tikhovsky water divider).

Depending on the method of managing releases at waterworks, calculations were made for three groups of scenarios, with priorities for different water users:

1. The management of the Krasnodar reservoir ($У_{кгу}$) is aimed at maintaining the required level, at which the necessary conditions for the operation of pumping stations that take water for water users are realized: —R1—R7

The management of releases of the Tikhovsky waterwork ($У_{тгу}$) is aimed at meeting the requirements of water users located on the branches Kuban and Protoka. The distribution of water flow between the arms of the Kuban and the Protoka was set in the required proportions (54% —Protoka, 46% —Kuban; 40% —Protoka, 60% —Kuban; 60% —Protoka, 40% —Kuban).

In the management scenario, the discharges of three water users —R1—R3 were cut by 50% in different variations, based on the consideration of reducing water supply by reducing acreage in farms.

Scenario 01: [Укгу-R1-R2]_[Уфгу-R1-R2]_[Утгу-54]_[R3-R7 – leftover principle], where: Укгу-R1-R2 – determines the Krasnodarsky waterwork management to provide required discharges, maintaining the required level of the main channel, the water level in which ensures the normal operation of irrigation systems R1 and R2; Уфгу-R1-R2 determines Fedorovsky waterwork management to maintain required water level in R1 and R2; and Утгу-54 determines the Tihovsky waterwork management so that the accepted ratio of the water distribution between the Kuban and the Protoka is maintained in the following proportions: 54% —the Protoka and 46% —the Kuban. The requirements of other irrigation systems R3–R7 are satisfied according to the leftover principle.

2. Krasnodarskoye reservoir management was carried out to maintain the specified discharges of 405, 340, 220, 175 и 90 m^3/s

Fedorovsky waterwork management was carried out to maintain the required upstream level (13,4 m), when normal water intake was provided for water users R1 and R2, with discharges of 32 m^3/s and 143 m^3/s.

The releases of the Tikhovsky waterwork were managed for discharge distribution between the Kuban and Protoka branches in the following proportions: 54% – the Protoka, 46% – the Kuban; 50% – the Protoka, 50% – the Kuban; 40% – the Protoka, 60% – the Kuban; 60% – the Protoka, 40% – the Kuban. The required distribution proportions were set for each discharge from the series (405, 340, 220, 175, and 90 m^3/s), for which, as well as for different proportions of water distribution between the branches, the water intakes of three water users —R1—R3 were also cut by 50% in different variations, based on considerations of reduction acreage in farms.

Scenario 57: [Укгу-405]_[Уфгу-R1-R2]_[Утгу-54]_[R3-R7 – leftover principle], where: Укгу-405 determines the Krasnodarsky waterwork management to maintain water discharge in controlled section line more ore equal to 405 m^3/s; Уфгу – R1-R2 determines Fedorovsky waterwork management to maintain required water level in R1 and R2; and Утгу-54 determines the Tihovsky waterwork management so that the accepted ratio of the water distribution between the Kuban and the Protoka is maintained in the following proportions: 54%—the Protoka and 46%—the Kuban. The requirements of other irrigation systems R3-R7 are satisfied according to the leftover principle.

3. Krasnodarskoye reservoir management was carried out to maintain the required upstream level in R6

Fedorovsky waterwork management was carried out to maintain the required upstream level (13,4 m), when normal water intake was provided for water users R1 and R2, with discharges of 32 m^3/s and 143 m^3/s.
Management of the Tikhovsky waterwork was carried out in the interests of water user R6.

Scenario 41: [Укгу-R6]_[Уфгу-((R1-R2)/2)]_[Утгу-R6]_[R3-R5, R7 – leftover principle], where: Укгу-R6 determines the Krasnodarsky waterwork management to provide required water level on R6; Уфгу-((R1-R2)/2) determines Fedorovsky waterwork management to maintain required water level in R1 and R-2, on R1 and R2 discharge provided half less than required; and Уттy-46 determines the Tihovsky waterwork management to maintain required water level in R6 and distribution between the Kuban and the Protoka in the following proportions: 46% и 54%, respectively. The requirements of other irrigation systems R3–R5 and R7 are satisfied according to the leftover principle.

One hundred and fifty-two scenarios of water distribution were developed, and the results of calculations on the hydrodynamic model were transferred to the Microsoft Excel. Using a computational scheme, the deficit was calculated (as a percentage of the nominal value) for water users (R1—fos; R2—k-mch-pk-os; R3—paos; R4—paos; R5—tos; R6—chos; R7—chos), the average deficit for each water user was found (AvDef-R1-R7), and the total deficit for all water users (SumDef-R1-R7). The percentage deficit was calculated based on the difference between the required water intake and, in fact, according to irrigation data correlated to the required nominal value. For reservoirs (R0-krs), the drawdown percentage was calculated.

The water deficit, expressed as a percentage for the entire vegetation period of j-th irrigation system, is calculated by formula 3:

$$D_j = \frac{W_j - W_j^*}{W_j} * 100\% = 1 - \frac{W_j^*}{W_j} \tag{3}$$

where W_j^* is the volume of water actually taken by the jth water intake of the jth irrigation system obtained as a result of modeling and W_j is the volume of water demand of water user.

The assessment of the availability of irrigation water during the vegetation period is calculated by formula 4:

$$C_j = \frac{W_j^*}{W_j} * 100\% \tag{4}$$

For reservoirs, the calculation of water scarcity was done according to the scheme: let V_i be the actual volume of the ith reservoir by the end of the growing season, obtained as a result of modeling in the MIKE11, then for the ith reservoir, the function setting the drawdown $E_i$as a percentage is determined by formula 5:

$$E_i = \frac{V_i^{\text{нпу}} - V_i}{V_i^{\text{нпу}} - V_i^{\text{умо}}} * 100\% \tag{5}$$

where $V_i^{нпу}$ is the reservoir volume at the normal retaining level and $V_i^{умо}$ is the reservoir volume at the dead storage level [9–11].

As an example, scenario 93 was calculated based on the initial inflow data for 2018, calculated according to the scheme [Укгу-340]_[Уфгу-R1-(R2/2)]_[Уттy-50]_[R3-R7 – leftover principle].

According to the calculations results (**Figure 5**), it can be seen that the water level at the water intakes of water users R3 and R4 in 2018 significantly exceeds the stated requirements, but there is a shortage for water user R6.

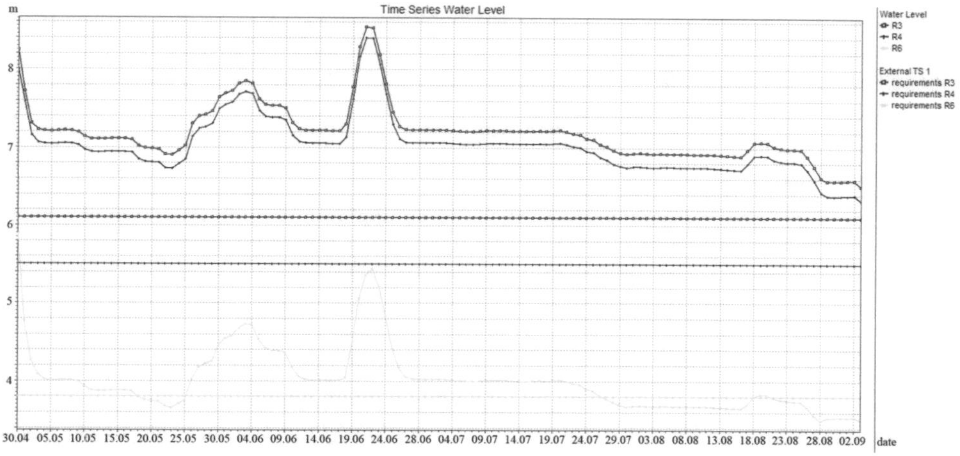

Figure 5.
Calculation results and requirements of water users.R3 (Petrovsko-Anastasievskaya irrigation system)—dark blue, R4 (Petrovsko-Anastasievskaya irrigation system)—blue, R6 (Chernoyerkovskaya irrigation system)—blue.

Scenarios	R0, %	R1, %	R2, %	R3, %	R4, %	R5, %	R6, %	R7, %	AvDef—R1–R7, %	SumDef-R1–R7, %
Scenario 1	8	0	0	64	37	67	87	86	49	23
Scenario 93	35	0	50	7	0	8	79	81	32	15
Scenario 69	40	0	0	9	0	7	94	96	29	20
Scenario 128	68	0	0	75	0	3	94	96	38	21
Scenario 149	65	0	0	35	4	56	38	50	26	7

Table 4.
Comparison of the calculations results of scenarios based on hydrology in 2018 (deficit, %).

Table 4 shows the results of calculations of water user deficits and reservoir drawdown for scenarios 1, 69, 128 and 149 according to the schemes, respectively:

[Укгу-R1-R2]_[Уфгу-R1-R2]_[Утгу-54]_[R3-R7- leftover principle];
[Укгу-220]_[Уфгу-R1-R2]_[Утгу-40]_[R3-R7- leftover principle];
[Укгу-340]_[Уфгу-R1-R2]_[Утгу-40]_[(R3/2), R4-R7- leftover principle];
[Укгу-410]_[Уфгу-R1-R2]_[Утгу-60]_[R3-R7- leftover principle].

Water management systems located on the water objects of river basins include a river network with a cascade of reservoirs and hydraulic structures designed to ensure the rational use and protection of water resources. Water management systems, as a rule, are multi-purpose and serve to provide water to various water users.

Rational use of water management systems involves the adoption of compromise management decisions that minimize damages from possible failures of water users' requirements. Decisions are made at various levels of planning and management. The ability to make informed decisions based on scientific, as well as social or political criteria is fundamental to the success of organizations involved in planning and managing water resources.

Efficiency criteria should be defined for significant water users, which provide a numerical assessment of the quality of decisions made on the water resources distribution. Some of these criteria may contradict each other (conflicting requirements of water users). In these cases, it is necessary to look for compromises between conflicting requirements of water users, which should be considered when searching for the "best" compromise solution.

Decisions made in the field of water resources management are inevitably associated with compromises among competing opportunities or goals. One of the tasks that are solved during planning is to evaluate alternative plans and identify compromises among competing opportunities, goals, or objectives. After that, the "best" compromise solution is worked out and adopted in the process of discussions and negotiations with the participation of all interested water users. The platform for such discussions in the Russian Federation is the Basin Council, and for negotiations – interdepartmental working groups, which in real time form the modes of operation of hydraulic structures included in the water management systems.

Professionals in the field of water resources who provide technically based options for compromise solutions to decision-makers should form a set of compromise options for water resources management. Their role is very important in the decision-making process since only professionals represent the full range and interrelation of problematic requirements and can form a set of compromise solutions agreed with all interested water users.

The task of regulating the water management systems' operation modes is a multi-criteria task with probabilistic initial information. A number of specific criteria should be considered, such as: the safety of hydraulic structures, protection of the downstream from floods, reliability of water supply, characterized by the issuance of guaranteed yield, etc.

When forming the water management systems modes operation, it is necessary to use multi-criteria optimization methods to form a set of alternative solutions in accordance with the list of different options for the water users' priorities hierarchy, to form a set of non-dominant solutions according to criteria (for example, according to the water users' supply) and to provide visualization tools for alternative plans for discussion and decision-making.

Multi-criteria methods or multi-purpose analysis methods are not intended to determine the best solution, they provide information about trade-offs between these sets of quantitative performance criteria [12]. Any final decision will be made in the process of discussion based on qualitative and quantitative information, and not in a computer. Determining acceptable and effective plans is a simpler task than deciding which of these effective plans is the best. The multi-criteria analysis was performed on the basis of the tools developed at the Dorodnitsyn Computing Centre of the Russian Academy of Science.

For the obtained matrix of solutions (deficits), using multi-criteria analysis methods based on the method of achievable goals, Pareto boundary curves were formed, which should be used in the process of negotiations between the decision-maker (Kuban Basin Water Management) and interested water users: the directorates of reservoirs and waterworks (Krasnodarsky waterwork, Fedorovsky waterwork, Tihovsky waterwork), the "Kubanmeliovodkhoz" management department and water users (departments of reclamation systems) that are directly subordinate to them – to choose the "optimal" in the sense of Edgeworth-Pareto trade-off solution. The Computer Expert Support Group also participates in the negotiations, and prepares several initial compromise scenarios and accompanies the decision-making process.

An example of such negotiation platforms for making decisions on the regulation of reservoirs of large river basins is the Interdepartmental Working Groups conducted by Rosvodresursy.

Before the start of negotiations, the Computer Expert Support Group prepares and sends compromise scenarios to interested water users, for example, with the lowest total and average deficits for all water users (**Figures 6** and 7).

The decision-maker conducts negotiations with interested water users, at which a decision is made: the reservoir will operate no more than 40% due to a small shipping operating in the water area of the Krasnodarskoye reservoir, increase water supply by R3, R4, R5 due to a slight reduction in the total deficit and "cutting" water users R6, R7.

Figure 8 shows scenario 69, in which the discharge from the Krasnodarskoye reservoir was 40%, the average deficit was 29%, and the total deficit was 20%. The deficit for other water users was, respectively: R1 = 0%, R2 = 0%, R3 = 9%, R4 = 0%, R5 = 7%, R6 = 94%, R7 = 96%.

This compromise scenario has been agreed with the majority of interested water users and approved by the decision-maker.

From the perspective of multi-criteria optimization, the inclusion of economic assessment in the scale of criteria "Methods of achievable goals" for each water distribution option allows you to significantly reduce the number of options and, most importantly, additionally focus on economic indicators: rice yield forecast, gross crop yield from irrigated lands of the irrigation system, guaranteed limited damage from irrigation water shortage.

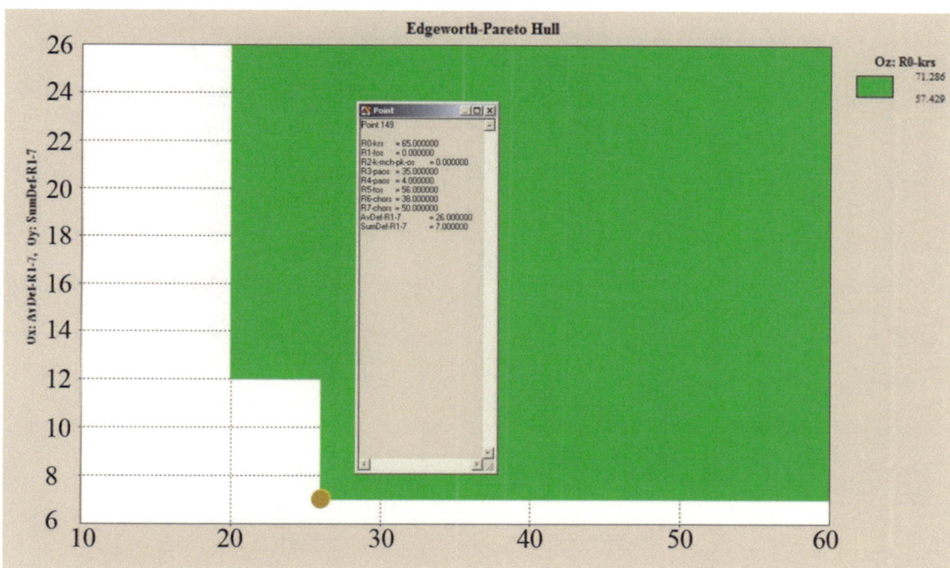

Figure 6.
Trade-off scenario 148.The drawdown of the Krasnodar reservoir (R0) is 65% and the total deficit is 7% (the smallest). The deficit for other water users is respectively: R1—Fedorovskaya irrigation system—0%, R2—Kubanskaya, Ponuro-Kalininskaya, and Maryano-Cheburgolskaya irrigation systems—0%, R3—Petrovsko-Anastasievskaya irrigation system—35%, R4—Pump stations No. 9 and No. 10 of Petrovsko-Anastasievskaya irrigation system—4%, R5—Temryukskaya irrigation system—56%, R6—Main canal of Chernoyerkovskaya irrigation system—38%, R7—Pump stations No. 1, No. 2, and No. 3 of Chernoyerkovskaya irrigation system—50%. For the following scenarios, the water users list is similar.

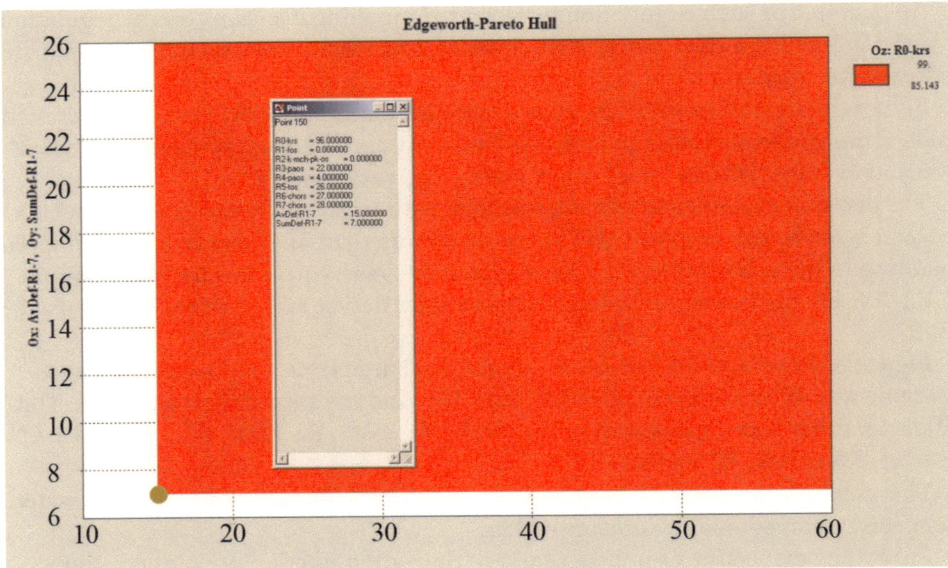

Figure 7.
Trade-off scenario 149.The drawdown of the Krasnodar reservoir is 96% and the average deficit is 15% (the smallest). Deficit for other water users is respectively: 0, 0, 22, 4, 26, 27, and 28%.

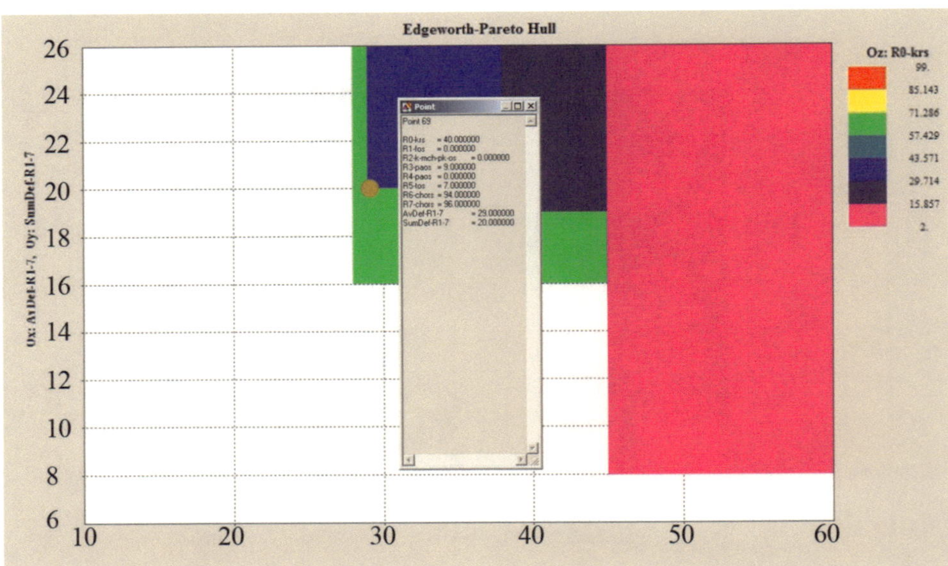

Figure 8.
Trade-off scenario 69.

Since it is not difficult for water users to include in the technology of periodic calculation of water supply modes with the same frequency calculations of the forecast yield of rice and other agricultural crops, as well as the calculation of guaranteed damage from the loss of crop production due to a shortage of irrigation water, the choice of the optimal option becomes more economically justified and can be selected as a compromise by the decision-maker, in the process of negotiations with interested water users (Basin Council, rice irrigation systems departments).

Forecast calculations of the expected yield are performed according to weather conditions and irrigation regimes for the past period of the growing season, which can be restored based on actual data on water intakes based on materials from dispatching services.

An example of such a calculation is the regression statistical model of rice yield (**Figure 9**), built on the production data of the branches of "Kubanmeliovodkhoz" for the last 10 years. Currently, field experimental research on this topic is actively developing, in the direction of creating a domestic dynamic model of rice, due to the need for a detailed description of the crop formation from natural and agrotechnical factors with reference to the vegetation phase.

Thus, according to the Kalininsky branch data, in 2018, there was a failure of water supply requirements in July, August, and September, which led to a decrease in rice yield from the maximum by 5.8 centner/ha. The actual irrigation rate was 21.9 thousand m^3/ha, which exceeded the declared volume for watering 2 thousand m^3/ha, while the actual yield in the branch was 71.5 centner/ha against the forecast 77.2 centner/ha (the forecast error was 7.5%).

Information materials on the technical characteristics and farm irrigation network state and the structure of the sown areas made it possible to detail the irrigation regime, considering the assumption that irrigation water is used in the mode specified by agricultural technology. The formulated restriction provides such an unambiguous assessment as "guaranteed damage"—the least damage caused by violation of the water regime under all other conditions being equal [13].

To perform such calculations, the system of optimal water distribution models described above on an inter-farm irrigation system is suitable, in which the hydraulic characteristics of the irrigation network are described by a small number of parameters: the network structure, the characteristics of water supply elements (channel length, discharge, efficiency, etc.). As previously stated, the model system allows you to determine optimal irrigation schedules, optimal water supply modes in the field, optimal irrigation standards, optimal operating modes of distributors with arbitrary

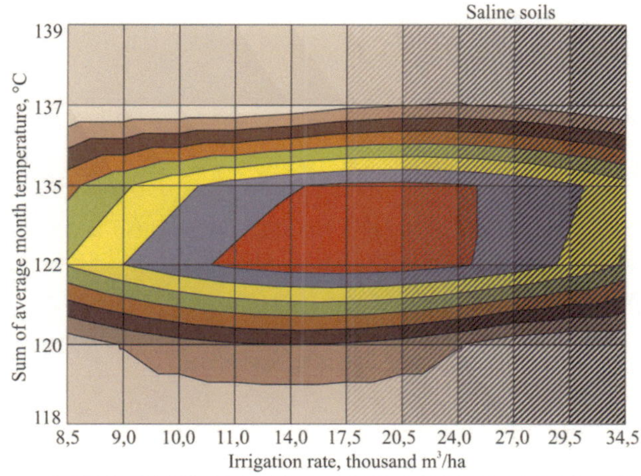

Figure 9.
Dependence of rice yield on irrigation rate and sum of average month temperature during vegetation period.

water intake operation and to develop a number of compromise solutions considering additional information about the forecast crop yields.

4. Conclusion

Water supply of rice irrigation systems in conditions of water scarcity is an extremely difficult task that requires improving the management of the entire structure of water use and irrigation technologies. The main direction in solving this problem is seen in the application of mathematical modeling and the latest digital technologies of communication and monitoring systems for collecting and processing information and integrating fundamental and applied disciplines.

The proposed approach to solving the problem of improving water supply and water use for rice irrigation systems is based on the development of an integrated system of tools for managing the distribution of water resources, considering the efficiency of water use by agricultural producers. In practical terms, a centralized water resources management system ensures high efficiency of their use in the implementation of the current scenario. On the other hand, the described approach makes it possible to identify the weakest links in water use, differentiated for each irrigation system, and to increase the efficiency of water use by agricultural producers. But the full solution of this set of tasks is still very far away, and we are still only at the very beginning of the planned path.

The performed studies have shown the possibility of studying the dynamics of hydrological processes under the influence of reclamation factors, analyzing water use in rice cultivation in detail, relying on models and calculation schemes. The main arrays of subject information obtained using numerical experiments on production data and standard hydrometeorological information were able to compensate for direct field measurements of water flows in the first approximation and allowed us to assess the state of water use in rice irrigation systems of the Lower Kuban.

Author details

Alina Buber*, Yuri Dobrachev, Alexander Buber and Evgenii Ratkovich
All-Russian Research Institute of Hydraulic Engineering and Land Reclamation named after A.N. Kostyakov, Moscow, Russia

*Address all correspondence to: buberalina@gmail.com

References

[1] Malysheva N, Kovaleva K, Hilko A. Development prospects of rice sowing. Scientist Notes. 2021;**2021**: 247-251

[2] Udalov A, Loginov V, editors. Estimating of Impact on Environment on Topic "Development of Project of Integrated Management and Water Object's Protection Scheme of the Kuban River basin". Krasnodar: Kuban'vodproekt; 2010. pp. 80-84

[3] Malysheva N, Yakuba S, Hadgidi A. To the question of reclamation development in steppe area of the Krasnodar Territory. Rice Sowing. 2021; **2021**:66-73

[4] Buber A, Bondarik I, Buber A. Development of approaches to water resources management in the lower Kuban to ensure water user requirements in low-water years. Irrigation and Drainage. 2020;**69**:3-10

[5] Borodychev V, Buber A, Dobrachev Yu: Calculation features of evaporation from the agrocoenosis soil surface at drip irrigation and fine dispersion sprinkling. In IOP Conference Series: Earth and Environmental Science. Mathematical Modeling of Technical and Economic Systems in Agriculture, 2020; Russia.

[6] Ivanov N. Determination of the evaporation value. News of the All-Union Geographical Society. 1954;**1954**: 189-196

[7] Berezkin VE, Kamenev GK, Lotov AV. A program for visualizing the multidimensional Pareto boundary in nonconvex multicriteria optimization problems (PFV-II). Certificate of registration of the computer program RU 2019664809, 13.11.2019. Application No. 2019663894 dated 29.10.2019

[8] Lotov A, Riabikov A, Buber A: A multi-criteria decision-making procedure with an inherited set of starting points of local optimization of the scalarizing functions. Scientific and Technical Information Processing. 2019; **46**:328-336

[9] Scheme of integrated management and water object's protection of the Kuban River basin' approved by Order No. 171 of the Kuban Basin Council dated 19.05.2014

[10] Main rules of water resources use of Krasnodarskoye reservoir on the Kuban River. 1980

[11] Operation rules of the Krasnodarskoye reservoir. 1982

[12] Loucks DP, van Beek E. Water Resource Systems Planning and Management An Introduction to Methods, Models, and Applications. Springer; 2017. p. 624. DOI: 10.1007/978-3-319-44234-1

[13] Vladimirov S, Hathohu E, Ponomarenko I. Reclamation state of the Lower Kuban irrigation systems. In: Scientific Mechanisms of Solving Problems of Innovation Development: Collection of Articles of International Practical Conference. Russia; 2017

Chapter 13

Smart Irrigation for Climate Change Adaptation and Improved Food Security

Erion Bwambale, Felix K. Abagale and Geophrey K. Anornu

Abstract

The global consequences of climate change cannot be ignored. The agriculture industry, in particular, has been harmed, resulting in poor production as a result of floods and droughts. One in every three people in the world's arid and semi-arid regions lacks access to healthy food and safe drinking water. Despite the fact that irrigation development is increasing in most developing nations, it still falls short of meeting current food demand, much alone predicted need by 2050. To feed the future population while combating climate change, agricultural practices must be precise. Scarce resources such as water, land, and energy will need to be exploited more efficiently in order to produce more with less. Smart irrigation is shaping up to bring answers to these twenty-first-century concerns. This chapter discusses improvements in smart irrigation monitoring and management systems that may be used to address climate, food, and population issues. It includes an overview of smart irrigation, smart irrigation monitoring, and smart irrigation management, as well as challenges and prospects related to climate change and food security. Smart irrigation may boost water savings and agricultural production, thereby improving food security.

Keywords: smart irrigation, water use efficiency, climate change adaptation, precision water management

1. Introduction

The world presently faces challenges ranging from extreme effects of climate change i.e. droughts and floods, to a rising population [1]. This has a huge impact on the present and future food security. Presently, 1.3 billion people are residents in water-stressed areas of the world, rendering water for agricultural production insufficient as competition from other sectors of the economy increases [2]. In countries with significant amounts of fresh and groundwater resources, irrigation has substantially contributed to sustainable food production all year round [3]. Presently only 24.1% of the total agricultural area is under irrigation, yielding about 40% of the total world food and fiber. Despite the benefits of irrigation reported so far, it has been regarded as a sector that uses a lot of water depriving other sectors of the economy. The FAO state of the food report posits that of the 70% freshwater abstracted for irrigation, only 50% is beneficial to plants [4]. The pressure on freshwater and groundwater resources due to the growing food and fiber demand will further exacerbate as agricultural production will need to expand by 1.7 times by 2050. This necessitates a 15% increase in freshwater withdrawals [3]. Agriculture will be required to re-allocate a fair share of the water abstracted to meet a 25–40% increase in future water demand from higher productive and employing sectors of the economy [5–7].

Given that irrigated agriculture has proven to deliver up to two-fold more food than rainfed agriculture [8], the demand for irrigation water will inevitably continue to increase as more land is secured for irrigation. Efficient land utilization with irrigation leads to crop diversification, which later buffers against climate variability. Climate change has already impacted agriculture to the extent that water for irrigation is becoming scarce in arid and semi-arid lands. Therefore, sustainable irrigation should aim at reducing water losses, meeting crop water requirements, and maintaining ecological flows in rivers and aquifers. However, improving water management in agriculture is typically constrained by inadequate policies, major institutional underperformance, lack of technology deployment, and financing limitations [9].

Sustainable smart irrigation is an essential step towards improving the state of food security and achieving sustainable development goal number 2. Smart irrigation ensures timely, real-time water application to the plant root zone, reducing losses associated with traditional irrigation systems like evaporation, seepage, and deep percolation [10]. With effective monitoring and control in smart irrigation, water, energy and labour are saved. As the notion of more crop per drop gains attention, smart irrigation is a potential climate change adaptation strategy and an effective way to ensure sustainable food supply all year round.

2. Current status of irrigation water application methods

In the past 50 years, irrigation has undergone tremendous changes driven by increased drought indices. From 1970 to 1985, automatic control of irrigation systems emerged in the United States of America as water scarcity became prevalent [11]. As a result, researchers were interested in developing ways of optimizing irrigation to achieve maximum yields. The advent of the Internet in 1989 spiked another interest in internet-based control of irrigation systems like data storage on the web became possible. And over the years, wireless sensor networks, sensors for monitoring and control, have since been developed to facilitate precision irrigation today. **Figure 1** depicts the evolution of irrigation from 1970 to 2020.

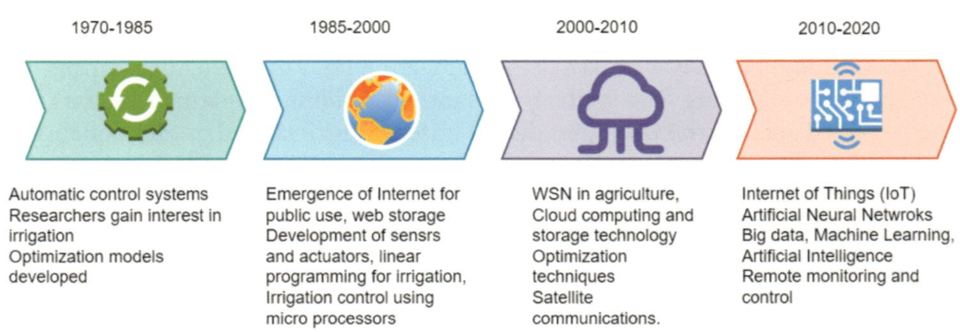

Figure 1.
Trends in irrigation since 1970s.

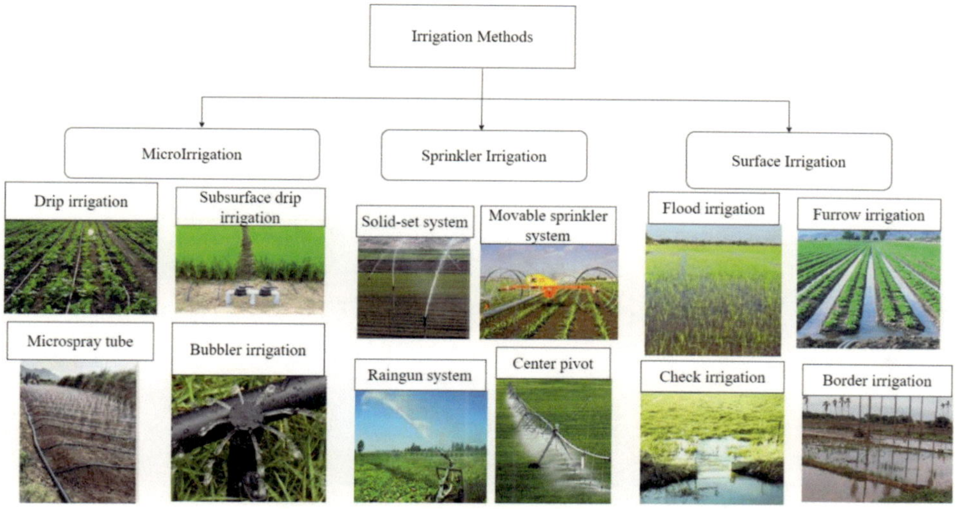

Figure 2.
Irrigation methods [12].

Eisenhauer and Martin [12] classified irrigation methods, as shown in **Figure 2**, ranging from micro-irrigation (surface drip irrigation, sub-surface drip irrigation, micro-spray tube irrigation, bubbler irrigation), sprinkler irrigation (solid-set system, portable sprinkler system, rain gun system, centre pivot system), surface irrigation (flood irrigation, furrow irrigation, check irrigation, border irrigation). Over the years, research has helped develop several tools and techniques for various production systems and ecologies that help save water and improve water productivity in agriculture. The advent of the Internet of Things (IoT) and wireless sensor networks has led to developing tools like soil, weather, and plant sensors that have improved monitoring that informs data-driven irrigation scheduling [13]. As depicted in **Figure 3**, the water use efficiency in these irrigation systems increases with the level of automation employed in the system.

2.1 Micro-irrigation

Water is delivered at low pressure through emitters by a distribution system located on the soil surface, beneath the surface, or suspended above the ground in micro-irrigation systems [12]. Irrigation water droplets are directed to the plant root zone via emitters, sprayers, or porous pipes, where it then infiltrates by gravity or capillary rise. The applicator's design reduces the water pressure within the delivery lines, resulting in a low discharge.

Micro-irrigation has received a lot of attention from agriculturalists, especially for high-value crops like vegetables, fruits, and nut trees. When determining whether to invest in a costly micro-irrigation system, a producer must consider whether the increase in crop production will be sufficient to pay for the system. The other concern is, can the system be built to filter the irrigation water to avoid the emitters from clogging? Another crucial choice is selecting emitters from the broad array available are suited for the specified function. Solving these difficulties will give an excellent irrigation system for decades. Micro-irrigation is defined by water being applied: at low rates, over longer durations, near the root zone of the plants, and at constant flow rate.

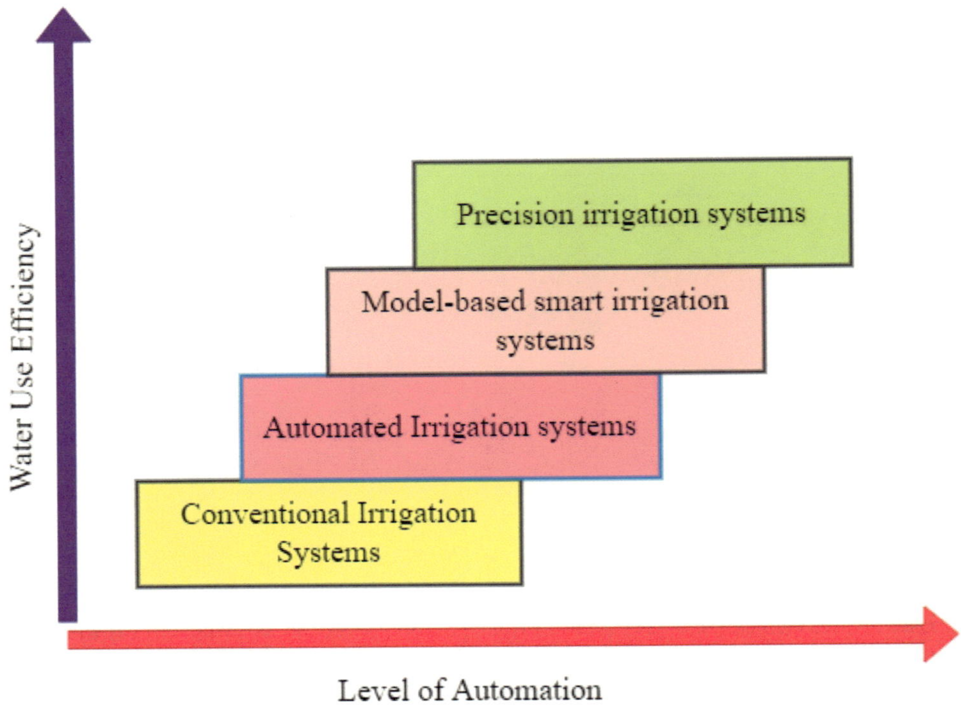

Figure 3.
Water use efficiency and level of automation Adapted from [14].

Precision water application with micro-irrigation systems has been reported to save up to 35–65% more water than standard flood irrigation systems, with a commensurate increase in the production of crops. Scholars [15–19] have all found consistent results confirming water savings and increased yields with micro-irrigation devices.

2.2 Sprinkler irrigation

Sprinkler irrigation is a system where water is uniformly applied over the crop canopy or soil surface identical to rainfall. With a sprinkler irrigation system, water is pumped and conveyed through high-density polyethylene (HDPE) pipes eliminating water losses through seepage and evaporation as in the case of surface canals under surface irrigation. Compared to conventional flood irrigation, sprinkler irrigation is more efficient, with irrigation efficiency of up to 80–90%. The performance of a sprinkler irrigation system solely depends on the design and selection of sprinklers. In irrigation design, it is recommended to select a sprinkler whose application rate is lower than the soil infiltration rate to prevent surface ponding and runoff.

The implementation of intelligent sprinkler irrigation systems involves high control precision, high intelligence, good dependability, simple operation, wired or wireless sensor network technology, and crop water demand data collection devices [20]. Fuzzy control logic, neural networks, and expert systems and machine learning, control technologies can be built for sprinkler irrigation [21]. Furthermore, intelligent precision irrigation systems are being built with remote transmission, monitoring, decision, and control functions. Therefore, it is vital to build automatic sprinkler irrigation equipment, encompassing flow meters, solenoid valves, precision control equipment, and robotic equipment.

2.3 Surface irrigation

Surface irrigation is the oldest irrigation application method in the world. It involves the application of water over the surface of the land to supply moisture to the plant. Surface Irrigation includes furrow, border, basin and, check irrigation. Surface irrigation requires less pressure than sprinkler or micro-irrigation systems. Under surface irrigation, irrigation water is applied at the inlet end, and the water subsequently flows to the downstream end. A part of the water infiltrates as it progresses over the field. Water is frequently applied by gated pipelines, siphons, or gates. Surface irrigation may be an effective application technique provided the soils and fields are well adapted to this approach. But, it may be exceedingly inefficient if the soils and other elements are not carefully addressed while constructing and administering the system. The soil infiltration rate is very crucial in the proper functioning of surface irrigation systems. If the soil's infiltration rate is excessively high, the depth of water that infiltrates at the entrance will be significantly bigger than the downstream end. The land slope and its regularity also considerably effect surface irrigation. Slopes that are overly steep create undue runoff and erosion. Acceptable slopes are usually less than 2%. The regularity of the slope is also crucial so that water does not gather in depressions on the surface.

To improve smart irrigation scheduling in surface irrigation systems, Supervisory Control and Data Acquisition (SCADA) software has been deployed in surface irrigation systems to improve the programming, monitoring and operation of an entire scheme from a central point [22]. Composed of field equipment, programmable logic controller (PLCs) and/or remote terminal units (RTUs) communication networks, SCADA host software and, third party systems, these components are connected to minimize human intervention but also ensure convenient operation and delivery of irrigation water with just a click of a button. SCADA systems provide real-time monitoring, remote supervisory or automatic control, troubleshooting, and automatic data reporting and archiving capabilities [22].

3. Smart irrigation scheduling approaches

Irrigation scheduling is a systematic process of determining when and how much to irrigate. This depends on various factors, including daily crop water requirement, the effective root zone, and the available soil moisture. Irrigation scheduling can be done using one or all of the following approaches: Plant-based, soil-based, and weather-based irrigation scheduling approaches. Each of these is shown in **Figure 4**.

3.1 Plant-based irrigation scheduling

Plant-based irrigation scheduling is based on the physiological and phenological status of the plant [16]. The physiological condition depicts the water stress level, which is estimated from canopy temperature depression relative to air temperature measured by infrared thermometry. The calculation of the cumulative stress degree days and crop water stress index can be used for scheduling irrigation. Phenological stages can also be used to determine when to irrigate. For example, in wheat

Figure 4.
Smart irrigation scheduling approaches. (a) soil sensors for irrigation scheduling, (b) plant sensor for detecting sap flow in plant stems, (c) ATMOS 41 for weather monitoring.

cultivation, crown root initiation (CRI), tillering, jointing, flowering, and the grain-filling stage are critical stages of growth that need irrigation [16]. Failure to supply irrigation water at these critical stages of growth leads to low yields as water stress becomes severe. The cumulative effect of water stress is determined with this method, making it effective as a water stress indicator. This helps to capture the moisture reduction in the soil through evapotranspiration [23]. Direct and indirect measurement techniques are used to determine plant water status. Direct plant water stress detection methods include using sap flow sensors, xylem sensors, leaf sensors, and others. On the other hand, indirect methods involve thermal sensing, near-infrared spectroscopy, and aerial imagery [24].

Several authors have used plant-based approaches for irrigation scheduling [23, 25–27]. For example, King et al. [27] used data-driven models for canopy temperature-based irrigation scheduling of sugar beet and winegrape. The data-driven models developed by the authors estimated reference temperatures enabling automatic calculation of the crop water stress index for crop water stress assessment. Similarly, Meeks et al. [25] used leaf water potential monitoring system for irrigation scheduling of winter rye cover crop. The authors reported significant water savings and an improvement in crop yields.

3.2 Soil moisture-based irrigation scheduling

Soil moisture-based irrigation scheduling involves determining the soil moisture status within the root zone, and knowing the permanent wilting point [28]. Soil moisture measurements are compared to moisture thresholds to trigger irrigation. Soil moisture monitoring is done by time-domain transmission sensors, neutron probes, capacitance sensors, or granular matrix sensors. Soil moisture-based irrigation scheduling allows variable rate irrigation scheduling due to its ability to measure spatiotemporal variability in the field.

The use of soil moisture-based irrigation scheduling has been reported in the literature. For example, Pramanik et al. [28] developed an automated basin irrigation system based on soil moisture sensors for irrigation scheduling. The authors highlighted that the ideal position of sensors for shutting the system would be at 37.5 cm depth put at 25% length from the intake in larger soil moisture deficit situations and at 7.5 cm depth set at 75% length in low moisture deficit conditions. Consequently, the irrigation application efficiency was enhanced up to 86.6% using automation.

Advances in geospatial technologies like remote sensing and geographical positioning systems have made it possible to determine soil moisture from space over large land areas. Satellites in space are able to predict soil moisture by taking images and using inbuilt algorithms to assess the soil moisture deficit. This is then used to inform irrigation scheduling. Recently, Kisekka et al. [29] compared in-situ soil moisture measurements and remotely sensed measurements. The authors concluded that remotely sensed soil moisture presents an effective means of soil-moisture-based irrigation scheduling in large agricultural fields.

3.3 Weather-based irrigation scheduling

Weather-based irrigation scheduling involves the use of weather sensors to monitor and measure the parameters that affect evapotranspiration. Automatic weather stations with temperature, humidity, wind speed, rainfall, and air pressure sensors are installed in the field to collect field data around the plant. The data from the weather sensors are then used to estimate water demand using evapotranspiration models. The Penman-Monteith evapotranspiration model is used to determine the daily water demand [30]. Irrigation is scheduled after a pre-determined amount of evapotranspiration has occurred and this threshold varies with soil type, crop type, and stage of growth.

4. Enabling technologies for smart irrigation

4.1 Communication technologies

4.1.1 Wireless sensor networks (WSN)

The advent of the industrial revolution and recent advances in electronics and wireless communications have led to the development of smart sensors with low power and cost solutions [31]. WSN provides a high spatio-temporal resolution for monitoring soil and crop parameters via wirelessly-connected sensor nodes installed across the field [32]. **Table 1** presents some of the common wireless sensor communication technologies used in smart irrigation.

WSNs allow the surveillance of plants and soil and may enhance production, efficiency, and profitability. Effective monitoring and communication helps to reduce risks due to climate aberrations, water shortages, insect infestation, and other factors unfriendly to agricultural growth and development. WSNs are helping to attain improved reaction times owing to real-time sensing and communication in agricultural contexts. There are numerous techniques for irrigation scheduling utilizing wireless sensors. Depending on the threshold levels of temperature and soil water content, the gateway permits automatic activation of the irrigation system. However, some of the sensors, modules, and valves that are commercially available for installation in an irrigation network are very sophisticated and costly to be implemented for managing stationary irrigation systems reference.

4.1.2 Internet of Things (IoT)

The IoT technology is the evolutionary phase of the Internet that builds a global infrastructure uniting devices and people [46]. IoT has its origins in numerous previous technologies: ubiquitous information systems, sensor networks, and embedded computers. Adelodun et al. [47] identified IoT as an interesting paradigm with seamless integration of smart capabilities into physical devices for context-oriented services. It

Technology	Description	References
Wi-Fi	Wi-Fi technology enables the wireless connection to a local area network through the Internet.	[33, 34]
ZigBee	Zigbee is an IEEE 802.15.4 standard-based wireless technology developed to enable personal wireless networks with low power radio signals	[35, 36]
Bluetooth	Bluetooth is a short-range wireless communication technology of standard IEEE 802.15.1 used to exchange data based on radiofrequency.	[37]
LoRa	LoRa is a long-range wireless technology that delivers low power and secure data transfer for machine-to-machine and Internet of Things applications. Founded by Semtech in 2012, LoRa is based on chirp spread spectrum (CSS) modulation, which has low power characteristics. LoRa supports the wireless connectivity of sensors, gateways, machines, devices, animals, humans etc., to the cloud across a range of 2–10 km.	[38, 39]
General Packet Radio Service (GPRS)	GPRS is a unique non-voice, rapid, packet-switching technology for worldwide system for mobile communications (GSM) networks. It permits the transmission of brief bursts and huge amounts of data, such as email and web surfing, via a mobile telephone network.	[39]
Ethernet	Ethernet technology allows devices to connect via a local or wide area network. It lets devices to interact with one other through a protocol, which is a set of rules or common network language.	[40, 41]
Fifth-generation wireless (5G)	5G is a fresh generation of cellular technology meant to speed up the responsiveness of wireless networks. It has the capacity to transport data at multigigabit rates hitting up to 20 Gb/s. 5G is effective in applications that demand real-time monitoring and control like precision irrigation.	[42]
Low-Power Wireless Personal Area Networks. (6LoWPAN)	6LoWPAN is a communication technology that offers a strategy for routing Internet Protocol version 6 (IPv6) across low-power wireless networks. It is created and supported by the Engineering Task Force (IETF), responsible for Internet standards.	[43, 44]
Message Queuing Telemetry Transport (MQTT)	MQTT is a lightweight, publish-subscribe network protocol for exchanging data between devices. It uses TCP/IP, although any network protocol that allows for ordered, lossless, bidirectional communications may support MQTT.	[45]

Table 1.
Most utilized communication technologies in smart irrigation.

has previously been effectively utilized in agricultural systems to monitor and manage environmental variables [48–51]. The Internet of Things provides a platform for precision farming, digitally integrating several soil sensing devices and, context-aware sensors, custom devices. Data analytical implementations enhance farmers' ability to resolve intricate agricultural issues such as soil preparation, water feed estimation, yield prediction, etc., throughout the whole growing and harvesting cycle. Several mechanistic irrigation scheduling systems have been presented employing most key farmed land parameters-soil moisture content and climatic data to predict the amount of water at certain time intervals. Campos et al. [51] built an IoT framework (**Figure 5**) to provide services for smart irrigation, such as data monitoring, pre-processing, fusion, synchronizing, storing, and irrigation management augmented by predicting soil moisture.

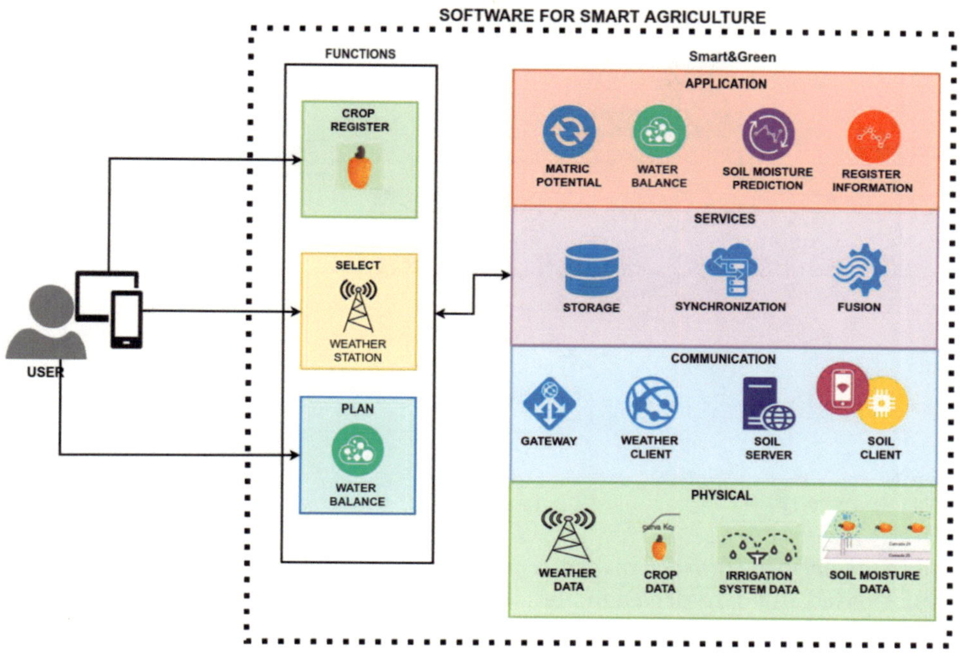

Figure 5.
IoT system for smart irrigation Adapted from [51].

4.2 Decision support systems

A decision support system for smart irrigation provides a framework for incorporating various tools and techniques for site-specific irrigation decisions. Commercial precision irrigation systems will thrive with improvement in robust and optimal decision support systems [52]. Decision support systems for irrigation scheduling/control can be categorized into two, namely: open-loop irrigation and closed-loop irrigation [14, 20].

4.2.1 Open-loop irrigation control

A couple of irrigation decision support systems schedule irrigation at predefined intervals and apply predefined irrigation volumes [52]. These decisions are not based on any sensor feedback on plant status, soil moisture status, or weather parameters. The decision to initiate an irrigation action is largely dependent on historical data and heuristics. This irrigation control strategy is inefficient which may lead to over or under irrigation, thereby wasting valuable water and fertilizer.

4.2.2 Closed-loop irrigation control

Closed-loop irrigation control is a control strategy where a mathematical model is used to make predictions about the future output. This is aided by real-time feedback from sensors that monitor the process. The model of the plant can either be a physics-based or a data-driven model. Whereas physics-based models are derived from the laws of physics, such as (conservation of mass, gravity, etc.), data-driven

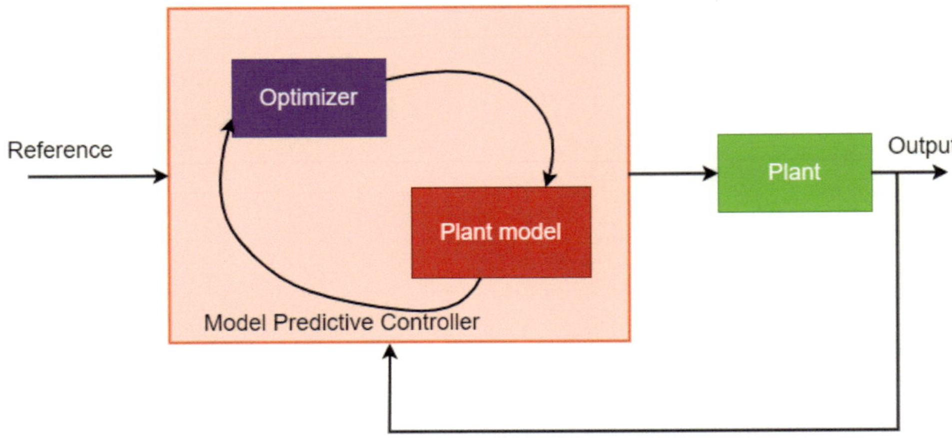

Figure 6.
Closed-loop control system.

models are derived from the real-time dynamics of a system. A combination of physics-based and data-driven models is a gray-box model. Here a mechanistic model is developed from physics and during operation data from the system is used to update the model depending on the dynamics in the plant environment. Under closed-loop irrigation, irrigation scheduling decisions are made by micro-controllers by comparing a current state with the desired state. A closed-loop control algorithm helps to initiate an action of the actuators. **Figure 6** is a schematic presentation of a closed-loop control system.

Closed-loop control is further divided into intelligent, optimal, and linear control strategies. With advances in computing power, smart irrigation systems are able to make decisions in real-time, depending on the prevailing environment of the plant. A detailed explanation of closed-loop irrigation strategies can be found in [14, 20, 53, 54].

4.3 Cloud platforms

Monitoring for smart irrigation results in enormous amounts of data that needs to be stored for processing. Data is generated by weather, soil, and plant sensors in real-time at every time step. This data is transferred to a cloud-based platform for storage and processing. Some of the common cloud platforms used in smart irrigation studies include Thinkspeak (**Figure** 7), (MATLAB), FIWARE, Dynamo, and MongoDB, among others [49]. Thinkspeak is a cloud-based data platform for storage, visualization, and retrieving of sensor data. A microcontroller can be connected to this platform for processing.

5. Smart irrigation and climate-risks

Deploying smart irrigation system may lower the overall irrigation water volume necessary to cultivate field crops in two ways. First, producers may eliminate non-cropped or marginal regions from water application, and second, producers can limit application rates in low-lying areas or soils with high water-holding capacity. Field

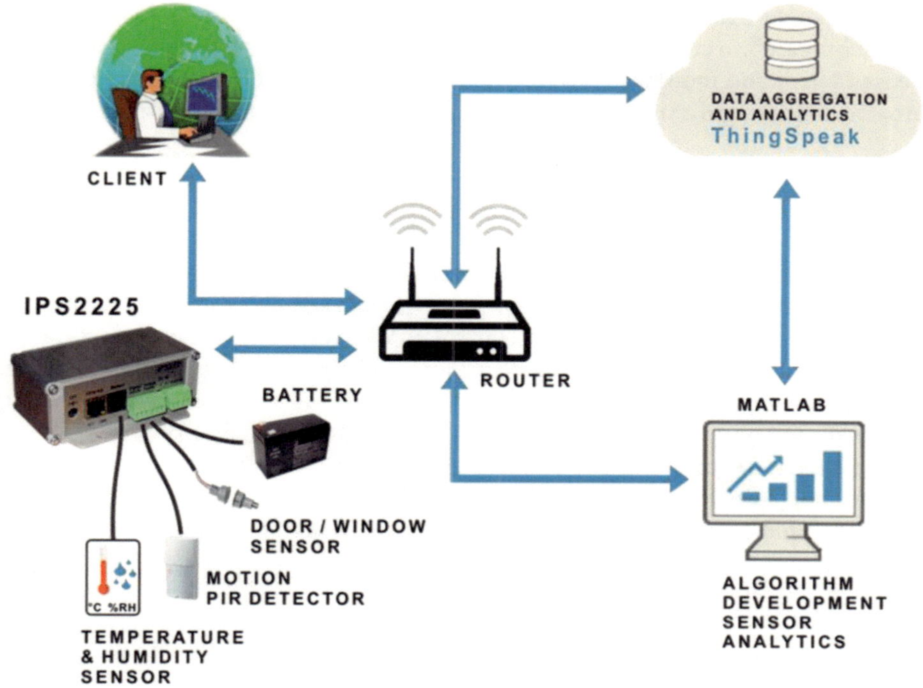

Figure 7.
Thinkspeak clould platform, Source [55].

installation of smart irrigation systems has demonstrated typical savings in irrigation water consumption of 8–20% [56] compared to uniform irrigation application. Using an intelligent irrigation system may assist minimize irrigation withdrawals while still maintaining a well-watered crop. This allows effective water usage and may lessen the likelihood of transient well failures during droughts. Having the machinery and availability to water for irrigation makes irrigated croplands less exposed to climate-related threats than their dryland equivalents. Smart irrigation decreases this danger by enhancing irrigation water-use efficiency and minimizing freshwater withdrawals to allow for more predictable water supply.

6. Smart irrigation and food security

Several studies carried out in different production systems and geographical areas worldwide have demonstrated the tangible benefits of smart irrigation systems over conventional irrigation practices. In the mild climatic conditions of Prince Edward Island, Afzaal et al. [57] measured crop productivity and water saving in potato production. The authors reported on the performance of a smart fertigation system and found significantly higher irrigation efficiency of the automated fertigation (1.42 kg/m^3) than for the traditional drip irrigation control system (1.19 kg/m^3). Thus, an automated drip irrigation system provides up to 26% water saving and high crop productivity compared to the conventional water application methods. Another study [58] compared a soil moisture sensor-based automated drip irrigation system with a non-automated drip irrigation system for nectarine that was irrigated when soil moisture content reached 70% of field capacity; the authors reported 43% more production compared to conventional irrigation methods. Belayneh et al. [59] reported increased yields and more water savings using a sensor-based irrigation scheduling approach compared to a time-based approach for old dogwood and redmaple trees.

7. Conclusion

Smart irrigation has gained significant attention as the need to improve water use efficiency increases. Smart irrigation involves data acquisition, interpretation, analysis, and control. Monitoring and control require sensors to collect real-time information for irrigation scheduling decisions. An irrigation control strategy has to be adopted for irrigation decisions. Closed-loop irrigation strategies with feedback from sensors are widely used for real-time irrigation decisions. Smart irrigation can save irrigation water and improve yield at the farm level, consequently leading to improved food security for the global population. In all the irrigation systems, it is possible to implement smart strategies that can help in saving irrigation water and improve yields.

Acknowledgements

This publication was made possible through support provided by the West African Centre for Water, Irrigation and Sustainable Agriculture (WACWISA), University for Development Studies, Ghana with funding support from the Government of Ghana and World Bank through the African Centers for Development Impact (ACE Impact) initiative. We would like to thank the unnamed reviewers and editor for their great contributions to this book chapter. Sincere gratitude to Rita Namoe Tabi for the helpful comments offered throughout the development of this work.

Author contributions

Conceptualization, EB; methodology, EB; formal analysis, EB; investigation, EB; writing-original draft preparation, EB; writing-review and editing, FKA, GKA YK; visualization, FKA; supervision, FKA, GKA YK; project administration, FKA; funding acquisition, FKA, GKA YK. All authors have read and agreed to publish the paper.

Funding

This study was financed by Institute de Recherché pour le Development (IRD), AFD and the West African Center for Water, Irrigation and Sustainable Agriculture (WACWISA), University for Development Studies, Ghana.

Conflict of interest

The authors declare no conflict of interest.

Author details

Erion Bwambale[1,2,3]*, Felix K. Abagale[1,2] and Geophrey K. Anornu[4]

1 West African Center for Water, Irrigation and Sustainable Agriculture (WACWISA), University for Development Studies, Tamale, Ghana

2 Department of Agricultural Engineering, University for Development Studies, Tamale, Ghana

3 Department of Agricultural and Biosystems Engineering, Makerere University, Kampala, Uganda

4 Civil Engineering Department, Regional Water and Environmental Sanitation Center Kumasi (RWESCK), Kwame Nkrumah University of Sciences and Technology, Kumasi, Ghana

*Address all correspondence to: erionbw209@uds.edu.gh

References

[1] IPCC. Summary for policymakers. In: Pörtner H-O, Roberts DC, Poloczanska ES, Mintenbeck K, AA MT, Craig M, et al., editors. Climate Change 2022: Impacts, Adaptation and Vulnerability. The Working Group II contribution to the IPCC Sixth Assessment Report. The Intergovernmental Panel on Climate Change. Geneva: Cambridge University Press; 2022. Available from: https://www.ipcc.ch/report/ar6/wg2/downloads/report/IPCC_AR6_WGII_SummaryForPolicymakers.pdf

[2] FAO, IFAD, UNICEF, WFP, WHO. In Brief to The State of Food Security and Nutrition in the World 2021 [Internet]. Rome, Italy: FAO, IFAD, UNICEF, WFP and WHO; 2021. Available from: http://www.fao.org/documents/card/en/c/cb5409en

[3] UN Water. Groundwater: Making the invisible visible. In: The United Nations World Water Development Report 2022. Paris, France: United Nations Educational, Scientific and Cultural Organization; 2022

[4] FAO. The State of Food and Agriculture 2020. Overcoming Water Challenges in Agriculture. Rome, Italy: FAO; 2020

[5] Bwambale E, Home P, Raude J, Wanyama J. Development of a water allocation model for equitable water distribution at doho rice irrigation scheme, Uganda. Hydrology. 2019;7:62. Available from: http://www.sciencepublishinggroup.com/journal/paperinfo?journalid=267&doi=10.11648/j.hyd.20190704.12

[6] Durodola OS, Bwambale J, Nabunya V. Using every drop: Rainwater harvesting for food security in Mbale, Uganda. Water Practice and Technology. 2020;**15**:295-310. DOI: 10.2166/wpt.2020.019

[7] Bwambale E, Home P, Raude J, Wanyama J. Hydraulic performance evaluation of the water conveyance system of Doho Rice Irrigation Scheme in Uganda. Journal of Sustainable Research in Engineering. 2019;**5**:101-112

[8] Li X, Troy TJ. Changes in rainfed and irrigated crop yield response to climate in the western US. Environmental Research Letters. 2018;**13**:064031. Available from: https://iopscience.iop.org/article/10.1088/1748-9326/aac4b1

[9] Sadeghi SH, Saedi SI, Peters RT, Stöckle C. Towards improving the global water application uniformity of centre pivots through lateral speed adjustment. Biosystems Engineering. 2022;**215**:215-227

[10] Bwambale E, Abagale FK, Anornu GK. Data-driven model predictive control for precision irrigation management. Smart Agricultural Technology. 2023;**3**:100074. DOI: 10.1016/j.atech.2022.100074

[11] Ramachandran V, Ramalakshmi R, Kavin BP, Hussain I. Exploiting IoT and its enabled technologies for irrigation needs in agriculture. Switzerland: Water, MDPI. 2022;**14**(5):1-20. DOI: 10.3390/w14050719

[12] Eisenhauer DE, Martin DL. Irrigation systems management. In: Heeren DM, Hoffman GJ, editors. American Society of Agricultural Engineers. USA: ASABE; 2021. Available from: https://elibrary.asabe.org/textbook.asp?confid=ism2021

[13] Abioye EA, Shukri M, Abidin Z, Saiful M, Mahmud A, Buyamin S, et al. Smart agricultural technology a data-driven Kalman filter-PID controller for fibrous capillary irrigation. Smart Agricultural Technology. 2023;**3**:100085. DOI: 10.1016/j.atech.2022.100085

[14] Abioye EA, Abidin MSZ, Mahmud MSA, Buyamin S, Ishak MHI, Rahman MKIA, et al. A review on monitoring and advanced control strategies for precision irrigation. Computers and Electronics in Agriculture. 2020;**173**:105441. DOI: 10.1016/j.compag.2020.105441

[15] Patle GT, Kumar M, Khanna M. Climate-smart water technologies for sustainable agriculture: A review. Journal of Water and Climate Change. 2020;**11**:1455-1466

[16] Sidhu RK, Kumar R, Rana PS, Jat ML. Automation in Drip Irrigation for Enhancing Water use Efficiency in Cereal Systems of South Asia: Status and Prospects. 1st ed. Amsterdam: Elsevier Inc.; 2021. DOI: 10.1016/bs.agron.2021.01.002

[17] Guan C, Ma X, Shi X. The impact of collective and individual drip irrigation systems on fertilizer use intensity and land productivity: Evidence from rural Xinjiang, China. Water Resources and Economics. 2022;**38**:100196. DOI: 10.1016/j.wre.2022.100196

[18] Wang F, Xue J, Xie R, Ming B, Wang K, Hou P, et al. Assessing growth and water productivity for drip-irrigated maize under high plant density in arid to semi-humid climates. Agriculture. 2022;**12**:97

[19] Yan H, Hui X, Li M, Xu Y. Development in sprinkler irrigation technology in China. Irrigation and Drainage. 2020;**69**:75-87

[20] Bwambale E, Abagale FK, Anornu GK. Smart irrigation monitoring and control strategies for improving water use efficiency in precision agriculture: A review. Agricultural Water Management. 2022;**260**:1-12. DOI: 10.1016/j.agwat.2021.107324

[21] Wahlin B, Zimbelman D. Canal automation for irrigation systems: american society of civil engineers manual of practice number 131. Irrigation and Drainage. 2018;**67**:22-28. DOI: http://doi.wiley.com/10.1002/ird.2140

[22] Brian W, Darell Z. Canal Automation for Irrigation Systems. Reston,Virginia: American Spciety of Civil Engineers; 2014. Available from: www.asce.org/bookstore

[23] Parkash V, Singh S. A review on potential plant-basedwater stress indicators for vegetable crops. Sustain. 2020;**12**:3945. DOI: 10.3390/su12103945

[24] Fernández JE. Plant-based methods for irrigation scheduling of woody crops. Horticulturae. 2017;**3**:35. DOI: 10.3390/horticulturae3020035

[25] Meeks CD, Snider JL, Culpepper S, Hawkins G. Applying plant-based irrigation scheduling to assess water use efficiency of cotton following a high-biomass rye cover crop. Journal of Cotton Research. 2020;**3**:1-12

[26] Gu Z, Qi Z, Burghate R, Yuan S, Jiao X, Xu J. Irrigation scheduling approaches and applications: A review. Journal of Irrigation and Drainage Engineering. 2020;**146**:04020007

[27] King BA, Shellie KC, Tarkalson DD, Levin AD, Sharma V, Bjorneberg DL. Data-driven models for canopy temperature-based irrigation scheduling. Transactions of the ASABE. 2020;**63**:1579-1592

[28] Pramanik M, Khanna M, Singh M, Singh DK, Sudhishri S, Bhatia A, et al. Automation of soil moisture sensor-based basin irrigation system. Smart Agricultural Technology. 2022;**2**:100032. DOI: 10.1016/j.atech.2021.100032

[29] Kisekka I, Peddinti SR, Kustas WP, McElrone AJ, Bambach-Ortiz N, McKee L, et al. Spatial–temporal modeling of root zone soil moisture dynamics in a vineyard using machine learning and remote sensing. Irrigation Science. 2022 under review. DOI: 10.1007/s00271-022-00775-1

[30] Allen RG, Pereira LS, Raes D, Smith M. Crop Evapotranspiration. Guidelines for Computing Crop Water Requirements. FAO Irrigation and Drainage paper 56. Rome, Italy: Food and Agricultural Organization of the United Nations; 1998

[31] Hamami L, Nassereddine B. Application of wireless sensor networks in the field of irrigation: A review. Computers and Electronics in Agriculture. 2020;**179**:105782. DOI: 10.1016/j.compag.2020.105782

[32] Aslan MF, Durdu A, Sabanci K, Ropelewska E, Gültekin SS. A comprehensive survey of the recent studies with uav for precision agriculture in open fields and greenhouses. Applied Sciences. 2022;**12**:1047. DOI: 10.3390/app12031047

[33] Srivastava P, Bajaj M, Rana AS. Irrigation system using IOT. In: 4th Int Conf Adv Electr Electron Information, Commun Bio-Informatics Overview. New York: IEEE; 2018. pp. 2-6

[34] Evans SR. Effects of Wi-Fi-enabled smart irrigation controllers on water use and plant health of residential landscapes in the intermountain west. [All graduate theses diss.]2020 [cited May 17, 2022]. p. 7920. Available from: https://digitalcommons.usu.edu/etd/7920

[35] Manjhi P, Sinha S, Vidyapeetham AV. Design of Automated Irrigation System using ZigBee. International Journal of Engineering Research and Advanced Development. 2020;**4**(4)

[36] Hamami L, Nassereddine B. Towards a smart irrigation system based on wireless sensor networks WSNs). In: Proc 1st Int Conf Comput Sci Renew Energies. ICCSRE 2018. SCITEPRESS – Science and Technology Publications, Lda. 2018:433-442

[37] Asadullah M, Ullah K. Smart home automation system using Bluetooth technology. In: 2017 International Conference on Innovations in Electrical Engineering and Computational Technologies. New York: Institute of Electrical and Electronics Engineers Inc.; 2017

[38] Kodali RK, Kuthada MS, Yogi Borra YK. LoRa based smart irrigation system. 2018 4th Int Conf Comput Commun Autom ICCCA 2018. New York: Institute of Electrical and Electronics Engineers Inc.; 2018

[39] Prathipa R, Kalaiarasi D, Raj PP, Sasidharan K, Kumar SS. LoRa based smart irrigation system for remote areas. International Research Journal of Engineering and Technology. 2021;**8**:160-164

[40] Mathematics A, Stellamercy M, Sathishkumar N. Automatic irrigation control system using embedded ethernet communication. International Journal of Pure and Applied Mathematics. 2018;**119**:127-132

[41] Li JL, Zheng WG, Shen CJ, Wang KW. Application of modbus

protocol based on μC /TCPIP in water saving irrigation in facility agricultural. IFIP Advances in Information and Communication Technology. 2014;**419**:281-288

[42] Boursianis AD, Papadopoulou MS, Damantoulakis P, Karampatea A, Doanis P, Geourgoulas D, et al. Advancing rational exploitation of water irrigation using 5G-IoT capabilities: The AREThOU5A project. In: 2019 IEEE 29th Int Symp Power Timing Model Optim Simulation, PATMOS 2019. New York: IEEE; 2019. pp. 127-132

[43] Abedin MZ, Chowdhury AS, Hossain MS, Andersson K, Karim R. An Interoperable IP based WSN for Smart Irrigation Systems. 2017 14th IEEE Annu Consum Commun Netw Conf CCNC 2017. Vol. 2017. New York: IEEE; 2017. DOI: 10.1109/CCNC.2017.8013434

[44] Abhinaya EV, Sudhakar B. Design and performance analysis of automatic irrigation system using 6LowPAN networks. Annals of the Romanian Society for Cell Biology. 2021;**25**:2836-2844

[45] Kodali RK, Sarjerao BS. A low cost smart irrigation system using MQTT protocol. In: TENSYMP 2017—IEEE Int Symp Technol Smart Cities. New York: IEEE; 2017. DOI: 10.1109/TENCONSpring.2017.8070095

[46] Bashir RN, Bajwa IS, Shahid MMA. Internet of Things and machine-learning-based leaching requirements estimation for saline soils. IEEE Internet of Things Journal. 2020;**7**:4464-4472

[47] Adelodun B, Mohammed AA, Adeniran KA, Akanbi SUO, Abdulkadir TS, Choi KS. Comparative assessment of technical efficiencies of irrigated crop production farms: A case study of the large-scale Kampe-Omi irrigation scheme, Nigeria. African Journal of Science, Technology, Innovation & Development. 2020;**0**:1-10. DOI: 10.1080/20421338.2020.1755111

[48] Bhattacharya M, Roy A, Pal J. Smart irrigation system using internet of things. In: Lecture Notes in Networks and Systems. Singapore: Springer; 2021. DOI: 10.1007/978-981-13-1217-5_20

[49] García L, Parra L, Jimenez JM, Lloret J, Lorenz P. IoT-based smart irrigation systems: An overview on the recent trends on sensors and iot systems for irrigation in precision agriculture. Sensors (Switzerland). 2020;**20**:1042. DOI: 10.3390/s20041042

[50] Nigussie E, Olwal T, Musumba G, Tegegne T, Lemma A, Mekuria F. IoT-based irrigation management for smallholder farmers in rural Sub-Saharan Africa. Procedia Computer Science. 2020;**177**:86-93. DOI: 10.1016/j.procs.2020.10.015

[51] Campos NGS, Rocha AR, Gondim R, da Silva TLC, Gomes DG. Smart & green: An internet-of-things framework for smart irrigation. Sensors (Switzerland). 2020;**20**:1-25

[52] Adeyemi O, Grove I, Peets S, Norton T. Advanced monitoring and management systems for improving sustainability in precision irrigation. Sustainability. 2017;**9**:1-29

[53] McCarthy AC, Hancock NH, Raine SR. Advanced process control of irrigation: The current state and an analysis to aid future development. Irrigation Science. 2013;**31**:183-192

[54] Moore KL. An introduction to iterative learning control. Csm Eges. 2006

[55] MathWorks. IoT analytics—ThingSpeak internet of things. Thingspeak. 2022. Available from:

https://thingspeak.com/ [cited May 25, 2022]

[56] Evans RG, Sadler EJ. Methods and technologies to improve efficiency of water use. Water Resources Research. 2008;**44**:1-15

[57] Afzaal H, Farooque AA, Abbas F, Acharya B, Esau T. Precision irrigation strategies for sustainable water budgeting of potato crop in prince Edward Island. Sustainabilty. 2020;**12**:2419. DOI: 10.3390/su12062419

[58] Conesa MR, Conejero W, Vera J, Ruiz-Sánchez MC. Soil-based automated irrigation for a nectarine orchard in two water availability scenarios. Irrigation Science. 2021;**39**:421-439. DOI: 10.1007/s00271-021-00736-0

[59] Belayneh BE, Lea-Cox JD, Lichtenberg E. Costs and benefits of implementing sensor-controlled irrigation in a commercial pot-in-pot container nursery. HortTechnology. 2013;**23**:760-769

Chapter 14

Resilience of Irrigated Agriculture to Face the Challenges in Mediterranean Climatic Conditions (Iberian Peninsula)

António Canatário Duarte, Amparo Melián-Navarro and Antonio Ruiz-Canales

Abstract

Climate change scenarios in Mediterranean basin point to a decrease in the amount of annual rainfall and the increased frequency of drought. In this framework of greater water scarcity, an increase in irrigation costs is expected, so its rational and efficient use is an unavoidable issue in modern irrigated agriculture. In the last 60 years in Portugal, it had a great increase in the efficiency of water use in agriculture, accompanied by a great increase in energy consumption, and the variation was 15,000 to 6000 m^3/ha.year and 200 to 1500 kWh/ha, respectively. The rational application of fertilizers is a priority, to prevent the contamination of superficial and subterraneous waters, and the process of soil salinization in semi-arid conditions. The pressure of water demand by agriculture implies the use of other water sources. For example, in 2010, the volume of unconventional water resources in Spain rose to 4.540 hm^3/year. Of the total used in agriculture, 450 hm^3 of water comes from the reuse of treated water, and 690 hm^3 comes from desalination. The use of modern/smart technologies in irrigated agriculture, like information and communication technologies, allows the rapid share of information between all the system components and can promote optimized answers at different scales.

Keywords: resilience of irrigated agriculture, Mediterranean climate change, water scarcity, water-efficient use

1. Introduction

Agriculture plays a vital economic role in the Mediterranean region. It employs more than a fifth of the population in 50% of the countries and contributes >10% of Gross Domestic Product (GDP) in eight countries alone [1]. The Mediterranean region in this study refers to the 21 countries surrounding the Mediterranean Sea in addition to Portugal. Mild winter temperatures and long hot dry summers that are characteristic of this region make it ideal for growing a diverse range of crops including olives, citrus, vineyards, and cereals, as well as high-value horticulture.

As precipitation across the region is subject to high inter-annual and seasonal variability, irrigation is an essential component of production for many farmers as it supports crop diversification, helps assure yield and quality, and helps to stabilize food supplies [2]. Water scarcity and droughts are frequent, widespread phenomena that affect more than 100 million people and around a third of the European territory. Global change (climate change and change in land use) is expected to aggravate this situation, especially in Mediterranean countries, such as Spain and Portugal [3].

Climate change is a diffuse and gradual phenomenon. Several factors intervene in it, although we can synthesize it in two in particular: that referred to the natural variability of the climate itself (with recurring events throughout history, famines with droughts due to temperature rises and a decrease in rainfall) and the actions of man who affect it (anthropic effects). Along these lines, the United Nations Framework Convention on Climate Change (UNFCCC) defines it as "a climate change attributed directly or indirectly to human activity that alters the composition of the global atmosphere and that is added to the natural variability of the climate observed during periods comparable time periods [4]. Another aspect of climate change that particularly affects irrigation is the increase in the frequency of extreme events: torrential rainfall and droughts. The recent DANAs (Isolated Depression in High Levels) have caused important floods in the Spanish Mediterranean in recent dates (like 2019), and in the Ebro, with social, economic, and environmental losses. At the same time, periods of drought have occurred with excessive frequency in both countries. Both phenomena entail significant social and economic costs to deal with them. Crops are essential to mitigate desertification and temper climate change. They contribute to avoiding or at least cushioning the succession of extreme phenomena. The concentration of CO_2 eq (CO_2 equivalent) in the atmosphere has not stopped increasing, since in 1880 it was about 280 ppm and today it reaches 408 ppm [5]. Irrigated crops are in a greater proportion than rainfed crops as they produce more biomass.

Agriculture defends itself from drought by improving the efficiency of water application in irrigation and with improved management. Obviously, irrigation requires significant energy consumption in most cases. For this reason, water stress reduces the productivity of crops, and avoiding water stress is the reason for the practice of irrigation, as old as agriculture itself [6].

Spanish irrigation has already begun to take measures to adapt to and mitigate climate change. The first of these has been related to the improvement of irrigation efficiency through the modernization processes of irrigated areas. This has meant that in the period 2002–2016 there have been significant savings both in total consumption (112.5%) and in unit consumption 118.24%). The installation of control networks has also contributed to this through the installation of agrometeorological stations, (e.g. SIARI Network to determine ETo values), such as that of irrigation water quality control networks (for example RECAREX in Extremadura to estimate the degree to which irrigation pollutes water masses). Irrigation is regarded as one of the main adaptations to support crop production in response to climate change and population growth [2]. However, any increase in irrigation demand will correspondingly impact energy consumption and greenhouse gas emissions suggesting potential conflicts in terms of mitigation and adaptation policies [7].

The availability and reliability of water resources is a limiting factor for economic development in many water-stressed countries. The Mediterranean region is one of the most water-scarce regions globally. Water is particularly scarce in Southern and Eastern countries and in some catchments in the North, such as Southeast Spain and the Ebro Depression, where the expansion of irrigated production, coupled with tourism and urbanization has created significant water supply challenges [8]. The improvement of agricultural water management, through irrigation systems modernization and application of information and communication technologies to increase crop productivity and reduce the influence of drought and promoting water conservation is one of the main objectives of current irrigated agriculture in Mediterranean countries [9]. Another challenge in intensive agricultural systems, such as irrigated agriculture, is the necessary compatibility of soil and water as natural resource and as well production factors [10].

2. Approaching the problem from a technical perspective

One of the most important problems brought by climate change in irrigated agriculture, particularly in Mediterranean basin, is the increased variability of weather patterns and a higher frequency of extreme weather events [11]. For the management of water resources, Water Authorities (WAs) have to make decisions before knowing the weather conditions they are going to face. So, every year, due Water Authorities are less able to make decisions consistent with the weather pattern of the following season due to the decreased predictability of events and the less relevant use of past records to make future decisions. As a consequence, current water management decisions are often a compromise between the outcome determined by all the weather states that could emerge. On the other hand, considering the necessary effort to reach convenient water use and management, the rehabilitation of the irrigation project structures, and the modernization of irrigation systems are a priority [12].

2.1 Modernization of the irrigation systems

In the first third of the 20th century, an ambitious plan of agricultural development works began in Portugal and Spain, which included some public irrigation schemes of considerable area, and which still are in operation as a result of adaptation and rehabilitation works of their infrastructures [13]. It should be noted that most of these irrigation schemes never had high levels of adherence, for various reasons, namely the incentive to farmers for the new reality of land use. The high cost of agricultural production factors (including fuel and electricity costs), combined with the volatility of agricultural product sales prices, has been a disincentive to investment in agricultural activity, in particular in irrigation. In recent years, we have witnessed a resurgence of interest on the part of farmers, mostly young people, in irrigated agriculture, greatly encouraged by the country's new irrigation schemes aided by modern technologies. In the last half-century, we can see a significant increase in efficiency in the use of water in agriculture, from 15.000 m^3/ha.year in 1960 to 6.600 m^3/ha.year in 2014 (**Figure 1**) [14], mainly due to the modernization of irrigation systems. This has corresponded to the replacement of traditional irrigation systems with water distribution by gravity, by automated systems equipped with pumping systems that require energy for their operation. Thus, energy consumption has increased sharply in the same period, from 200 kW.h/ha in 1960 to 1.534 kW.h/ha in 2014 (**Figure 1**) [14]. On the other hand, the economic productivity of irrigation water (Gross Added Value/m^3 of water, calculated at constant 2006 prices) has increased in the last decade by more than 30% [15].

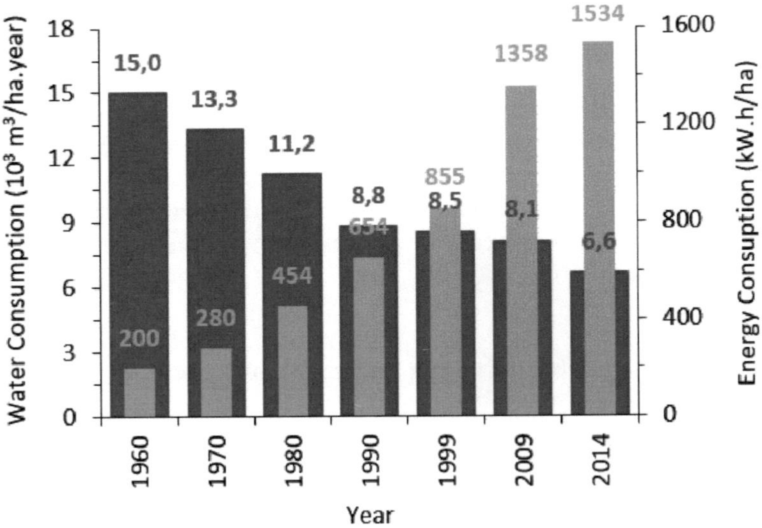

Figure 1.
Evolution of water consumption and energy in Portugal, in irrigated agriculture (Adapted of [14]).

It is foreseeable that in the coming years the efficiency of water use in irrigation will continue to increase, driven by operational programs, financial incentives, and the use of innovative technologies [16]. As mentioned above, irrigation farming is unavoidable in the summer months in most of the Iberian Peninsula, especially in the south of this territory, where there are climatic conditions to enhance the productive capacity of the crops. The climatic variables that influence evapotranspiration (solar radiation, temperature, relative humidity, and wind) reach comparatively more favorable values in the south of the Iberian Peninsula, which determine high rates of that indicator of the physiological activity of plants [17]. It is evident that the productive potential of the crops is not only expressed through the convenient supply of water but also through other cropping practices allowed to obtain good production, such as adequate levels of fertilization, the effective fight against pests and diseases, among others [9]. In the context of climate change, in many aspects already confirmed, the Intergovernmental Panel on Climate Change (IPCC) has been alerting to the scenario of a drier climate determining less available water resources, less production of hydropower energy and losses of 25% of volume water intended for agriculture activity [18]. In this context of greater water scarcity, the greater pressure in its demand for irrigation and other activities should lead to an increase in the costs of irrigated agriculture [19].

2.2 Information and communications technologies

Adaptation and mitigation are the two pillars facing climate change. In this regard, weather and climate services can help decision-makers in making informed decisions to improve adaptation capacity by assessing and forecasting existing and emerging risks. Since all adaptation actions depend on the availability of adequate information, the rapid diffusion of Information and Communication Technologies (ICTs), such as mobile phones and the internet, poses new opportunities to face climate change by improving access to information and consequently by improving the information environment under which water suppliers and water users operate [20]. The availability of ICTs might contribute to mitigating the moving-target problem by providing timely information on future climate and weather conditions, thereby reducing uncertainty before and during the irrigating season. Overall, the ICT-informed decision process of water management could help irrigated agriculture by reducing losses from climate shocks and taking advantage of favorable years. The potentialities of ICTs for the management of water resources in agriculture motivate many studies normally with the objective of quantitatively estimating economic benefits from the ICT-informed decision process of water management in agriculture at the Water Authorities (WAs) level. Some authors developed theoretical models based on insights from the Bayesian Decision Theory [21–23]. It assesses the economic benefits brought by new pieces of information, influencing WA's perception of uncertain events with direct consequences on its strategic decisions.

Background in agriculture, some models using ICT investigate the role played by information in supporting WAs to rationalize the management of water resources and the prevention of extreme weather event impacts. Because decisions on land and water allocation are sequential across the season and influenced by one another, the methodology accounts for the passing of time in the decision process to assess how the time of information provision affects its usability [21]. Great potential is found for such technologies in contributing to food security and climate change adaptation in the agricultural sector. Qualitative studies showed their benefits for both developed and developing countries. Among these, [24] identified the following: (i) promoting economic performance, (ii) raising efficiency, and (iii) fostering innovation. Nevertheless, reference [25] suggested that ICTs impacts on decision outcomes are highly variable. One reason for this variability lies in the findings of the authors in [26]; according to them, ICTs are successful only when key information needs are addressed. In addition, many ICTs projects do not reach the expected success because developers take useful information for granted. As a consequence, ICTs developers tend to poorly consult end users on their information requirements and the resulting ICTs may turn out to be inapplicable in their decision process.

The elements characterizing information and determining its value are: (a) content of information: the WA must be able to implement the additional information in the decision process; if the WA is not able to act upon information, it has no value for it; (b) accuracy of information: the more accurate the information is, the smaller the risk of failures and the higher the VOI (Value of Information); imprecise information is not capable of inducing any change in WA beliefs; (c) timing of information provision: information must be provided at the right time in the decision process; late messages have no value. The timing factor (c) plays a key role in influencing the accuracy of information (b). Usually, information provided well in advance of the occurrence of an event might condition strategic decisions but it will not be so accurate. If information is provided in a short advance, the decisions influenced by the information will not be so strategic, but the information will be likely more accurate. This is typically the case with emerging information, such as weather forecasts.

Developing and applying a method to assess the economic value of ICTs seems to be an interesting topic for agricultural and resource economists. Moreover, considering the growing societal demand for climate services, together with the limited budget available, this topic is of high policy relevance.

3. Approaching the problem from policy, economic and institutional perspective

Since antiquity, the use of water in agriculture has been a basic element for the survival and economic and social progress of humanity. This explains why, since then, the irrigated area of the world has not ceased to grow [27], making irrigation a key element for feeding the planet. All over the world, Asia is irrigating 41% of the total cultivated surface, America 13%, Europe covering 9%, Oceania is irrigating the 7%, and Africa the 5%. The authors in [28] estimate a total irrigation potential of some 402 million hectares in developing countries, of which half is currently in use. In this way today, although the extension of global irrigation is limited since it represents only 7% of the world's useful agricultural area, this type of agriculture is a key element for feeding the planet [29]. Irrigation contributes exceptionally to social cohesion and stability since it generates a strong demand for labor and favors the commercial exchange of products and supplies, with the consequent economic flows (of consumption and savings) [30]. From north to south (Mediterranean area) the natural scarcity increases, being greater in the area of the Region of Murcia and Almería (Andalusia). They are the most productive irrigation but also the most deficient. In addition, this famine is aggravated by recurrent extreme events (droughts and floods) [31], that force us to think about management measures to prevent their effects and consequences [32]. Climate change projects a decrease for the year 2027 between 2 and 11% in the average contributions of the peninsular basins, whose effects will be greater in the most limiting or sensitive areas. The need for irrigation and its efficiency is indisputable [33].

3.1 Politic, economic and institutional aspects

The role of water in the economy covers various aspects from the analysis of the uses of water and its impact on the different economic sectors to economic instruments in the planning and management of water resources. In the first approach, the economics of water advocates determining the volume that is used, the forecast of future demand, its elasticity in relation to variations in other parameters, and the productivity of the different uses. The second focuses on the instruments of economic content that public authorities use in the planning and management of water resources. The National Agrarian Accounting Network of Spain indicates that for the year 2017 [34] the annual value of the production of an average hectare of irrigated land in Spain is 5.4 times higher than one of rainfed land (€5.576 compared to €1.030). Irrigation, in addition to promoting higher income, also makes it more secure by favoring the diversification of production and not depending on rainfall. In line with the greater profitability of the irrigation activity is the land value of the irrigated land. The transformation into irrigation on average multiplies by three the value of the land. In fact, in 2018 the average price of rainfed land for the whole of Spain stood at €9.447/ha, while irrigated land reached €28.444/ha [34]. Irrigated agriculture contributes more than half (64%, 16,000 million euros) to the value of the Spanish Final Agricultural Production, using less than a quarter (22.5%) of the national cultivated area (**Table 1**). The profitability and economic productivity of water are also different depending on the type of crop. 80% of irrigation water has returned between 0.02 and 0.60 €/m^3. It stands out that 19% of the volume used in irrigation is used for crops with very low profitability (less than €0.02/m^3. And only 1% for crops with returns greater than €3.00/m^3 [36].

Contribution of irrigation in Spain	Irrigated crops	Rainfed crops
Economic contribution		
Production value (€/ha)	5.576	1.030
Agriculture net income (€/ha)	2.328	484
Land value (€/ha)	28.444	9.447
Aid CAP/Agriculture net income (%)	24.2%	47.3%
Final Agricultural Production Contribution (%)	64%	36%
Social contribution		
Employment generation (Agricultural Work Unit (AWU/ha)	0.109	0.024
Net Value Added/AWU (€/AWU)	31.782	25.516

Table 1.
Economic and social contribution of irrigation in Spain (year 2017) [35].

In Spain and Portugal, an important part of irrigation is supplied by surface water, coming from the reservoirs of the great Iberian rivers. Therefore, it is from these river valleys, both those that flow into the Mediterranean and, mainly, those that flow into the Atlantic, that supply three-quarters of the Iberian irrigation. Irrigation is associated with high energy consumption. This makes irrigation communities especially vulnerable to energy prices with consequences on their economic situation [37–39]. Some prices are appraised and have a political component. Until a few years ago, there was a subsidized electricity rate for irrigation in Spain. For example, in the Segura Hydrographic Confederation for 2018, the different prices paid by the final user of water for irrigation are barely €0.033/m^3 for river water, €0.10/m^3 when water comes from the transfer (between river valleys), or €0.5/m^3 if it comes from a water treatment plant [40]. In Portugal, great variability is also observed in the form of water pricing. The cost of water in most irrigation communities is based on area, and soil quality, because in some areas the hydrants do not have water meters; the most modern irrigable areas already have them and in these cases the water will be paid for the volume spent. Among those communities that charge water considering the irrigation area, they also differentiate their price considering the crops, and supposedly the water will be more expensive for the most profitable crops and with better soils. Thus, for example, the cost can vary between €77.5/ha (Cova da Beira Irrigation District) and €556.8/ha (Alqueva Irrigation District) on land cultivated with corn [41]. These prices include utilization fees and maintenance fees.

Special mention in Spain requires the cost of the Tajo Segura Transfer water. It is an aqueduct that transfers water from the Tagus River in an amount of 421 hm^3 per year to the Segura River and therefore allows to reduce the scarcity of water in the Spanish southeast, where it is used with great efficiency. As a note to indicate that a few years the agreed total has been transferred and the transferred water covers not only irrigation needs but also other uses (urban). For some provinces of the Spanish southeast, the impact of the transfer is very high. It is enough, for example, to indicate that 62% of the agricultural area of the province of Alicante corresponds to areas that can be irrigated by transfer, while in Murcia it is 55% [42]. Transfers, like any other water policy measure, present economic costs, as well as environmental ones, which must be compensated. The total invoiced for the transfer represents 20% of all the income from fees and tariffs that are produced in all the Hydrographic Confederations of Spain, despite the fact that the transferred water is only 3% of the total water used for irrigation in Spain [43]. As of 2017, a modification has been introduced in the rate that obliges water users to cover the construction costs, as well as the fixed costs of the infrastructure (of the entire infrastructure despite not transferring all the water agreed), increasing average costs.

3.2 Unconventional waters use

The scarcity of water determines and even forces the search for other non-conventional alternatives such as desalinated water or water treatment effluents so that water from treatment plants or desalination plants is used for irrigation, mainly mixed with well or surface water. Agriculture is the main user of reclaimed water and is reported to be used for this purpose in around 50 countries, on 10% of all irrigated land [44]. The main difficulty lies in the price, too high to be supported by farmers, and in the characteristics of the water obtained, due to the excess of salts, with the problems involved by the aquifers and soils. However, improving the quality of effluents is essential for their subsequent use, since water is a reusable resource that can have an almost unlimited life if it's well managed [45].

In 2010, the volume of unconventional resources in Spain rose to 4.540 hm^3/year, of which, those with effective use would add 450 hm^3 of water coming from the reuse of treated water and 690 hm^3 coming from desalination [46]. This amount has only grown and there are currently some 1.000 desalination plants in Spain with an installed capacity of 1.205 hm^3/year and 2.530 treatment plants that treat a flow of 3.375 hm^3/year, although their capacity is 30% higher [3]. Reclaimed water is one more resource within water management, and although it cannot be considered a conventional resource, it does have a key role in comprehensive water planning [47]. Despite this, its importance, at a quantitative level, is still very low, representing around 3% of the total available water resources. Although in practice it's found that the use of treated water is limited in agriculture basically because of its price ranges between €0.6/m^3 and €0.8/m^3.

4. Approaching the problem from an environmental perspective

The rational use of water should imply its moderate and efficient consumption, and the conservation of its quality after use and release back into the water environment. This idea is a fundamental concern in the Water Framework Directive (Directive 2000/60/EC of October 23rd, 2000). Several authors point out that improving water management at the fields level is the most effective tool in reducing the impacts of irrigation on the environment [48, 49], is the achievement of this objective, inseparable from convenient management of water resources at the scale of irrigation district [16, 50]. Agricultural and livestock activity is one of the primary sources of non-point source pollution, occurring in extensive areas and highly dependent on its hydrological behavior, which leads to the water bodies substances previously deposited in the soil (fertilizers, phytosanitary products, organic matter) [51]. For this reason, the focus on controlling this type of pollution has been more on indirect instruments such as codes of good agricultural practice embodied in agri-environmental measures complemented with monetary incentives to farmers [52]. In the climatic scenario of water scarcity, a considerable increase in irrigation costs is expected due to the pressure of water demand, so its rational and efficient use, combined with environmental concerns, is an unavoidable issue in modern irrigated agriculture.

4.1 Rational application of fertilizers

The amount of nitrogen fertilizers incorporated into the soil should be in proportion to its removal by the crops, safeguarding unnecessary spending with excess fertilizer and situations of contamination of soil, water, and atmosphere [53], since, in more oxidized forms, nitrogen is very soluble and mobile, presenting itself as one of the most problematic substances in water pollution [54]. However, this desideratum has not been achieved in recent years in Portugal, and there has even been an increase in the excess of nitrogen fertilizers between 2000 and 2017 from 144.7 to 153.1 thousand tons, corresponding to an increase of 5.8% (**Figure 2**). It should also be noted the importance of gaseous nitrogen emissions from the application of fertilizers in agricultural activity, which in 2017 amounted to 36.3 thousand tons, with a considerable impact on air quality and the water cycle. Nitrogen losses to the atmosphere are volatilized in the form of elemental nitrogen, ammonia ammonium ion, and nitrous oxide, the latter having a very harmful influence on the greenhouse effect in the atmosphere [56].

Concerning the nutrient phosphorus, the amounts applied in fertilizations are much lower, and its mobility in the soil is lower than that of nitrogen and is preferentially loaded outside of agricultural fields together with sediments [57, 58]. Even considering the improvements that have been seen in the characteristics of fertilizers, particularly in the efficiency of absorption of this nutrient by plants, the excess phosphorus applied still has a very high value, contributing to the progressive saturation of the soil with this element. We can verify by reading the graph in **Figure 2** that the phosphorus balance had a significant decrease between 2000 and 2017 (35%), passing the surplus amount of this element from 36.6 to 238 thousand tonnes. In the same period, the situation regarding the nitrogen balance presents an opposite trend.

4.2 Irrigation water quality and the problem of salinization

The high concentration of salts in the various compartments of its cycle, depending on its nature, may cause inconveniences to economic, environmental, and social order. This problem usually occurs under particular climatic conditions, being typical in irrigated areas where water with high salinity is used [59], frequently by downstream reuse of water that has already been used once or several times in irrigation [60, 61]. In a given agricultural system, the balance of salts is considered adequate when the level of soil salinization is compatible with the expected crop yield [62]. In a scenario of climate

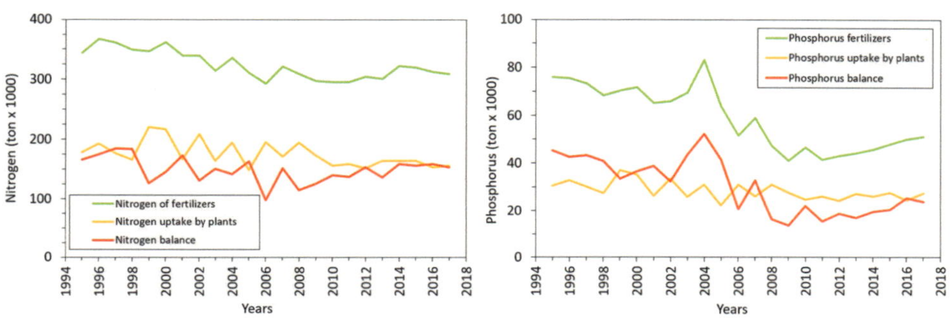

Figure 2.
Gross balance (incorporation – Removal by plants) of nitrogen and phosphorus from fertilizers applied in agricultural activity in Portugal (Adapted of [55]).

change for Portugal and Spain, where an increase in temperature is predicted, with a greater incidence of drought phenomena, and a decrease in total annual precipitation, especially in the south of the Iberian Peninsula, the problem of salinization may become an agro-environmental problem with some acuity, due to the decrease in the washing of salts from the soil and its evapoconcentration in the plant root zone [41]. Of particular concern are the areas of more intensive agriculture, with massive applications of fertilizers, namely the areas of irrigation perimeters in southern Portugal and Spain, and, for similitude of climatic conditions, the southern European countries [63]. In Portugal the estuarine zones of some rivers and other coastal areas, and in Spain, especially in the Ebro River valley and large areas between Almeria and Valencia, the soils are potentially affected by salinization. A significant problem is the particular case of salinization called sodization, currently affecting more than 50% of areas near the coast of Cadiz in Spain. Areas near the coast may be affected by the advancement of the seawater interface due to the intensification of groundwater abstractions [64].

In addition to the excessive application of fertilizers to crops, especially irrigated crops, the quality of the water derived for irrigated areas is also a determining factor in inducing salinity in soils [65]. This aspect is crucial when the water abstracted for an irrigation district already has return flows enriched with nutrients (salts) from several irrigation zones and portends a deteriorated quality [66]. If the irrigation water has of good quality, it is usually not degraded after being used and returned to the natural drainage network. To exemplify this idea, the results of two irrigation seasons observed in an experimental basin located at the Irrigation District of Campina da Idanha, in the center-eastern region of Portugal, are presented below. Indeed, it is possible to verify the low mineralization of the irrigation water derived from a distribution channel, whose quality refers to a category of excellent, and slight variation throughout the irrigation season, rarely exceeding the limit of 100 S/cm (**Figure 3**). On the other hand, the return fluxes from this basin also present a good quality regarding this parameter, not compromising its use downstream [68].

Once the problem of soil salinization is installed, some corrective measures can be adopted to improve crop yields, such as (i) adoption of rotations that include crops more tolerant to salinity and that ensure acceptable yields for farmers; (ii) paying particular attention to the crops tolerance to salinity in the emergence phase, practicing

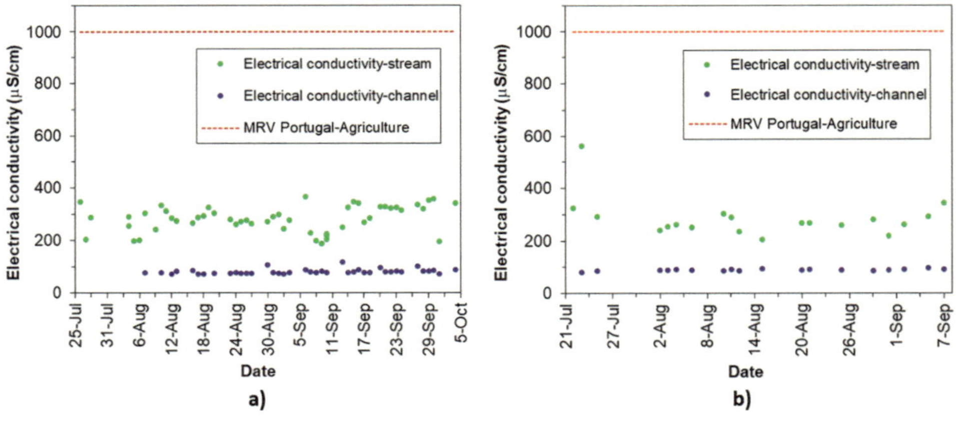

Figure 3.
Evolution of water quality (electrical conductivity) in a distribution channel of Idanha Irrigation District (Portugal), and the return fluxes of an irrigated small basin, in irrigation seasons 2004 (a) and 2005 (b), and the maximum recommended value (MRV) of irrigation water quality in Portugal (Adapted of [67]).

one or more irrigations to decrease the osmotic potential in the soil; (iii) more frequent irrigations than contemplate an adequate leaching fraction for the maintenance of an acceptable salinity level; (iv) adoption of localized irrigation systems, allowing to have high moisture contents in the humidified soil volume mitigating the salinity effect [67].

5. Conclusions

The edaphoclimatic particularities of the Mediterranean basin, namely in the Iberian Peninsula, determine that irrigated agriculture is unavoidable to guarantee the quality of agricultural products and levels of production to the necessary economic viability. However, this goal implies strong challenges, like the compatibility between agricultural activity and the conservation of natural resources.

The great increase in the efficiency of water use in agriculture was accompanied by a great increase in energy consumption, due the modernization of irrigation systems. In the last 60 years in Portugal, the variation was 15,000 to 6000 m^3/ha.year, and 200 to 1500 kWh/ha, respectively. The use of modern/smart technologies in irrigated agriculture, increasingly widespread in the last time, like Information and Communication Technologies, allow the rapid share of information between all the system components, according to the actual or forecasted situation, and using models and artificial intelligence, can promote optimized answers at different scales (irrigation systems in field, distribution water network in the irrigation project, water storage).

The alternative/unconventional sources of water to supply the necessity in agriculture, already quite contribute to the resilience of this activity in Spain. For example, in 2010, the volume of unconventional water resources in Spain rose to 4.540 hm^3/year. Of the total used in agriculture, 450 hm^3 of water comes from the reuse of treated water, and 690 hm^3 comes from desalination.

Nevertheless, the main difficulty lies in the price, too high to be supported by farmers, and, many times, in the characteristics of the water obtained. Special mention must be referred about the water transfer between regions under rules of equity, and periodically reviewed. An example in Spain is the water transfer from the Tagus River in a volume of 421 hm^3/year to the Segura River.

The rational application of fertilizers, in line with the crops needs, is a priority, to prevent the contamination of superficial and subterraneous waters, and to prevent the gradual process of soil salinization in semi-arid regions. Equally important is the evaluation of the water quality to irrigation and accounting the amount of nutrients that already contain.

Acknowledgements

This work is supported with Portuguese national funds by FCT - Foundation for Science and Technology, I.P., within the following projects:

GEOBIOTEC—Project UIDB/04035/2020;

CERNAS-IPCB—Project UIDB/00681/2020.

Conflict of interest

The authors declare no conflict of interest.

Author details

António Canatário Duarte[1]*, Amparo Melián-Navarro[2] and Antonio Ruiz-Canales[2]

1 Scholl of Agriculture/Polytechnic Institute of Castelo Branco, Castelo Branco, Portugal

2 Higher Polytechnic School of Orihuela/University of Miguel Hernández, Orihuela, Spain

*Address all correspondence to: acduarte@ipcb.pt

References

[1] FAO. Rethinking rural development in the mediterranean. Paris, France: International Centre for Advanced Mediterranean Agronomic Studies (CIHEAM). 2009. p. 3761

[2] Hanjra MA, Qureshi ME. Global water crisis and future food security in an era of climate change. Food Policy. 2010;**35**(5):365-377

[3] Jódar A. Hydrological Impacts of Global Change: Adaptation Alternatives in Complex Water Systems with Water Resources Scarcity and Severe Natural Risks (Southeast of Spain). Alicante, Spain: Universidad de Alicante; 2021

[4] UNFCCC. Kyoto protocol to the United Nations framework. Review in European Community International Environmental Law. 1998;**7**:214-217. DOI: 10.1111/1467-9388.00150

[5] Sánchez-Mora JI. Los regadíos ante el cambio climático. In: Berbel-Vecino J, editor. Externalidades positivas del regadío. Madrid: Federación Nacional de Comunidades de Regantes de España (FENACORE).; 2020. pp. 89-102

[6] Testi L, López-Bernal A. El regadío y la mitigación del cambio climático (captura de CO2). In: Berbel Vecino J, editor. Externalidades positivas del regadío. Madrid: Federación Nacional de Comunidades de Regantes de España (FENACORE); 2020. pp. 80-88

[7] Carrillo-Cobo MT, Camacho-Poyato E, Montesinos P, Díaz JAR. Assessing the potential of solar energy in pressurized irrigation networks. The case of Bembézar MI irrigation district (Spain). Spanish Journal of Agricultural Research. 2014;**12**(3):838-849

[8] García-Ruiz JM, López-Moreno JI, Vicente-Serrano SM, Lasanta–Martínez T, Beguería S. Mediterranean water resources in a global change scenario. Earth-Science Reviews. 2011;**105**(3-4):121-139

[9] Duarte AC, Mateos L. How changes in cropping intensity affect water usage in an irrigated Mediterranean catchment. Agricultural Water Management. 2021;**260**:107274

[10] Duarte AC. Soil erosion by water and climate change, how can reaching the sustainability in European Mediterranean Basin. International Journal of Environmental Science and Natural Research. 2021;**29**:556269

[11] Lemus-Canovas M. Changes in compound monthly precipitation and temperature extremes and their relationship with teleconnection patterns in the Mediterranean. Journal of Hydrology. 2022;**608**:127580. DOI: 10.1016/j.jhydrol.2022.127580

[12] Playán E, Mateos L. Modernization and optimization of irrigation systems to increase water productivity. Agricultural Water Management. 2006;**80**(1-3):100-116. DOI: 10.1016/j.agwat.2005.07.007

[13] Raposo JR. História da rega em Portugal. Lisboa: Ministério do Ambiente e Recursos Naturais; 1994. p. 272

[14] Duarte AC. O uso da água na agricultura em Portugal Parte I - Evolução da agricultura de regadio; aspetos sociais e institucionais. Revista EGITANEA SCIENCIA, Instituto Politécnico da Guarda. 2017;**1**(20):7-28

[15] Silva F. Agricultura: o grande utilizador de água em Portugal – Realidades e desafios. In: Boletim

Informativo n°145 da Associação Portuguesa dos Recursos Hídricos. Lisbon, Portugal; 2012

[16] Lorite IJ, Mateos L, Fereres E. Evaluating irrigation performance in a Mediterranean environment. I. Model and general assessment of an irrigation scheme. Irrigation Science. 2004;**23**:77-84

[17] Allen RG, Pereira LS, Raes D, Smith M. Crop Evapotranspiration: Guidelines for Computing Crop Water Requirements. Rome, Italy: FAO; 1998

[18] IPCC. Climate Change 2007: Synthesis Report. In: Core Writing Team, Pachauri RK, Reisinger A, editors. Contribution of Working Groups I, II and III to the Fourth Assessment Report of the Intergovernmental Panel on Climate Change. Geneva, Switzerland: IPCC; 2007

[19] Fereres E, Connor DJ. Sustainable Water Management in Agriculture. Challenges of the New Water Policies for the XXI Century. Lisse, The Netherlands: AA Balkema; 2004. pp. 157-170

[20] Sørensen C, Kateris D, Bochtis D. ICT innovations and smart farming. In: International Conference on Information and Communication Technologies in Agriculture, Food & Environment. Cham: Springer; 2017. pp. 1-19

[21] Perea RG, Camacho E, Montesinos P, Rodríguez-Díaz JA. Optimisation of water demand forecasting by artificial intelligence with short data sets. Biosystems Engineering. 2019;**177**:59-66. DOI: 10.1016/j.biosystemseng.2018.03.011

[22] Liu W, Shao X, Wu C, Qiao P. A systematic literature review on applications of information and communication technologies and blockchain technologies for precision agriculture development. Journal of Cleaner Production. 2021;**298**:126763. DOI: 10.1016/j.jclepro.2021.126763

[23] Pulido-Calvo I, Gutierrez-Estrada JC. Improved irrigation water demand forecasting using a soft-computing hybrid model. Biosystems Engineering. 2009;**102**(2):202-218. DOI: 10.1016/j.biosystemseng.2008.09.032

[24] Deichmann U, Goyal A, Mishra D. Will digital technologies transform agriculture in developing countries? Agricultural Economics. 2016;**47**:21-33

[25] Aker JC, Ghosh I, Burrell J. The promise (and pitfalls) of ICT for agriculture initiatives. Agricultural Economics. 2016;**47**:35-48

[26] Nakasone E, Torero M. A text message away: ICTs as a tool to improve food security. Agricultural Economics. 2016;**47**(S1):49-59

[27] Arcas N, Alcón F, Gómez-Limón JA, De Miguel MD, Review. The evolution of research regarding the economics of irrigarion water. Spanish Journal of Agricultural Research. 2010;**8**(2):172-186

[28] Faurès JM, Hoogeveen J, Bruinsma J. The FAO irrigated area forecast for 2030. Rome, Italy: FAO; 2002. pp. 1-14

[29] FAO. The State of the World's Land and Water Resources for Food and Agriculture: Managing Systems at Risk. Abingdon (UK): Food and Agriculture Organization (FAO) & Earthscan-FAO; 2011

[30] Melián A, Melgarejo J. El regadío y el desarrollo regional. In: Editado por la Federación Nacional de Comunidades de Regantes de España (FENACORE). Madrid, Spain; 2020. pp. 47-58

[31] Prólogo OJ. Convivir con la inundación. In: López Ortiz MI, Melgarejo-Moreno J, editors. Riesgo de inundación en España: análisis y soluciones para la generación de territorios resilientes. Alicante, Spain: Universitat d'Alacant; 2020. pp. 21-26

[32] Molina A. Instrumentos jurídicos para la regulación de la sequía. In: Melgarejo- Moreno J, López-Ortíz M, Fernández-Aracil P, editors. Inundaciones y sequías. Análisis multidisciplinar para mitigar el impacto de los fenómenos climáticos extremos. Alicante, Spain: Universidad de Alicante; 2021. pp. 1115-1138

[33] Vargas J, Cánovas-García F. Ordenación del territorio y análisis de vulnerabilidad para la prevención del riesgo de inundación. In: Melgarejo-Moreno J, López-Ortíz MI, Fernández-Aracil P, editors. Inundaciones y sequías. Análisis multidisciplinar para mitigar el impacto de los fenómenos climáticos extremos. Alicante, Spain: Universidad de Alicante; 2021. pp. 101-126

[34] MAPA. Metodología y resultados empresariales 2017. Madrid: Ministerio de Agricultura Pesca y Alimentación; 2019

[35] Gómez-Limón JA, Gutiérrez-Martín C. El regadío y la mejora socioeconómica. In: Berbel Vecino J, editor. Externalidades positivas del regadío. Madrid: Federación Nacional de Comunidades de Regantes de España (FENACORE); 2020. pp. 59-68

[36] Puxeu J. El agua como palanca de desarrollo: un recurso global de utilización local. In: Jordana-Butticaz J, editor. El sector agro-mar-alimentario español. Una visión renovada. Cajamar: Mediterráneo Económico; 2015. pp. 185-213

[37] Langarita R, Sarasa C, Jiménez S. Los costes energéticos en la agricultura de regadío. Alternativas para su reducción y efectos de la implantación de una tarifa verde en España. 2016

[38] Corominas J. Agua y energía en el riego, en la época de la sostenibilidad. Ingeniería del agua. 2010;**17**(3):219-233

[39] Braz-Tangerino F, Ferreira MI, Moreno-Hidalgo MA, Playán E, Pulido-Calvo I, Rodríguez-Sinobas L, et al. Visión del regadío. Ingeniería del Agua. 2014;**18**(1):39-53

[40] López-Peñalver FJ, Melián-Navarro A, Ruiz-Canales A, Piccinetti L, Egea-Pérez R, Chazarra- Zapata JP. Uso del agua regenerada en la Europa Mediterránea. Beneficio hídrico, perjuicio económico. In: Actas Congreso Nacional del Agua Orihuela. Innovación y Sostenibilidad. Orihuela: Universidad de Alicante; 2019. pp. 913-923

[41] Duarte AC. O usa da água na agricultura em Portugal-Parte I. Evoluçao da agriultura de regadio. Aspectos sociais e institucionais. Egitania Sciencia. 2017;**1**(20):7-28

[42] Melgarejo P, Martínez-Nicolás JJ, Hernández-García F, Legua P, Martínez-Font R. Necesidades hídricas de la agricultura en la provincia de Alicante. In: Melgarejo J, Fernández M, editors. El agua en la Provincia de Alicante. Diputación Provincial de Alicante y Universidad de Alicante. Alicante, Spain; 2020. pp. 281-296

[43] Chamber of Alicante. El Trasvase Tajo-Segura: Una Infraestructura Necesaria. Cámara Oficial de Comercio, Industria, Servicios y Navegación de Alicante. 2021. Available from: https://www.camarali-cante.com/estudios-economicos/analisiseconomico/infraestructuras/trasvase-tajo-segura/

[44] Winpenny J, Heinz I, Koo-Oshima S. Reutilización del agua en la agricultura:

¿Beneficios para todos?. FAO. Informe sobre temas hídricos. Cuaderno. 2013;**35**:124

[45] Melián A, Fernández-Zamudio MA. Reutilización de agua para la agricultura y el medioambiente. Revista Agua y Territorio. 2016;**8**:80-92

[46] Olcina J, Moltó E. Recursos de agua no convencionales en España: Estado de la cuestión. Universidad de Alicante; 2010. pp. 131-163

[47] Costa-Botella D, Melián-Navarro A. Los retos del uso del agua depurada y desalada en el sureste español: aplicación a la agricultura. In: Navarro Caballero TM, editor. Desafíos del Derecho de Aguas. Variables jurídicas, económicas, ambientales y de Derecho comparado. Thompson Reuters ARANZADI; 2015. pp. 451-460

[48] Pereira LS. Water and agriculture: Facing environment and scarcity challenges. Agricultural Engineering International. CIGR Journal. 2005

[49] Mateos L. Irrigation management and fertlizer/pesticide leaching. In: Trimble SW, Stewart BA, Howell TA, editors. Encyclopedia of Water Science. Boca Raton, FL, US: Marcel Dekker Inc.; pp. 282-284

[50] Mateos L, López-Cortijo I, Sagardoy JA. 2002SIMIS: The FAO decision support system for irrigation scheme management. Agricultural Water Management. 2002;**56**:193-206

[51] Braden JB, Shortle JS. Agricultural Sources of Water Pollution. In: Shogren JF, editor. Encyclopedia of Energy, Natural Resource, and Environmental Economics. Amsterdam, Netherlands: Elsevier; 2013

[52] Evans A, Mateo-Sagasta J, Qadir M, Boelee E, Ippolito A. Agricultural water pollution: Key knowledge gaps and research needs. Current Opinion in Environmental Sustainability. 2019;**36**:20-27

[53] Kopittke PM, Menzies NW, Wang P, McKenna BA, Lombi E. Soil and the intensification of agriculture for global food security. Environment International. 2019;**132**:105078

[54] Hatch DJ, Goulding KW, Murphy DV. Nitrogen as a pollutant from agriculture in surface and groundwater. In: Haygarth PM, Jarvis SC, editors. Agriculture, Hydrology and Water Quality. Wallingford: CABI Publishing; 2002. pp. 7-27

[55] Duarte, A. C. Melián-Navarro, A. 2021. O uso da água na agricultura em Portugal Parte II – Aspectos ambientais da agricultura de regadio. Revista EGITANEA SCIENCIA, N°27, 185-214.

[56] Sun S, Liu J, Zhang M, He S. Simultaneous improving nitrogen removal and decreasing greenhouse gas emission with biofilm carriers addition in ecological floating bed. Bioresource Technology. 2019;**292**:121944

[57] Krasa J, Dostal T, Jachymova B, Bauer M, Devaty J. Soil erosion as a source of sediment and phosphorus in rivers and reservoirs – Watershed analyses using WaTEM/SEDEM. Environmental Research. 2019;**171**:470-483. DOI: 10.1016/j.envres.2019.01.044

[58] Horta C, Roboredo M, Carneiro JP, Duarte AC, Torrent J, Sharpley A. Organic amendments as a source of phosphorus: Agronomic and environmental impact of different animal manures applied to an acid soil. Archives of Agronomy and Soil Science. 2017;**64**(2):257-271. DOI: 10.1080/03650340.2017.1346372

[59] Umali DL. Irrigation-Induced Salinity – A Growing Problem for

Development and the Environment. Washington: World Bank; 1993

[60] Mateos L, Young CA, Wallender WW, Carlson HL. Simulating spatially distributed water and salt balances. Journal of Irrigation and Drainage Engineering. 2000;**126**(5):288-295

[61] Aragués R, Tanji KK. Water quality of irrigation return flows. In: Trimble SW, Stewart BA, Howell TA, editors. Encyclopedia of Water Science. Boca Raton, FL, USA: Marcel Dekker Inc.: 2003. pp. 502-506

[62] Aragués R, Cerdá AC. Salinidad de aguas y suelos en la agricultura de regadío. In: Jiménez-Días RM, Espinosa JL, editors. Agricultura Sostenible. Madrid: Ediciones Mundi-Prensa; 1998

[63] Vargas-Amelin E, Pindado P. The challenge of climate change in Spain: Water resources, agriculture and land. Journal of Hydrology. 2014;**518**:243-249

[64] Argamasilla M, Barberá JA, Andreo B. Factors controlling groundwater salinisation and hydrogeochemical processes in coastal aquifers from southern Spain. Science of the Total Environment. 2017;**580**:50-68. DOI: 10.1016/j.scitotenv.2016.11.173

[65] Chen L, Li C, Feng Q, Wei Y, Zhao Y, Zhu M, et al. Direct and indirect impacts of ionic components of saline water on irrigated soil chemical and microbial processes. Catena. 2019;**172**:581-589. DOI: 10.1016/j.catena.2018.09.030

[66] Minhas PS, Ramos TB, Ben-Gal A, Pereira LS. Coping with salinity in irrigated agriculture: Crop evapotranspiration and water management issues. Agricultural Water Management. 2020;**227**:105832. DOI: 10.1016/j.agwat.2019.105832

[67] Duarte AC. Water pollution induced by rainfed and irrigated agriculture in Mediterranean environment, at basin scale. Ecohydrology and Hydrobiology Journal. 2011;**11**(1-2):35-46

[68] Duarte AC. Non-point source pollution in an agricultural catchment and the quality of return flows, under Mediterranean conditions. Fresenius Environmental Bulletin. 2014;**23**:3237-3247

Chapter 15

Seed Soaking Times and Irrigation Frequencies Affected the Nutrient Quality and Growth Parameters of *Hordeum vulgare* L. Cultivated in Hydroponics

Ryan Anthony Smith, Muhali Olaide Jimoh and Charles Petrus Laubscher

Abstract

The choice of hydroponic systems for fodder production is of great importance to Sub-Saharan Africa and specifically South Africa, considering the current water crisis. This study investigated the impacts of seed soaking times and irrigation frequency on the vegetative and nutritive properties of *Hordeum vulgare* grown in a hydroponic room. *H. vulgare* seeds were weighed and soaked in sterile containers filled with 500 mL solution of 20% solution of sodium hypochlorite for 1, 3, 8, 16 and 24 h at room temperature. Once soaked, the seeds were transferred to a hydroponic system and irrigated using flood irrigation. After the 8-day growing period, growth parameters were measured, and samples were oven-dried, pulverized and then subjected to nitrogen and protein analysis. It was observed that shorter soaking time with varied irrigation frequencies had the highest impact on the weight, and nutrient yield of *H. vulgare* although other growth parameters investigated such as leaf length and root map expansion deviated from this trend under different soaking times and irrigation frequencies. This study revealed that a 1-h pre-soaked treatment was the best for cultivating barley hydroponically. This treatment is recommended for the cultivation of barley as it proved to be beneficial to the farmer in terms of quality yield.

Keywords: barley, fodder feed, forage crop, organic fodder, subsistence farming

1. Introduction

Hordeum vulgare (barley) was one of the first crops to be domesticated for human consumption and is still one of the most important cereal crops grown worldwide [1, 2]. Due to its high nutritional value and palatability, barley is the fourth-largest cereal in terms of grain production worldwide, with nearly 60% used as animal feed, around 30% used for malt production, 7% used for seed production, and only 3%

 IntechOpen

used for human food [3]. The South African Department of Agriculture, Forestry, and Fisheries reported 421,800 tons of barley production in 2018. According to the report, the country's high maize production hampers the feed market for barley, so barley is grown specifically for malting. Furthermore, because barley production is limited to winter rainfall areas, there is a need to optimize cultivation techniques so that it can be produced year-round [4].

Globally, the demand for feeds and forage has increased due to the increase in the livestock population [5]. Hydroponically sprouted cultivars are used as a dietary supplement for animals in South Africa, the United Kingdom, the United States of America, Australia, and other parts of the world due to their ease of germination and growth [6, 7]. Besides, various authors have reported different harvesting and growth cycles for a barley fodder mat, ranging from a 6-day harvest cycle to a 10-day harvest cycle, implying that hydroponically produced fodder has a short growth period [8, 9]. The quality and quantity of fodder produced by hydroponics are also of interest. Farmers in India, for example, discovered that feeding their dairy cattle hydroponically grown barley increased milk yield from 0.5 to 2.5 l/day. They discovered an improvement in animal health as well as the fat content of their cow's milk in addition to the increased yield [10].

Climate change is affecting agriculture and natural water resources all over the world, which has an impact on the sustainability of food and water resources [11]. Because water and agriculture are inextricably linked and vital to most societies' economies and security, hydroponic cultivation ensures year-round production while consuming less water [9, 12], as opposed to run-to-waste systems in field production [13]. It was reported earlier that hydroponic production used only about 2% of the water required for field production of the same crop [14] thus, the introduction of hydroponic systems for fodder production can help overcome the challenges encountered in conventional production.

Previous research has shown that soaking barley seed before sowing increases the rate of germination, softens the seed coat, and breaks seed dormancy, though the number of hours recommended for soaking barley seed ranged from 3 to 28 h [15–18]. Similarly, the importance of using clean seeds for cultivation and seed sterilization during the soaking procedure cannot be overstated [9, 19]. Soaking seeds in a solution of 20% (bleach) for 30 min was recommended to prevent the formation of any fungal contamination [20, 21]. However, [22] using Mercuric chloride to prevent the proliferation of fungal contaminants while [23] tested both the effects of sodium hypochlorite (NaOCl) and Mercuric chloride ($HgCl_2$) on a range of pathogens and proposed that surface sterilization of the seed is important to remove unwanted fungal growth.

Although earlier studies showed a wide range of water types, including both mist and flood irrigation [14, 24, 25], there is a scarcity of data on the amount and frequency of water/irrigation used in a hydroponic chamber to germinate barley seed. This was necessary to address the challenges of nutrient imbalance in hydroponic systems, soil quality and productivity, and soil-based ecosystem services [26–28] affecting barley biomass production. In this study, an 8-day harvest cycle was used in conjunction with post-germination irrigation frequency to determine the most effective method to break seed dormancy, cause germination, and grow into a seedling mat and see the impacts of these treatments on the nitrogen value, protein content, fresh weight, and dry weight of the seedling forage mat post-harvest compared to the original untreated seeds. Because soil-based farming is facing serious challenges in South Africa and other developing countries, fodder crop producers can use the study's findings to obtain critical information that will aid in the optimization of inputs and the efficient utilization of resources in hydroponic fodder production.

2. Materials and methods

2.1 Source of seeds

Viable seeds of *H. vulgare* sv13 were obtained from Kaap Agri Bedryf Ltd. located in Malmsbury, Western Cape. The seeds used originated from the Swartland District of the Western Cape. The seeds were first weighed on a balance that measures in 100 g increments.

2.2 Experimental design and hydroponic setup

The experiment was carried out in the plant tissue culture laboratory at the Cape Peninsula University of Technology's Bellville Campus. A 230 cm × 450 cm growing room was used to control light and temperature and determine the best growing conditions. Shelving units measuring 200 cm in height, 127 cm in length, and 40 cm in depth were installed in the growing room. The shelving unit had six shelves that were 37 cm apart and measured 120 cm × 40 cm. Two fluorescent light bulbs were installed on each shelf. For drainage, a corrugated fiberglass sheet was cut to the size of the shelf below and positioned at a 55-degree angle. The front, bottom end was fitted with a D-shaped gutter. This was used to collect the runoff from the fiberglass sheets. The run-off was then directed back to a sump via the gutter, resulting in an ebb and flow closed watering system. After cleaning and soaking the seeds, they were placed in perforated aluminum containers measuring 10 cm × 20 cm. The perforations were evenly spaced across the tray's bottom surface, with approximately 2 cm between each perforation. There was no need for a medium because the seeds germinated and formed a root mat that held the seedlings in place. The seed trays were then placed on the fiberglass sheeting, and each tray was fitted with an irrigation tube. Irrigation water was delivered to the seeds in their respective trays using a pump (HJ 1542 submersible), which delivered 622.5 mL/min to each tray for 2 min, for a total of 1245 mL. The pump was linked to a timer (MajorTech model MTD7), which controlled the amount of water delivered to each tray [29, 30]. Before the treated seeds were placed in the growing system, the entire setup, including the sump, Perspex shelves, and seed containers, was thoroughly cleaned and disinfected. The sump was filled with deionized water containing a 20% sodium hypochlorite solution, and the system was flushed to disinfect all surfaces [13].

The temperature of the room was kept at 23°C, as it was found that a temperature range of 20–30°C did not have a significant impact on growth [25, 31]. Two Samsung Smart InverterTM air conditioners were used to regulate the temperature. Fresh air was brought into the growing chamber through heap filters from outside the building. Lighting was provided with fluorescent tubes [32, 33]. The fluorescent bulbs used were Osram (L36/640) cool white fluorescent tubes with a light output of 5.96 kilo lux. The ExTech—Heavy Duty Digital Light Meter, model number HD 400, was used to measure the intensity of the light. A Panasonic TB178K timer control unit was used to set the lighting system to provide a photoperiod of 16 h day/8 h night [34, 35].

Treatment code	Description	Treatment code	Description	Treatment code	Description	Treatment code	Description	Treatment code	Description
T1	1 h soak–2 h irrigation	T6	3 h soak–2 h irrigation	T11	8 h soak–2 h irrigation	T17	16 h soak–2 h irrigation	T21	24 h soak–2 h irrigation
T2	1 h soak–4 h irrigation	T7	3 h soak–4 h irrigation	T12	8 h soak–4 h irrigation	T18	16 h soak–4 h irrigation	T22	24 h soak–4 h irrigation
T3	1 h soak–8 h irrigation	T8	3 h soak–8 h irrigation	T13	8 h soak–8 h irrigation	T19	16 h soak–8 h irrigation	T23	24 h soak–8 h irrigation
T4	1 h soak–10 h irrigation	T9	3 h soak–10 h irrigation	T14	8 h soak–10 h irrigation	T19	16 h soak–10 h irrigation	T24	24 h soak–10 h irrigation
T5	1 h soak–12 h irrigation	T10	3 h soak–12 h irrigation	T15	8 h soak–12 h irrigation	T20	16 h soak–12 h irrigation	T25	24 h soak–12 h irrigation

Table 1.
Treatments of H. vulgare *seeds soaked in 500 mL distilled water diluted with a 20% solution of sodium hypochlorite under different irrigation frequencies*

2.3 Treatment preparation

There were 25 treatments, each with 10 repetitions. Each treatment included a pre-soaking period followed by a post-soaking irrigation period (**Table 1**). Each repetition began with 100 g of viable seeds placed in a sterile plastic container containing 500 mL of distilled water containing a 20% solution of sodium hypochlorite (bleach) at room temperature [8, 14]. It was decided to test a range of seed soaking times, namely: 1, 3, 8, 16 and 24 h, which was compared against the control of 16 h. After the allotted soaking time, the seeds were washed in running, deionized water and placed in their respective growing trays without being exposed to darkness. Each tray was 10 cm × 20 cm in size. This ensured that the washed seeds had a depth of 1 cm. After that, the containers were placed in the hydroponic system to germinate (**Figures 1** and **2**). The seeds were allowed to germinate and grow into a forage mat for 8 days at 23°C under a photoperiod of 16-h day/8-h darkness.

Drip irrigation tubes were used to flood each seed tray with 1245 mL of water, with the excess running off through drainage holes in the seed container. The runoff was collected and channeled back into the sump of the hydroponic system for reuse. When necessary, the sump was refilled with distilled water mixed with a 20% bleach solution to ensure disinfection. The five previously mentioned treatments were subjected to five different irrigation intervals, each with 10 repetitions. Flood irrigation was used to fill each seed tray with water every 2; 4; 8; 10; and 12 h, with the control being a 2 hourly water interval [24, 36].

2.4 Data collection

Before removing the seedlings from their trays at the end of the 8-day growing cycle, a grid of 2 cm × 2 cm blocks was placed over the surface of the container, dividing the space into 50 blocks. This was used to determine the average leaf height per block by measuring the height of each leaf in the respective 2 cm × 2 cm block. The average leaf height of the sample plants for each block was then measured to determine the container's overall average leaf height. The longest leaf in each tray was

Figure 1.
Photograph showing the hydroponic setup and irrigation supplied to each tray.

Figure 2.
Photograph of barley seedlings at harvest (photo by R.A. Smith).

also measured and recorded. Thereafter the seedling mat was removed from its tray and the depth of the root mat was recorded to determine whether the initial 1 cm of soaked seed had expanded over the 8-day growing period.

For nutrient analysis, the trays were removed from the experiment and all excess remaining surface water was allowed to drain away after the allotted growth period of 8 days. The seedlings in their respective trays (**Figures 1** and **2**) were then weighed using a Kern KB 360-3 N scale that measures up to 0.01 g to determine their fresh weight. The weight of the seedling mat was calculated by subtracting the weight of the container from this measurement. Once the fresh weight of the plant material was determined, the seedling mat was removed from its tray and placed in brown paper bags before being dried in an oven (Labtech LDO-150F) at 60–70°C for 36–48 h. The plant material was weighed again after it had completely dried to determine its dry weight. Using a Culatti Typ MFC CZ13 mill, the dried plant material was ground and sieved after being weighed. Following that, samples of dried plant material were sent to the Agrifood Technology Station for protein and nitrogen analysis [37].

2.5 Statistical analysis

Data collected were analyzed using a two-way analysis of variance (ANOVA). The analysis was performed using Minitab 19.2.0/October 2, 2019, a stable release developed at the Pennsylvania State University, USA by Minitab LLC. Where F-value was found to be significant ($P \leq 0.05$), Tukey honest significant difference (HSD) was used to compare the interaction between soaking time and irrigation interval at $P \leq 0.05$ level of significance [38].

3. Results

3.1 Average leaf length and root mat expansion

When comparing all soak treatments in conjunction with all irrigation intervals on average leaf length, Treatment 20 with 16-h soaking time and 12-h irrigation duration produced the longest leaf of 14.33 0.9 cm, while Treatment 9 produced the shortest leaf (6.23 0.40 cm) when compared to the control treatment. Both the independent and combined effects of soaking time and irrigation interval on leaf length were significant at $P \leq 0.05$ (**Figure 3**). Additionally, the root mat expansion was most

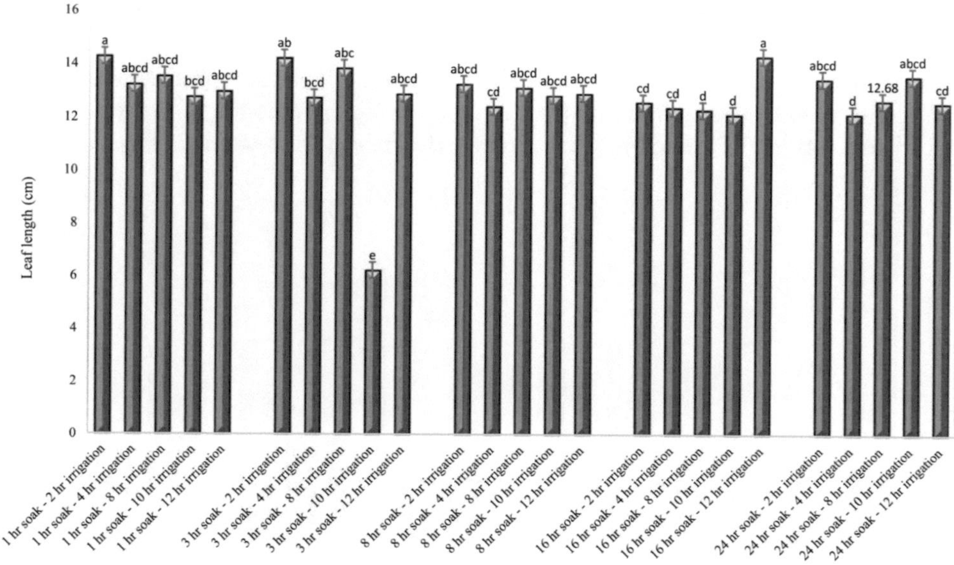

Figure 3.
Effect of soaking time and irrigation interval on average leaf length of H. vulgare. *Means that do not share the same letters are significantly different.*

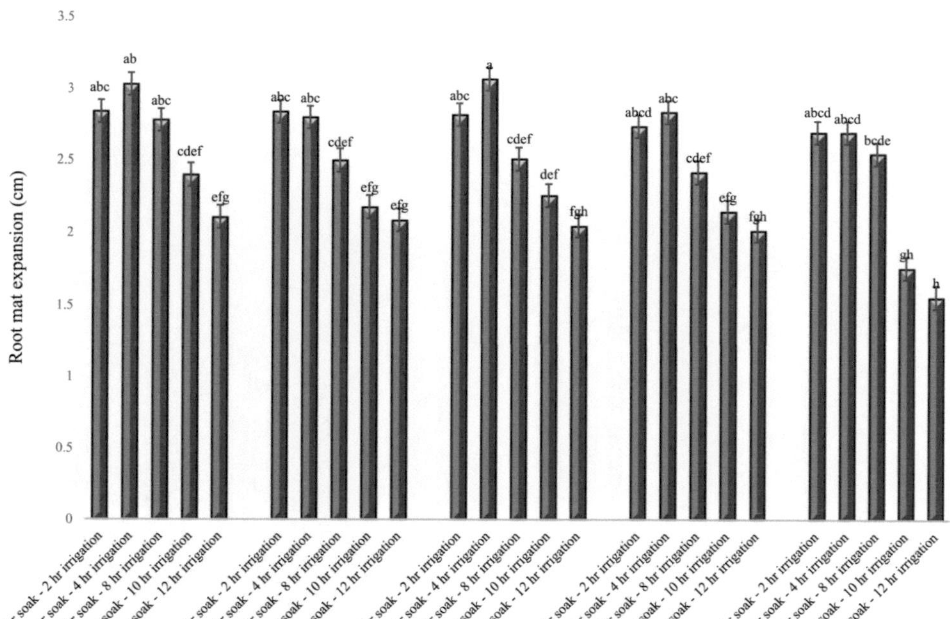

Figure 4.
Effect of soaking time and irrigation interval on root mat expansion of H. vulgare. *Means that do not share the same letters are significantly different.*

significant in Treatment 12 (8 h soak–4 h irrigation) and less significant in Treatment 25 (24 h soak–12 h irrigation) at a 95% confidence limit (**Figure 4**).

3.2 Nitrogen, protein, fresh weight and dry weight analyses

The highest nitrogen yield was obtained from seeds that had been soaked for 1 h and irrigated for 12 h (Treatment 5), while the lowest yield was obtained from Treatment 17 (**Figure 5**). Similarly, the protein content of the samples follows the same pattern, with the highest and lowest nitrogen yields recorded in Treatments 5 and 17, respectively (**Figure 6**). Furthermore, the freshly harvested *H. vulgare* sample

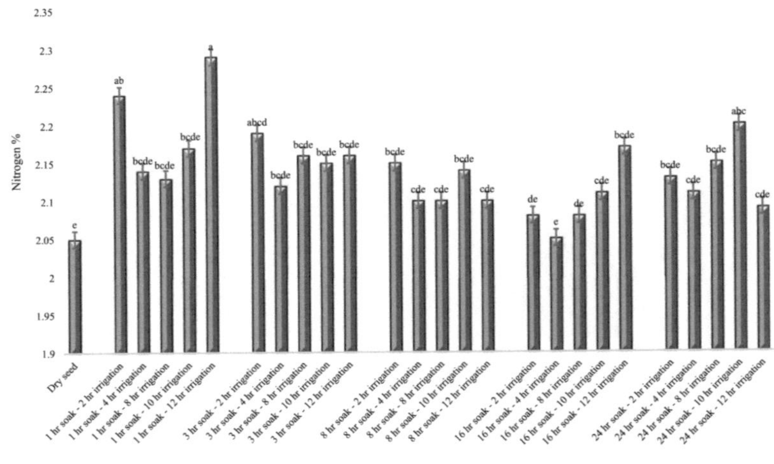

Figure 5.
Effect of soaking time and irrigation interval on the nitrogen content of H. vulgare. *Tukey pairwise comparisons was used to compare means of combined effects of soaking time and irrigation interval at* $P \leq 0.05$. *Means that do not share the same letters are significantly different.*

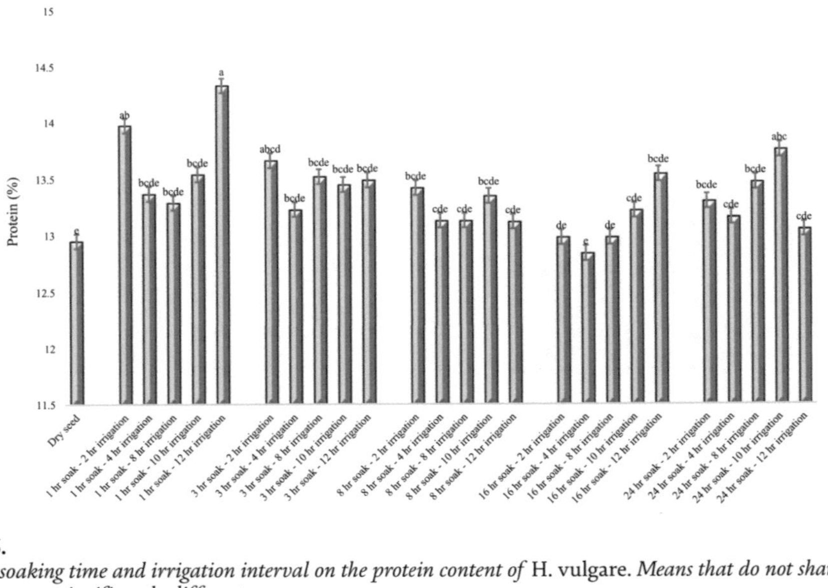

Figure 6.
Effect of soaking time and irrigation interval on the protein content of H. vulgare. *Means that do not share the same letters are significantly different.*

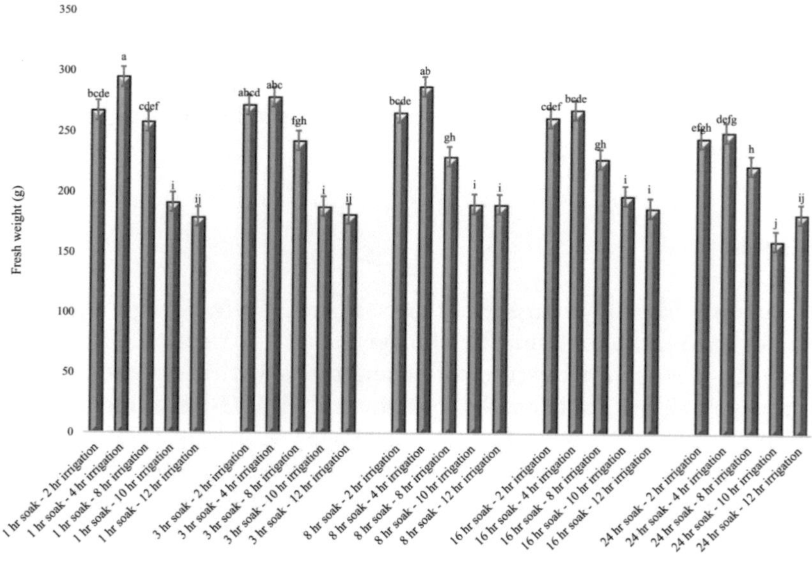

Figure 7.
Effect of soaking time and irrigation interval on fresh weight of H. vulgare. *Means that do not share the same letters are significantly different.*

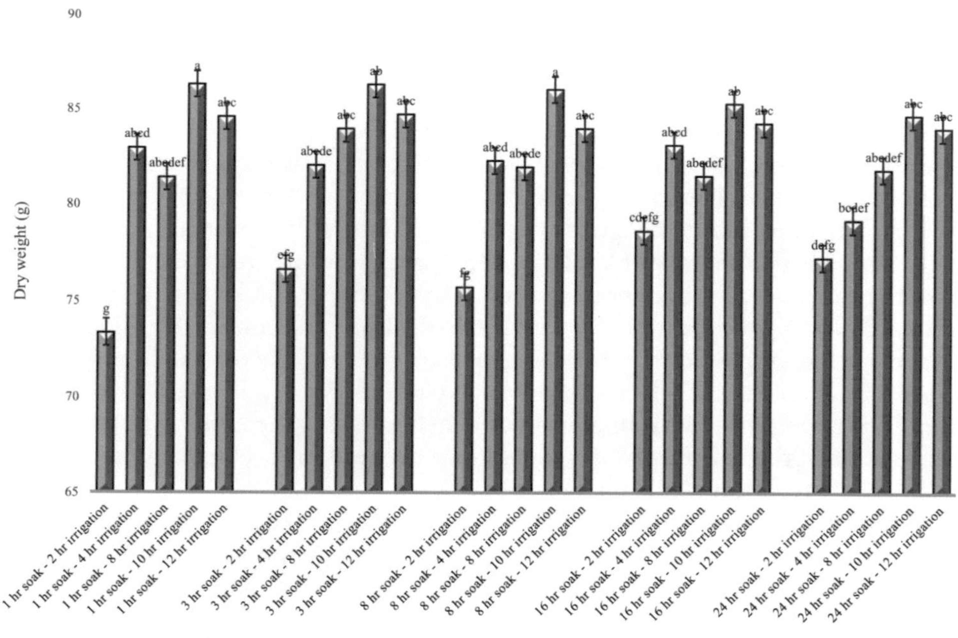

Figure 8.
Effect of soaking time and irrigation interval on the dry weight of H. vulgare. *Means that do not share the same letters are significantly different.*

weighed the most in Treatment 2 (1 h soak–4 h irrigation) and the least in Treatment 24 (**Figure 7**). The dry weight of *H. vulgare*, on the other hand, was highest in Treatments 4 and 14, and lowest in Treatment 1 (**Figure 8**).

Furthermore, the two-way ANOVA revealed that soaking time and irrigation interval had no independent effect on nitrogen and protein yield of *H. vulgare*, as well as the interaction of the two factors. Similarly, the fresh weight was not significantly affected by the two factors independently, but soaking time significantly affected the dry weight of the harvested plant samples. The interaction of the two factors had a significant effect on the dry weight of the sample, although the irrigation interval indicated otherwise.

4. Discussion

Findings from this study suggest that the seeds of *H. vulgare* responded more favorably to a shorter soaking time for a longer period of irrigation. The average leaf height increases as the irrigation period lengthens, with mean average heights ranging from 8 to 12 cm. For example, Treatment 6 with a 3-h soaking time and a 2-h irrigation interval had the greatest effect on average leaf height. Treatments 1 and 3 (**Figures 3–5**) were of slightly less significance because they both had a 1-h soaking treatment but were subjected to 2 hourly and 8 hourly irrigation intervals. These treatments differed greatly from the soaking control of 16 h as reported in [33, 39] but in agreement with [24] with an irrigation control of 2 h. Although Treatment 20 had the highest mean length value of 14.33 cm, it appeared that the seedlings and their corresponding lengths responded more to the increased irrigation frequency of 2 h, confirming the irrigation control used in this study (**Figure 4**). This was consistent with the soak control of 16 h, but it differed significantly from the irrigation control of 2 h [33, 40]. Also, the average height of barley recorded in this experiment differed significantly from previous studies where heights of 14.0 cm and 6.2 cm were respectively recorded by [20, 41].

Treatment 12 (8-h soak with 4 hourly irrigation) had the highest mean root expansion (3.07 cm) after the 8-day growing cycle, which was significantly different from the control. Treatment 2 followed with a mean of 3.03 cm (1-h soak with a 4-hourly irrigation interval; **Figure 5**), which was less significant than Treatment 12 due to a marginal difference of 0.04 cm, which is significant. Treatment 2 would thus allow the cultivator to reduce soaking time again to achieve similar results, with both treatments having the greatest effect with an irrigation frequency of 4 h, though other researchers have reported that transient exposure of *H. vulgare* roots to heavy metals may also have an adverse effect on root mat expansion [35, 42]. It would be interesting to see if changing the irrigation type, from drip to spray, or the mineral/trace metals composition of the irrigation water, would improve seedling root expansion, and if adding liquid fertilizer would do the same.

Furthermore, the highest nitrogen and protein concentrations were obtained in Treatment 5 with a 1-h soaking treatment and a 12-h irrigation interval (**Figures 6** and 7). These results did not agree with the controls of 16 h of soaking and 2 hourly irrigation intervals. This indicated that the seed requires a shorter soaking treatment (1 h) as well as a longer (12 h) irrigation interval to achieve the highest level of nutrients in the seedling at harvest time. However, as discovered during the growth experiment, a 12-h irrigation interval is not beneficial to seedling growth. It was interesting to note that the next highest statistical mean belonged to treatment 1 (1-h soak with a 2-h irrigation interval), which was only marginally less than the highest mean achieved in treatment 5, which also had a 1-h soak time but a 12-h irrigation interval. This was consistent with the control and resulted in seedlings that were stronger and healthier. The shorter soaking time benefits the grower by reducing the time spent pre-soaking the seed, but it does not help with water conservation because it is irrigated every 2 h. However, as the salinity of the hydroponic medium changes, this may change. However, this might change as the salinity of the hydroponic medium changes [43, 44]. Only crude nitrogen and protein were tested in this experiment, and further research into trace element levels would be required to determine the seedlings' full nutrient spectrum post-harvest. Other studies found that shortening the growing period from 8 to 4 days resulted in higher nutrient levels, which could be investigated further.

Furthermore, all of the highest dry weight means were obtained using a 10-h irrigation interval. The most significant results were obtained with soaking treatments lasting 1 and 3 h, as shown in Treatments 4 and 9, respectively. Although less significant, Treatment 14 with an 8-h soaking time was only 0.24 g lighter than treatments 4 and 9. This indicated that the seed pre-soaking time before germination could be reduced to 1 h. Under a 10 hourly irrigation interval, all three treatments (Treatments 4, 9, and 13) produced the highest dry weight. The highest fresh weight recorded came from that of Treatment 12, with a soaking time of 8 h and an irrigation interval of 4 h. It can be deduced that the seedlings benefitted from longer pre-soaking treatment. They still required moderate watering as the irrigation interval was every 4 h. Neither of the treatments agreed with the controls for soaking nor irrigation. The second highest mean value was achieved with Treatment 2 with a 1-h soak and 4 hourly irrigations. This proved that the pre-soaking time could be reduced without it affecting the total weight of the seedling post-harvest, however, the water consumption required to enable growth remained relatively high. Therefore, the disparity recorded in the weight of fresh samples compared to dry samples agrees with the results of the previous study as reported by Emam [20]. An investigation into the use of a nutrient solution to the irrigation water would be required to establish if this would improve overall fresh weight, post-harvest and if the introduction of a nutrient solution would allow the irrigation interval to be decreased thereby saving water.

5. Conclusion

This study reveals that a 1-h pre-soaked treatment, under 4 or 12 hourly irrigation intervals (T2 and T5) was the best treatment for cultivating barley hydroponically to achieve better yield for optimal fresh fodder production. For dry fodder weight, the highest yields were obtained with irrigation intervals of 10–12 h. Even though other growth parameters investigated such as length and root map expansion deviated from this trend, shorter soaking time at increased irrigation frequencies, proved to be beneficial to the farmer in terms of weight, nutrient yield, and height of *H. vulgare*. In the future, it would be helpful to ascertain if a change in irrigation type, from drip to spray, or increasing the mineral/trace metals composition of irrigation water would improve the root expansion of the seedling or bring about a marked decrease in irrigation water, thereby saving water.

Authors' contributions

Ryan Anthony Smith: Designed and performed the experiments; collated and analyzed data; wrote the original paper.

Muhali Olaide Jimoh: Analyzed and interpreted collated data; revised the manuscript with technical inputs.

Charles Petrus Laubscher: Conceived and designed the experiments; sourced for funding; supervised the experiments; edited the manuscript.

Funding

The authors wish to acknowledge the financial assistance received from the Cape Peninsula University of Technology through the University Research Fund.

Competing interests

Authors declare no conflict of interest.

Data availability statement

All data associated with this research are available on reasonable request from the corresponding author.

Author details

Ryan Anthony Smith, Muhali Olaide Jimoh and Charles Petrus Laubscher*
Faculty of Applied Sciences, Department of Horticultural Sciences, Cape Peninsula University of Technology, City of Cape Town, South Africa

*Address all correspondence to: laubscherc@cput.ac.za

References

[1] Zhou X, Yu J, Spengler RN, et al. 5,200-year-old cereal grains from the eastern Altai Mountains redate the trans-Eurasian crop exchange. Nature Plants. 2020;**6**:78-87

[2] Sato K, Shin-I T, Seki M, et al. Development of 5006 full-length CDNAs in barley: A tool for accessing cereal genomics resources. DNA Research. 2009;**16**:81-89

[3] Baik BK, Ullrich SE. Barley for food: Characteristics, improvement, and renewed interest. Journal of Cereal Science. 2008;**48**:233-242

[4] DAFF. A Profile of the South African Barley Market Value Chain. Department of Agriculture, Forestry and Fisheries. Arcadia 0007, South Africa; 2019

[5] Pattanaik AK, Jadhav SE, Dutta N, et al. Eco-responsive feeding and nutrition: Linking livestock and livelihood. Thematic papers. In: Proceeding of 9th Biennial Animal Nutrition Association Conference; Guwahati, India. 2015. p. 354

[6] Sneath R, McIntosh F. Review of hydroponic fodder production for beef cattle. Department of Primary Industries: Queensland Australia. 2003;**84**:54

[7] Agius A, Pastorelli G, Attard E. Cows fed hydroponic fodder and conventional diet: Effects on milk quality. Archives Animal Breeding. 2019;**62**:517-525

[8] Mooney J. Growing cattle feed hydroponically. Meat and livestock Australia. 2005. p. 30

[9] Cuddeford D. Hydroponic grass. Farm Animal Practice. 1989;**11**:211-214

[10] Naik P, Swain B, Singh N. Production and utilisation of hydroponics fodder. Indian Journal of Animal Nutrition. 2015;**32**:1-9

[11] Rockstrom J, Lannerstad M, Falkenmark M. Assessing the water challenge of a new green revolution in developing countries. Proceedings of the National Academy of Sciences of the United States of America. 2007;**104**:6253-6260

[12] Ngxabi S, Jimoh MO, Kambizi L, et al. Growth Characteristics, Phytochemical Contents, and Antioxidant Capacity of Trachyandra ciliata (Lf) Kunth Grown in Hydroponics under Varying Degrees of Salinity. Horticulturae. 2021;7(8):244

[13] Al-Karaki GN, Al-Hashimi M. Green fodder production and water use efficiency of some forage crops under hydroponic conditions. Agronomy. 2012. Article ID 924672;1-5

[14] Al-Karaki GN, Al-Momani N. Evaluation of some barley cultivars for green fodder production and water use efficiency under hydroponic conditions. Jordan Journal of Agricultural Sciences. 2011;7:448-457

[15] Pettersson S. Low root zone temperature effects on net mineral nutrient uptake and distribution in barley (*Hordeum vulgare*). Journal of Plant Physiology. 1995;**145**:459-464

[16] Sang N, Li G, Xin X. Municipal landfill leachate induces cytogenetic damage in root tips of *Hordeum vulgare*. Ecotoxicology and Environmental Safety. 2006;**63**:469-473

[17] Singh B, Singh R, Singh R, et al. Effect of insecticides on germination, early growth and cytogenetic

behavior of barley (*Hordeum vulgare*). Environmental and Experimental Botany. 1979;**19**:127-132

[18] Van Campenhout L, Shen HY, Iserentant D, et al. The gas environment of germinating barley in various microbial states during malting. Process Biochemistry. 1999;**34**:929-937

[19] Luna T, Wilkinson KM, Dumroese RK. Seed germination and sowing options [Chapter 9]. In: Wilkinson KM, Landis TD, Haase DL, Daley BF, Dumroese RK, editors. Tropical Nursery Manual: A guide to starting and operating a nursery for native and traditional plants. Agriculture Handbook 732. Washington, DC: US Department of Agriculture, Forest Service. 2014;**732**:163-183

[20] Emam MSA. The sprout production and water use efficiency of some barley cultivars under intensive hydroponic system. Middle East Journal of Agriculture Research. 2016;**05**:161-171

[21] Jimoh MO, Afolayan AJ, Lewu FB. Germination response of *Amaranthus caudatus* L. to soil types and environmental conditions. Thaiszia—Journal of Botany. 2019;**29**:085-100

[22] Jagatheeswari D, Ranganathan P. Influence of mercuric chloride on seed germination, seedling growth and biochemical analysis of green gram (*Vigna radiata* (L.) Wilczek. var. Vamban-3). International Journal of Pharmaceutical and Biological Archive. 2012;**3**:291-295

[23] Ramakrishna N, Lacey J, Smith JE. Effect of surface sterilization, fumigation and gamma irradiation on the microflora and germination of barley seeds. International Journal of Food Microbiology. 1991;**13**:47-54

[24] Dung DD, Godwin IR, Nolan JV. Nutrient content and in sacco degradation of hydroponic barley sprouts grown using nutrient solution or tap water. Journal of Animal and Veterinary Advances. 2010;**9**:2432-2436

[25] Nxawe S, Ndakidemi PA, Laubscher CP. Possible effects of regulating hydroponic water temperature on plant growth, accumulation of nutrients and other metabolites. African Journal of Biotechnology. 2010;**9**:9128-9134

[26] Bünemann EK, Bongiorno G, Bai Z, et al. Soil quality—A critical review. Soil Biology and Biochemistry. 2018;**120**:105-125

[27] Teixeira HM, Bianchi FJJA, Cardoso IM, et al. Impact of agroecological management on plant diversity and soil-based ecosystem services in pasture and coffee systems in the Atlantic forest of Brazil. Agriculture, Ecosystems and Environment. 2021;**305**:107171

[28] Zikeli S, Deil L, Möller K. The challenge of imbalanced nutrient flows in organic farming systems: A study of organic greenhouses in southern Germany. Agriculture, Ecosystems and Environment. 2017;**244**:1-13

[29] Faber RJ, Laubscher CP, Rautenbach F, et al. Variabilities in alkaloid concentration of *Sceletium tortuosum* (L.) N.E. Br in response to different soilless growing media and fertigation regimes in hydroponics. Heliyon. 2020;**6**:e05479

[30] Butcher JD, Laubscher CP, Coetzee JC. A study of oxygenation techniques and the chlorophyll responses of *Pelargonium tomentosum* grown in deep water culture hydroponics. HortScience. 2017;**52**:952-957

[31] Pardo E, Malet M, Marín S, et al. Effects of water activity and temperature on germination and growth profiles of ochratoxigenic *Penicillium verrucosum* isolates on barley meal extract agar. International Journal of Food Microbiology. 2006;**106**:25-31

[32] Frossard R, Oertli JJ. Growth and germination of fungal spores in guttation fluids of barley grown with different nitrogen sources. Transactions of the British Mycological Society. 1982;**78**:239-245

[33] Hafsi C, Russo MA, Sgherri C, et al. Implication of phospholipase D in response of *Hordeum vulgare* root to short-term potassium deprivation. Journal of Plant Physiology. 2009;**166**:499-506

[34] Ali S, Farooq MA, Yasmeen T, et al. The influence of silicon on barley growth, photosynthesis and ultra-structure under chromium stress. Ecotoxicology and Environmental Safety. 2013;**89**:66-72

[35] Venter G. Successful Hydroponics: 21st Century Technology for Commercial and Home Applications: A Comprehensive Practical Guide to Scientifically Based Hydroponic Crop. Bloomington, Indiana, United States: Xlibris Corporation; 2010

[36] Fazaeli H, Golmohammadi HA, Shoayee AA, et al. Performance of feedlot calves fed hydroponics fodder barley. Journal of Agricultural Science and Technology. 2011;**13**:367-375

[37] Jimoh MO, Afolayan AJ, Lewu FB. Nutrients and antinutrient constituents of *Amaranthus caudatus* L. cultivated on different soils. Saudi Journal of Biological Sciences. 2020;**27**:3570-3580

[38] Khan RM. Problem Solving and Data Analysis Using Minitab: A Clear and Easy Guide to Six Sigma Methodology. New York: Wiley; 2013

[39] Chung HJ, Jang SH, Cho HY, et al. Effects of steeping and anaerobic treatment on GABA (γ-aminobutyric acid) content in germinated waxy hull-less barley. LWT—Food Science and Technology. 2009;**42**:1712-1716

[40] Ella ES, Dionisio-Sese ML, Ismail AM. Seed pre-treatment in rice reduces damage, enhances carbohydrate mobilization and improves emergence and seedling establishment under flooded conditions. AoB Plants. 2011;**11**:1-11

[41] Snow AM, Ghaly AE, Snow A. A comparative assessment of hydroponically grown cereal crops for the purification of aquaculture wastewater and the production of fish feed. American Journal of Agricultural and Biological Sciences. 2008;**3**:364-378

[42] Zelinová V, Halušková L, Huttová J, et al. Short-term aluminium-induced changes in barley root tips. Protoplasma. 2011;**248**:523-530

[43] Mahjoor F, Ghaemi AA, Golabi MH. Interaction effects of water salinity and hydroponic growth medium on eggplant yield, water-use efficiency, and evapotranspiration. International Soil and Water Conservation Research. 2016;**4**:99-107

[44] Sogoni A, Jimoh M, Kambizi L, et al. The impact of salt stress on plant growth, mineral composition, and antioxidant activity in *Tetragonia decumbens* Mill.: An underutilized edible halophyte in South Africa. Horticulturae. 2021;**7**:140